Organization of Neural Networks

VCH

Distribution:

VCH Verlagsgesellschaft, P.O. Box 1260/1280, D-6940 Weinheim (Federal Republic of Germany)

Switzerland: VCH Verlags-AG, P.O. Box, CH-4020 Basel (Switzerland)

Great Britain and Ireland: VCH Publishers (UK) Ltd., 8 Wellington Court, Wellington Street, Cambridge CB1 1HW (Great Britain)

USA and Canada: VCH Publishers, Suite 909, 220 East 23rd Street, New York, NY 10010-4606 (USA)

ISBN 3-527-26715-8 (VCH Verlagsgesellschaft)

ISBN 0-89573-652-7 (VCH Publishers)

Organization of Neural Networks

Structures and Models

Edited by
W. von Seelen, G. Shaw, U. M. Leinhos

VCH

Editors:
Prof. Dr. Werner von Seelen
and
Ulrich M. Leinhos
Johannes-Gutenberg-Universität
Institut für Zoologie III
Saarstraße 21
D-6500 Mainz 1
Federal Republic of Germany

Prof. Dr. Gordon Shaw
Department of Physics
University of California
Irvine, CA 92717
USA

Cover illustration: Pyramidal cells and axonal network in the visual cortex of *Macaca mulatta* (layer 3, area 18). Bielschowsky silver stain. Width of the picture: 290 μm. Photo and preparation: Claudia Martin-Schubert, Max-Planck-Institut für Biologische Kybernetik, Tübingen, Federal Republic of Germany.

Editorial director: Dr. H.-J. Kraus
Prodution manager: Max Denk

British Library Cataloguing in Publication Data:

Organization of neural networks: structures and models.
1. Nervous system – Vertebrates
I. Seelen, W. von; II. Shaw, G. (Gordon); III. Leinhos, U.M.
596'.0188 QP361
ISBN 3-527-26715-8

Library of Congress Card. No. 87-31606

Deutsche Bibliothek Cataloguing-in-Publication Data:

Organization of neural networks: structures and models / ed.
by W. von Seelen ... – Weinheim; Basel (Switzerland); Cambridge; New York, NY: VCH, 1988
ISBN 3-527-26715-8 (Weinheim ...) Pp.
ISBN 0-89573-652-7 (Cambridge ...) Pp.
NE: Seelen, Werner von [Hrsg.]

Reproduction: Mitterweger Werksatz GmbH, D-6831Plankstadt
Printing: Graphischer Betrieb Konrad Triltsch, D-8700 Würzburg
Bookbinding: Graphischer Betrieb Konrad Triltsch, D-8700 Würzburg
Printed in the Federal Republic of Germany

Preface

Decoding the way brains function is one of the great scientific challenges of our time, requiring the cooperation of scientists from a great variety of disciplines. A key role is played by theoretical neuroscience, which uses methods from diverse scientific fields in order to develop formal concepts on the strategies applied by brains in processing information. Knowledge on the implementation of intelligent behavior in living systems, which autonomously produce purposeful behavior, is of great significance for basic research in the natural sciences as well as for technical applications in the fields of computer science and artificial intelligence.

This volume contains papers on the theoretical analysis and modeling of brain functions, mainly of higher vertebrates. Furthermore, reviews, which provide the necessary background information on the models described, are given so as to allow a larger circle of readers insight into theoretical neuroscience. An introductory article facilitates classification of the contributions for the reader less familiar with this fascinating scientific discipline. A "summary statement" sketches a scientific-historical map of theoretical neuroscience, assigning each author of a paper to a "site" and describing future trends of development of this discipline.

The theoretical analysis and models in this book were presented and discussed at a meeting on theoretical neuroscience held in the Werner-Reimers-Stiftung in Bad Homburg, Federal Republic of Germany, on September 16 – 19, 1986. This meeting with the working title "Organization of Structure and Function in the Brain" was the second in a series of planned meetings that began in Trieste in 1984 and that is to be continued every two years. The aim of these international meetings is to provide a selected interdisciplinary circle with an opportunity for intensive exchange of ideas on theoretical problems of neural information processing.

The contributions as a whole do not lay claim to complete coverage of the field, but rather correspond to the idea of the meeting as a forum for detailed discussions. The papers given have been revised and supplemented with edited versions of the discussions or comments subsequently sent to the editors. The diversity in the form of these supplements reflects the open nature of the meeting itself.

The editors would like to thank the authors of the reviews for their supplementary contributions. We are also grateful to Dr. Ad Aertsen, who contributed considerably to the preparation of this volume, as well as to Ms. Käthe Rehbinder, Ms. Ingrid Krämer and Mr. Neil Beckhaus for their assistance.

The meeting was made possible by the generous support of the Werner-Reimers-Stiftung, to which the editors, on behalf of all participants, would like to express their thanks once more.

Finally, we are especially thankful to Dr. Hans-Joachim Kraus und VCH Verlagsgesellschaft for their patience and generous support.

September 1987 The Editors

Contents

List of Contributors

Edward H. Adelson
Center for Biological
Information Processing
Whitaker College E25 – 201
Massachusetts Institute of Technology
Cambridge, MA 02139
USA

Ad Aertsen
Max-Planck-Institut
für Biologische Kybernetik
Spemannstraße 38
D-7400 Tübingen 1
Federal Republic of Germany

Terry Allard
Coleman Laboratory
Departments of Otolaryngology
and Physiology
871 HSE
University of California at San Francisco
San Francisco, CA 94143
USA

Shun-ichi Amari
Faculty of Engineering
Department of Mathematical Engineering
and Instrumentation Physics
University of Tokyo
Bunkyo-ku
Tokyo 113
Japan

James A. Anderson
Department of Psychology
Center for Neural Science and
Department of Cognitive
and Linguistic Sciences
Brown University
Providence, RI 02912
USA

Elie Bienenstock
Laboratoire de Neurobiologie
du Développement
Centre d'Orsay
Bâtiment 440
Université de Paris-Sud
91405 Orsay Cedex
France

Egbert de Boer
Audiologisch Centrum
Abt. KNO,
RM D2-210
Academisch Medisch Centrum
Universiteit van Amsterdam
Meibergdreef 9
1105 AZ Amsterdam zuidoost
The Netherlands

Jean-Pierre Changeux
Institut Pasteur
25 rue du Docteur Roux
75724 Paris
France

Stanislas Dehaene
École normale supérieure
45 rue d' Ulm
75005 Paris
France

Rolf Eckmiller
Institut für Physikalische Biologie
Abteilung Biokybernetik
Universität Düsseldorf
Universitätsstraße 1
D-4000 Düsseldorf
Federal Republic of Germany

Enis Ersü
ISRA Systemtechnik GmbH
Schöfferstraße 15
D-6100 Darmstadt
Federal Republic of Germany

Walter J. Freeman
Department of Physiology and Anatomy
2549 Life Sciences Building
University of California
Berkeley, CA 94720
USA

George L. Gerstein
The School of Medicine
Department of Physiology G4
University of Pennsylvania
Philadelphia, PA 19104
USA

Alfred Gierer
Max-Planck-Institut
für Entwicklungsbiologie
Molekularbiologische Abteilung
Spemannstraße 35/IV
D-7400 Tübingen 1
Federal Republic of Germany

Norberto M. Grzywacz
Center for
Biological Information Processing
Whitaker College E25-201
Massachusetts Institute of Technology
Cambridge, MA 02139
USA

Gert Hauske
Lehrstuhl für Nachrichtentechnik
Technische Universität München
Arcisstraße 21
D-8000 München 2
Federal Republic of Germany

Ellen C. Hildreth
Center for
Biological Information Processing
Whitaker College E25-201
Massachusetts Institute of Technology
Cambridge, MA 02139
USA

Klaus-Peter Hoffmann
Allgemeine Zoologie und Neurobiologie
Ruhr-Universität Bochum
Universitätsstraße 150
D-4630 Bochum 1
Federal Republic of Germany

Victor K. Inada
Center for
Biological Information Processing
Whitaker College E25-201
Massachusetts Institute of Technology
Cambridge, MA 02139
USA

William Jenkins
Coleman Laboratory
Departments of
Otolaryngology and Physiology
871 HSE
University of California at San Francisco
San Francisco, CA 94143
USA

Peter I. M. Johannesma
Lab. voor med. fysica en biofysica
Katholieke Universiteit Nijmegen
Geert Grooteplein noord 21
Postbus 9101
6525 EZ Nijmegen
The Netherlands

Jan J. Koenderink
Fysisch laboratorium
Rijksuniversiteit Utrecht
Postbus 80 000
3508 TA Utrecht
The Netherlands

Ulrich Michael Leinhos
Abteilung Biophysik
Institut für Zoologie III
Johannes-Gutenberg-Universität
Saarstraße 21
D-6500 Mainz 1
Federal Republic of Germany

Hanspeter Mallot
Abteilung Biophysik
Institut für Zoologie III
Johannes-Gutenberg-Universität
Saarstraße 21
D-6500 Mainz 1
Federal Republic of Germany

Christoph von der Malsburg
Max-Planck-Institut für
Biophysikalische Chemie
Postfach 28 41
D-3400 Göttingen
Federal Republic of Germany

Michael M. Merzenich
Coleman Laboratory
Departments of
Otolaryngology and Physiology
871 HSE
University of California at San Francisco
San Francisco, CA 94143
USA

Graeme Mitchison
Kenneth Craik Laboratory
Department of Physiology
University of Cambridge
Downing Street
Cambridge, CB2 3EG
UK

Jean-Pierre Nadal
École normale supérieure
24 rue Lhomond
75231 Paris
France

Günther Palm
Max-Planck-Institut
für Biologische Kybernetik
Spemannstraße 38
D-7400 Tübingen 1
Federal Republic of Germany

Gregg Recanzone
Coleman Laboratory
Departments of
Otolaryngology and Physiology
871 HSE
University of California at San Francisco
San Francisco, CA 94143
USA

Christoph E. Schreiner
Coleman Laboratory
Departments of
Otolaryngology and Physiology
871 HSE
University of California at San Francisco
San Francisco, CA 94143
USA

Werner von Seelen
Abteilung Biophysik
Institut für Zoologie III
Johannes-Gutenberg-Universität
Saarstraße 21
D-6500 Mainz 1
Federal Republic of Germany

Martin I. Sereno
Division of Biology 216 – 76
California Institute of Technology
Pasadena, CA 91125
USA

Gordon Shaw
Department of Physics
University of California
Irvine, CA 92717
USA

Arthur P. Shimamura
Veterans Administration Medical Center
San Diego, CA 92161
USA

Wolf Singer
Max-Planck-Institut für Hirnforschung
Deutschordenstraße 46
D-6000 Frankfurt/M 71
Federal Republic of Germany

Larry R. Squire
Department of Psychiatry
University of California
San Diego, CA 92093
USA

Henning Tolle
Department Control Systems Theory
Institute of Control Engineering
Technical University Darmstadt
D-6100 Darmstadt
Federal Republic of Germany

Gerard Toulouse
École normale supérieure
Département de physique
24 rue Lhomond
75231 Paris Cedex 05
France

Christian Wehrhahn
Max-Planck-Institut
für Biologische Kybernetik
Spemannstraße 38
D-7400 Tübingen 1
Federal Republic of Germany

An Introduction to Theoretical Neuroscience as Portrayed in this Book

by *Ulrich Michael Leinhos* and *Werner von Seelen*

SUMMARY

This chapter introduces the reader who is less familiar with theoretical neuroscience to this discipline as it is depicted in this book to facilitate classification of the individual contributions. The problems, subject matter, methods, and the relations to related branches of science will be sketched. Associative memory, distributed systems and self-organization are the focal points on which the articles in this book concentrate. Multi-unit recordings mediate between theoretical and experimental neuroscience in a special way.

The individual contributions will be introduced briefly against this background. For a more detailed classification of the contributions based on scientific theory, the reader is referred to the summary statement by Walter J. Freeman.

1 PROBLEMS TREATED IN THEORETICAL NEUROSCIENCE

The contributions in this volume deal with the theoretical analysis and modeling of brain functions in higher vertebrates, i.e., with theoretical neuroscience. Using the methods of various disciplines, they attempt to extend our knowledge on the strategies employed by biological brains in information processing. Subsuming these attempts under the frequently encountered concept "*brain theory*" might give the impression that a uniform theory on biological information processing systems already exists. In reality, so far only a number of tentative steps starting from familiar scientific ground have been taken, and the theoretical connections between these steps result from the problems with which brains are confronted and which require the analysis of brain functions. Instances of such problems, which define the subject matter of theoretical neuroscience, are:

- What are the basic components of neural systems?
- How do brains implement dynamic self-organization for specific tasks?
- How are variable hierarchies structured in brains, and which strategies are used in the various subsystems?
- How can brains interpret scenes in the environment?
- How do brains construct models for controlling behavior?
- Which learning strategies are implemented in brains?
- How is knowledge represented in brains; are there modality specific forms of this representation?
- What part does time play in the internal organization of brains?
- How have brains evolved phylogenetically and how do they develop ontogenetically?
- How do brains solve multidimensional optimization tasks?
- What is intelligence?

It is extremely difficult to define such questions more precisely as this would presuppose an hypothesis on the structure of the system. Reasons for this difficulty are to be found

a) in the high complexity of the systems under investigation,
b) in the fact that the subsystems are in principle inseparable,
c) in the nature of the empirical data obtainable.

On a). Since expanding knowledge in the natural sciences presupposes a verifiable model of the real system, the *complexity* of the structure to be investigated is the highest barrier to be overcome and the differences in complexity of the problems lead to models of different levels of complexity. So, one of the central tasks is to design experiments adequate for investigating these models and which pose answerable questions.

On b). Experimental *decoupling* of subsystems that are in principle non-separable has often been practised successfully in physics for purposes of analysis. The same method is often used in brain research. Decoupling can be achieved by operations on the anatomical structures, or - more elegantly - by using specific, mostly particularly simple signals. In questions concerning the preprocessing of sensory signals, separation is successful but it always leads to difficulties in problems involving integrative performances: in most cases, anatomical decoupling involves loss of performance in the whole system while decoupling by means of signals presupposes that the organization of the brain is task specific in each case with the result that the degree of organization becomes a limiting condition rather than an experimental parameter.

On c). The third complicating factor is the *structure of the data* to be ascertained experimentally. Neuroanatomy, neurophysiology and behavioral physiology provide results that exclude complete observability and control of the systems under investigation. The dynamic weighting of the synapses are not determinable in neuroanatomy, nor are the codes of the spike sequences in neurophysiology, nor the involvement of the coupled subsystems (nonlinear as a rule) in behavioral physiology. Consequently, the description is in principle incomplete. None of the methods available at present is able to solve the problem of analysis directly. The only feasible way seems to be to estimate model parameters. The efficiency of this method depends essentially on the quality of the models which means that the discovery of concepts, the determination of their efficiency and the calculation of experimentally verifiable predictions is decisive. So the theoretical study and modeling of brains assumes a key role in neuroscience.

2 SUBJECT MATTER AND METHODS OF THEORETICAL NEUROSCIENCE

Although defining a scientific subject is always difficult, the questions mentioned at the beginning of the previous section do indicate certain boundaries. Accordingly, theoretical neuroscience is the *formal study of the strategies of the intelligent activities of living systems*, systems that autonomously produce purposeful behavior in a physical world and so react successfully to changes in that physical world as well as producing successful changes in it themselves.

As an empirical science, theoretical neuroscience deals with the structures implemented in living creatures such as sensors and effectors, and central nervous systems. Furthermore, the physical environment is an indispensable part of the system description insofar as it combines sensors, brain and effectors into a closed loop of information flow because firstly, effector activity changes the environment and the changes lead to direct feedback to the system, and secondly, phylogenetic and ontogenetic adaptations to the environment occur in the organisms.

On the other hand, theoretical neuroscience abstracts from the actual material implementation of these structures so far that the strategies for information processing that have proved successful in the course of phylogeny and so are common to all or many central nervous systems become clear. Consequently, the descriptive categories can be applied to non-living information processing systems although the connection with defined neural structures is lost as a rule.

Corresponding to the various complexity levels of the systems, such abstractions can be made at various logical levels. The subdivision of the area of scientific interest that results is an arbitrary one that is practically motivated. The same can be said of the separation into sensory and motor subsystems, central nervous system and physical environment, which reflects the necessity of interrupting the feedback loop of interactions between organism and environment at a certain point according to specifically defined criteria for purposes of observation and description. As mentioned above, this always leads to a qualitative loss of information in the case of complicated functions and simplified signals due to the fact that the subsystems are inseparable in principle. For instance, it is hardly possible to draw a dividing line between the sensory organs and the more centrally situated information processing systems of an organism since, even at the lowest levels of processing, central nervous, top-down influences have to be reckoned with. In the case of such influences, intermodal interactions can, for example, play a part, and investigating them presupposes the integration of the sensory organs and central nervous system. Consequently, one important problem for theoretical neuroscience is how to enhance the level of formal description so far that these integrating performances can be included.

The inventory of methods in theoretical neuroscience is the same as that in other theoretical natural sciences; however, in addition to the mathematical formalization of deterministic models, algorithmic descriptions acquire particular significance in theoretical neuroscience due to the distinct complexity of the systems to be modelled.

3 THEORETICAL NEUROSCIENCE IN THE SYSTEM OF SCIENCES

Theoretically oriented brain research has only been recognized as an independent scientific discipline since the seventies. It is a highly interdisciplinary research field with contacts to numerous other sciences. As a natural science dealing with living systems in their environment, it is founded on the other natural sciences, in particular *physics*, *physiology*, *anatomy* and *ethology*, and has to refer back to the knowledge they provide in describing biological information processing systems. It need hardly be mentioned that this gives rise to very close interactive contact to *neurophysiology*. The scientific subject of theoretical neuroscience also suggests exchanges with branches of science that deal with the performance of biological brains at the uppermost level of complexity, i.e. *psychophysics*, *cognitive psychology* and *psycholinguistics*. The overlaps with *mathematics*, *system theory* and *information theory and science*, from which theoretical neuroscience draws its repertoire of methods, are obvious. But there are also very close cross-connections with the more technologically oriented *computer sciences*, *artificial intelligence* and *robotics*.

Despite these close interconnections in the network of sciences, the development in the spectrum of methods of theoretical neuroscience has successively led to a certain independence of the field as a scientific discipline.

3.1 Theoretical neuroscience and neurobiology

The relation of theoretical science to the experimental basis of neurobiology is similar to that of theoretical to experimental physics. The level of abstraction is chosen such that the focal point of considerations lies not in the individually implemented anatomical and physiological details but in the strategies biological systems use to carry out certain tasks. But the abstract description and modeling in terms of mathematical and system theoretical procedures can still be evaluated by referring to the experimental data. This kind of division of labor between theoretical and experimental disciplines is becoming increasingly significant in neuroscience.

This development in neuroscience is proceeding more slowly than in other natural sciences because of the complexity of living, and in particular of neural systems. The consequence is that, in many cases, our inventory of methods for system theoretical description still trails behind technological innovations in experimental methods and the enormous increase in empirical data they have produced so that a central task of theoretical science, viz. definition of problems, provision of concepts for experimental neuroscience and support for the interpretation of experimental findings, is often still not sufficiently fulfilled.

The complexity of biological systems makes such system theoretical descriptions as attempt to explain these systems in detail in terms of analytic or combinatorial mathematics seem inappropriate. Instead, theoretical neuroscience must employ such mathematical methods as, through the combination of algorithmic formalization on the one hand, and computational capacity made available by innovations in computer technology on the other, supply, with respect to prediction capability, more and more exact simulations and models of the way central nervous systems operate, and, in this way, lead to qualitative statements on the subject being studied.

Statements formulated in this way, influence the traditional empirical methods of working to the extent that their theoretical approaches in their turn make increasing use of ways of explanation that move away from pure phenomenology and make exchange with theoretical neuroscience possible. If, despite division of labor within neuroscience, intensive exchange of ideas between empiricists and theorists continues, the mutual insemination of such a 'computational neuroscience' and experimental neuroscience will lead to distinct progress in our understanding of biological brains in the predictable future.

3.2 Theoretical neuroscience and the cognitive sciences

Compared with scientific disciplines like psychology, psycholinguistics or other cognitive sciences that, although they are also interested in a more abstract understanding of the performances of central nervous systems, regard the brain primarily as an integrating unit, theoretical neuroscience is less able to dispense with continually referring to the material substrate with its physiological and anatomical structures so as to attain system theoretical statements on all the various levels of complexity dictated by the subject matter.

3.3 Theoretical neuroscience and information science

Compared with mathematics, systems theory and information science, theoretical neuroscience, as a natural science, is not only concerned with synthetic, formal symbolic systems, but also with the flow of information in organisms and between organisms and the environment, i.e. with systems that are given in the physical world. Consequently, it has to take the material substrate and its physical laws into account in its system description also. Moreover, organisms are systems that, compared with physical systems in a narrower sense, have a higher, biological degree of complexity with a phylogenetic and ontogenetic history. Partial knowledge of this history makes problems and insights possible that deal with the constructibility of such systems under defined conditions and are extraordinarily important for analyzing them.

Nevertheless, there is a close connection between the two fields, from which theoretical neuroscience profits because information science provides possibly more useful methods for describing complex biological systems than the traditional tools of analysis and discrete mathematics. Information science, in turn, profits from theoretical neuroscience since it directs attention towards performances of brains that are possible in principle and so can be implemented, and towards the solutions that nature has found.

The case is similar for the applied branches, microelectronics, computer technology, artificial intelligence and robotics. Right from the beginning, developments in computer technology have been heavily stimulated, practically if not explicitly, both by the experimental findings of neuroscience as well as by its theories. Although theoretical neuroscience is rather a matter of basic research where the relation to applications is not so direct as in engineering, the level of abstraction of theories on neural information processing and of the methods makes it possible to apply them not only to biological but also to artificial brain systems. Since the structure of the latter is known a priori, they, in turn, can be used as models and criteria for the quality of theoretical predictions.

Both these directions are at present converging more and more: it has already been mentioned that brain theorists are making use of the processing capacities made available by information science and computer technology to an ever larger extent. Correspondingly, they are modeling increasingly larger functional neural assemblies even of histologically inhomogeneous layers and nuclei in the brain, and more extensive distributed networks. In doing so, they have to rely on computer technology to develop systems with architectures that anticipate as far as possible the non-serial structures to be simulated.

The questions formulated in Section 1 demonstrate the close proximity to applications problems in information science and artificial intelligence. Their difficulties in constructing intelligent automata must be considered as still largely unsolved at present. Biological brains, however, are the living proof that these problems can be solved.

In addition to this long-term goal, there is a large number of problems of detail that result in close interaction. For instance, computer technologists are displaying increasing interest in fields of brain theory such as pattern and speech recognition, sensory multiprocessing, structuring of processing hierarchies by delegation of command, associative memory, neural parallel processing systems and the principles of self-organization.*

4 FOCAL POINTS OF RESEARCH: ASSOCIATIVE MEMORY, PARALLEL PROCESSING AND SELF-ORGANIZATION

The assistance theoretical neuroscience lends development in the field of *associative memory* seems to be the most advanced. These structures for representing information, which are probably implemented by synaptic plasticity in biological systems, are an indispensible condition for flexible autonomic behavior due to indirect addressing and high tolerance to errors. In addition to applying the fundamental concepts in reconstructing data distorted by

---

* That a connection between neurobiology and information science can have far-reaching political and social implications ought to be mentioned at least in passing. So, neurobiology - like the other natural sciences - also ought to make its competence clear to the general public, and initiate and maintain contact with the social sciences so as to guarantee that new scientific knowledge and its technological application is subject to social control and to prevent harmful consequences.

noise in technical systems for instance, the memory performances of biological systems are also of great general significance for artificial information processing systems. However, the code used to represent knowledge is still unclear.

Interest in *neural distributed and parallel processing systems* is also directly connected with application. Conventional computers use sequential algorithms that can merely simulate distributed processing. It seems that the technical limits of the processing capacity of serial computing systems has been reached in the current generation of computers. Further increases in performance will require parallel computers in which the capacity will be extended by increasing the number of processors, high connectivity and the capability to modify on-line connectivity rapidly. Processing is based on the activity pattern of a matrix of neuron-like elements, in which the activity level of each unit depends on the current degree of excitation or inhibition. This means that the units are activated in parallel, and the flow and storage of information is distributed over a matrix of such units. However, parallel computers involve problems in the communication between the processors as well as in the structure of task-sharing and interactive control that have to be solved when programming such machines. This requires the generalization of problem solving strategies that are converted into program structures by the computer system after the goal has been set by man. This means that parallel processing systems must have a non-specific data format and a certain degree of error tolerance, and be capable of self-organization so as to be able to put themselves into functional states in which they can autonomously evaluate situations, extract the relevant information, store it as knowledge and react flexibly to new, unpredictable situations.

An obvious way of solving these exacting problems is to recur to the strategies of biological brains, which demonstrate that all these performances can be implemented. The necessary structures are evolved in the course of ontogenesis not only through direct genetic determination but also through *self-organization*. Global mappings of connectivity organize themselves through local cooperative mechanisms based on the activity patterns of participating neurons and through synaptic plasticity. Theoretical neuroscience and computer technology pay particular attention to research into the developmental physiology of self-organization since it can lead to models for self-programming. -

At the same time, these problems indicate the great distance between current knowledge and the goal to be attained, and circumscribe the difficulties involved in theoretical neuroscience. On the one hand, the present state of research in theoretical neuroscience is based either on extremely abstract proposed solutions to problems that have arisen experimentally, or on the system description of very small, delimited neuron populations. This means that theoretical neuroscience still has to find a bridge between deterministic description of relatively simple neural arrangements and statistical description of large neural assemblies. As already mentioned, advances in computer technology have made it possible to collect such large amounts of data that we are not able to make full use of these at present. So a central problem is how to deal with *multi-unit recordings* conceptually. Since biological systems are distributed parallel processors, the question arises of how the complex sets of data acquired in these experiments can be managed and depicted. - On the other hand, it means that the barrier confronting the acquisition of knowledge in theoretical neuroscience is primarily conceptual, i.e. it cannot be broken through simply with technological improvements since one can assume that the exact description of a performance is the essential presupposition for constructing a system that can execute it. What we still lack for constructing such systems is not the necessary technology but the conceptual penetration that would enable us to describe the performance of biological brains with sufficient accuracy.

5 THE ARTICLES

The brain theoretical models presented in this book are mainly based on experimental data from the auditory, visual, somatosensory and memory system of primates (including humans) and the cat, but also of birds and lower vertebrates. The sequence of the articles is fairly arbitrary; it is based on the experimental database. A review of the particular experimental field precedes each group. The focal points of research described above are mentioned in one way or another in all contributions but problems of self-organization of central nervous networks play a central part and so are dealt with in a separate review.

The review "*Elements of Signal Coding in the Auditory Nervous System*" describes some fundamental anatomical and physiological features of the central nervous auditory system and stresses the essential differences from the visual and sensomotory system such as result, for instance, from the necessity of an extremely high temporal resolution. - In "*Pattern Recognition of Spectral Information in Hearing*", a possible neural wiring schema for the perception of pitch based on physiological measurements and signal detection theory is proposed and the potential anatomical localization of such a circuit indicated. Proposals for experimentally verifying the suggested mechanism are made. - The significance of multi-unit recordings has already been pointed out. Such experiments concentrate on the observation of dynamic interaction between larger groups of neurons rather than on single neurons. Although, in the meantime, several laboratories around the world carry out such simultaneous recordings, the relevant theoretical tools have not yet been developed very far. For this reason, a special discussion of this topic was held at the meeting, which, due to its detailed nature, is published word for word to a large extent. First, "*Information Flow and State in Cortical Neural Networks: Interpreting Multi-Neuron Experiments*" introduces an experimental and analytic method for investigating the simultaneously recorded firing patterns of an assembly of neighboring neurons (in the auditory cortex of the cat in this case) so as to be able to describe not only pairs of neurons but the whole assembly of neurons and extract its effective connectivity. The following "*Discourse on Current Issues in Multi-Neuron Studies*" discusses the relation between the current status of theory and experimental research and stresses the need for theories with variables that can be qualitatively verified. The field of multi-unit recordings is thought to be particularly well-suited for promoting discussions between the different approaches to brain theory and so for supporting such theoretical developments.

The review "*Principles of Cortical Network Organization*" takes recourse to paradigms of pattern formation that are wellknown from other natural sciences to describe important characteristics of self-organization in physical systems in general and in neural networks in particular. Some examples from the visual system referring to experimental findings from deprived animals, mathematical analyses and computer simulation are discussed: in addition to ocularity domains and orientation specificity, retinotopic mapping such as plays a part for example in the connection between the retina and the tectum opticum is treated as a paradigm for self-organization. The latter is a well-studied system that is also the center point of the following contribution. Plasticity of mapping alterations in late ontogenesis as described in "Self-Organizing Processes in Adult Neo-Cortex" (see below) are also mentioned. - "*Generation of Spatial Order in the*

Developing Nervous System" deals with the mechanisms of topographic self-organization during early brain ontogenesis, presents a model based on studies of retino-tectal mapping and mathematical simulation, and goes into the possible relations between the principles of ontogenesis and the function of brain structures. - "*Co-Operation Between Retinal and Cortical Input in the Midbrain of Mammals*" is based on a combination of measurements of response behavior of neurons, and sensory deprivation and brain lesions, as well as interspecific comparisons, and describes an example of the specification of the topology of a neural network during a learning phase in early ontogenesis before connectivity is permanently fixed which leads to dynamic self-organization of topographical mapping through co-operation and competition of retinal and cortical inputs to midbrain nuclei. The subsequently submitted comment "*Concepts for Changing Neural Network Topology*" points out a further possibility for controlling dynamic changes in network topology, the so-called command neurons, and discusses their significance for the development of new parallel and self-organizing computer architectures with neuron-like networks.

The practice of first producing a mathematical analysis and only then looking for its possible relevance for existing neural networks is criticized in the review "*The Visual System*". The concern of the review is that abstract mathematical descriptions should not depart too far from the complexity of real networks so as to guarantee the connection between the two. For this purpose, it summarizes anatomical and physiological data, and quantitative details for modeling the existing circuitry of the visual system. According to the review it is necessary to indicate more exactly than has so far been the practice how they are related to real neurons and neural assemblies. With respect to the required data for doing this, the responses of single neurons are still of great significance in contrast to global approaches. A combination of both approaches such as is provided in the techniques of multi-unit recordings (cf. above) seems necessary for better data and better theories. - "*Design for a Sensorium*" demonstrates in terms of differential geometry the feasibility of exploiting whole mathematical fields for neurophysiology. Problems of the structure of sensory systems are treated with the visual system as an example and with reference to physiological data. Questions concerning the position of the sensory systems between the environment and sensomotory loops, and the creation of the topology of the visual field are discussed. - An application of graph theory to the problem of recognizing objects is considered in "*Neural-like Graph-Matching*

Techniques for Image-Processing". This problem requires generation of invariance with respect to various image transformations, image segmentation and separation of figure from background. Possibilities for implementation in biological systems with rapidly changeable synapses or in neurally oriented artificial systems with parallel algorithms for digital image-processing are suggested. At present, the application is limited due to the above-mentioned sequential algorithms of the current computer generation. - "*The Temporal Integration of 3-D Structure from Motion: A Computational and Psychophysical Study*" investigates important features of a model for the ability of the visual system to ascertain the three-dimensional form of an object that appears two-dimensionally on the retina from the relative motion of the image points by referring to an internal model of the object that is constantly up-dated. The method used is a combination of computer simulations and psychophysical measurements.

The review "*Some Basic Organizational Features of the Somatosensory Nervous System*" summarizes anatomical, physiological and behavioral data on the essential somatosensory pathway from the peripheral receptors in the skin via the afferent pathways to the primary somatosensory cortical areas. - These data provide the background for the contribution "*Self-Organizing Processes in Adult Neo-Cortex*". While it has so far been assumed that cortical neurons attain their functional specificity mainly due to the maturation of their anatomical connections during ontogenesis, it is here shown for primates with reference to observations in human medicine that functional specificity can be radically altered during the whole life-span. Such adaptive, intrinsic self-organization may possibly be the constitutive basic operation of the neo-cortex. This work demonstrates the plastic re-organization of topographic representations of the skin surface within the somatosensory cortex areas, but similar results for the auditory and visual areas are also mentioned. Cortical neural assemblies obviously have a larger inventory of anatomical afferences than can be inferred from physiological studies of coupling width. The choice from the inventory of inputs for the current form of the local mapping is made in dependence on the use of afferences as caused by behavior. The proposed choice mechanism is the coincidence of afferences so that the cortical representations do not reflect a static anatomy but rather the probability of coincidental neural events.

Proceeding from the premise that explanations of neural plasticity as foundation of learning and memory have to take the various abilities observed in human behavior into account, the results of neuropsychological studies on disturbances in this behavior due to brain damage insofar as they are relevant to biological models are summarized in the review "*Aspects of Human Memory and Amnesia Relevant to Neurobiological and Computational Models of Memory*". The article also suggests a conceptual framework for interpreting these results, which concern the anatomical substrate of memory and memory control, separation of various memory dimensions and the distinction between memory and other cognitive performances. There is a discussion of the extent to which connectionist models can provide a bridge between the neural and behavioral levels. - "*Categorization in a Neural Network*" presents a parallel distributed associative model with categorization ability - one of the main performances of the human cognitive system. Simulations check the correctness of the model's categorization. These simulations imply that a categorization system can indeed be developed from a small amount of abstract pre-knowledge but that it requires high connectivity and parallel processing to be implemented. As has been mentioned, such neuron-like, connectionist networks will probably be implemented in future generations of computers. - Human memory performances partly consist in learning sequences of actions or events. In "*The Organization of Sequential Memory: Sparse Representations and the Targetting Problem*" an explanation is sought for the way in which any element in such a sequence elicits recall of the whole sequence. Quantitative estimations based on anatomical data imply that the enormous memory capacity is only possible if one assumes that only a small number of neurons code each element of a sequence. The flexibility in recalling sequences means that the neurons of each coded element of a sequence must be able in principle to activate the neural representation of every other element in the sequence, which leads to the necessity for special topographical connectivity patterns. Accordingly, a model is described in which elements of a sequence are represented by smaller topographically ordered subnetworks and the connections are arranged so that the axons of the subnetworks form their synapses with the dendrites of other subnetworks. - There are limits to the memory capacity and so to the ability to acquire information in non-selective memory models. "*Neural Networks and Models for Learning and Memory*" presents a simulation model that keeps its memory capacity constant by preventing memory overloading through overwriting older information. The model, which is based on the possibility of synaptic modifications in the seconds range and assumes saturation characteristics for synaptic efficiency, has parallels to human short-term memory the stationary capacity of which is made clear by the 'magical number seven'. Simulations indicate analogies to classical findings in experimental learning psychology. - "*Mathematical Theory of Self-Organization in Neural Nets*" offers a mathematical model for the analysis of learning and the ability of neural nets to organize themselves into smaller subnetworks. The article describes simulation studies of the topological features of the model of a self-organizing neural encoder, such as participate in retino-tectal mapping as already mentioned several times, and in the self-organizing modifications to neural representation in the somatosensory cortex described in "Self-Organizing Processes in Adult Neo-Cortex", and compares these features with physiological data. Neuron-like encoders could serve as preprocessors for complex associative networks or parallel computers, to which the method of analysis described would also be applicable. - Principles for the construction of control loops for complex nonlinear technical processes in industry can be derived from studies on the human brain. To bridge the lack of mathematical predictability in these processes, such control structures would have to be adaptive and capable of learning. In the model presented in "*Learning Control Structures with Neuron-like Associative Memory Systems*", strategies used in human problem solving such as are analysed in cognitive psychology are combined with neuron-like associative memories. The implementation of this concept based on associative memory in a massively parallel distributed computer architecture would provide extensive possibilities for industrial application.

Finally, the summary statement "*An Epigenetic Landscape of Brain Theory*" classifies the papers given at the meeting into four traditional research fields in neuroscience by assigning each to an 'epigenetic landscape' on the basis of its genealogy and so provides an overall view of the context and points out the relationships between the individual contributions. Future trends in the development of theoretical neuroscience are discussed in terms of the papers presented here.

Review:

Elements of Signal Coding in the Auditory Nervous System

by *Christoph E. Schreiner* and *Michael M. Merzenich*

SUMMARY

Some basic anatomical and functional features oforganization of the central auditory system are summarized. After a description of special anatomical features of organization of the system, some aspects of the selective response properties of central auditory neurons defined with the use of simple sound stimuli are described. Receptive fields of auditory neurons can be constructed for spectral, temporal and spatial aspects of sound stimuli. This response selectivity underlies the encoding and representation of the spectral content, temporal fluctuations and locations of sounds.

1 INTRODUCTION

Species-specific communication sounds like speech, and other natural (almost invariably spectrally complex) sounds consist of a stream of many acoustical features definable in terms of frequency, intensity and duration. The main perceptual attributes of acoustical signals are a) pitch, b) loudness, c) spatial location, and d) temporal characteristics. The acoustic correlates of these basic perceptual qualities have been used to study and model the principles of the analysis and representation of complex signals within the central nervous system. The parametric exploration of these general acoustic properties (e.g., frequencies, intensities, between-ear differences, fluctuations over time, etc.) and their representation at different levels of the auditory system has proven to be a useful analytic approach (1).

Historically, the analysis of the auditory system focused on spectral parameters of pure tones , e.g. frequency, intensity and phase representation (2). From the nonlinearities in the coding of these parameters as well as from apparent difficulties in the predictibility of responses to complex signals it became evident that the description of the response characteristics of single neurons had to be based on a larger set of parameters (3,4). Among the additional response parameters that were investigated in more detail were two-tone interactions, adaptation, binaural interactions, and coding of envelope changes. Some attempts were made to develop more complete descriptions of response properties that combine several of those aspects. An example for such an integrative approach is the construction of 'spectro-temporal receptive fields' (5,6) that combine aspects of frequency coding and temporal coding. However, the complexity in descriptions of response properties already apparent for simple signals renders a 'complete' characterization of auditory neurons for complex signals difficult for most practical purposes.

A compromise has to be found between the need for complete response characterization and practical limitations imposed by the large number of possible stimulus continua and time constraints. Evidently, for each neuron, each of a number of acoustical parameters assert some influence on the activity of that neuron. Input/output relationships or filter characteristics based on simple stimulus continua can be easily defined for every auditory neuron. A basic set of these 'simple' filter characteristics can be used to derive a 'compound' receptive field consisting, for example, of an estimate of the frequency response, the temporal response and the spatial response. Such analyses can provide i) a preliminary picture of signal processing by single neurons or populations of neurons, and establishes ii) a common framework for the comparison of particular response properties obtained in more detail. In addition, 'compound' receptive fields derived with the use of simple stimuli allow some aspects of the processing of more complex stimuli to be deduced.

In this brief review, we describe some basic features of the anatomical organization of the auditory nervous system, and then describe some functional features deduced from neuronal responses to simple stimulus continua at different neural levels that are essential for the processing of complex auditory stimuli.

2 BASIC ANATOMICAL ORGANIZATION OF THE CENTRAL AUDITORY SYSTEM

Information from the auditory receptors in the organ of Corti of the inner ear is conveyed to the cochlear nuclear complex (CN) by the axons of spiral ganglion cells, which form the 8th cranial (auditory) nerve. The projection of the auditory nerve to the three large divisions of the CN (antereroventral, posteroventral and dorsal division) is highly ordered geometrically. As described below, this highly ordered projection conserves an orderly representation of the sensory epithelium of the inner ear, and thereby, of represented sound frequency. Indeed, projections all across the main-line auditory system are cochleotopic, preserving an orderly representation of cochlear location (or "best" sound frequency) at all system levels (7,8).

From the CN, major anatomical projections terminate within ipsilateral and contralateral superior olivary complexes (SOC) and within the ipsilateral and contralateral nuclei of the lateral lemniscus (NLL) in the pontine brain stem; and within the contralateral central nucleus of the inferior colliculus (ICC) at the level of the midbrain (see Figure 1). The CN, SOC and NLL on each side of the brain stem have 3, 2 and 3 major anatomically and functionally distinct subdivisions, respectively. Outputs from most of these functionally and anatomically distinct sources on the two sides of the brain stem (12-13 sources overall) converge in a broadly overlapping projections

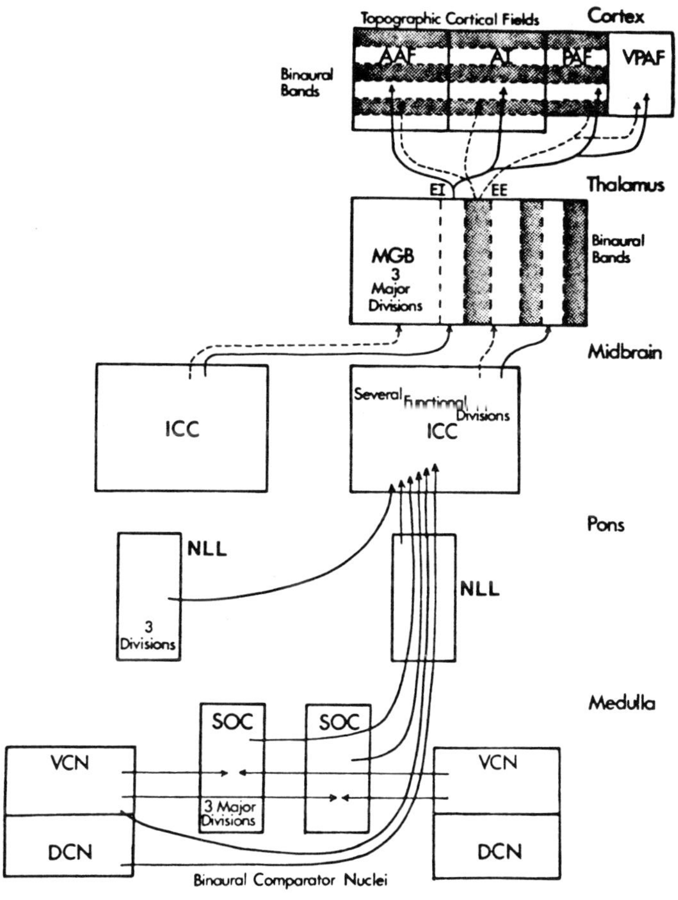

Figure 1: Basic organization of the main-line ascending auditory pathway in the cat (from 9). Only projections to one cerebral hemisphere are shown. The ventral and dorsal cochlear nuclei (VCN and DCN, respectively) receive ascending projections from the spiral ganglion of one cochlea through the fibers of the auditory nerve. In the superior olivary complex (SOC), inputs from both ears converge, and interaural intensity and time difference cues for sound location representation are processed. From the superior olivary complex , there are major projections to the central nucleus of the inferior colliculus (ICC) via the lateral lemniscus. Other projections from the VCN and SOC terminate within the nuclei of the lateral lemniscus (NLL), which also, in turn, project forward to the ICC. Projections from the inferior colliculus to the medial geniculate body (MGB) maintain a spatial segregation of neurons with different binaural response characteristics, i. e., of neurons with excitatory and inhibitory interaction between inputs from the two ears (EI, white area in MGB) or excitatory and excitatory interaction (EE, shaded area) . A similar segregation of the neurons is maintained in the projection of the MGB to the primary auditory cortical field. Several other cortical fields also have similar functionally segregated 'binaural bands'.

to the principal midbrain auditory nucleus, the ICC.

The main projection from the contralateral midbrain ICC terminates is a projection that, again, conserves an orderly representation of cochlear place or sound frequency; however, in the extended isofrequency dimension of the projection, there is a highly convergent-divergent projection to the different subnuclei of the contralateral medial geniculate body (MGB), the thalamic auditory nucleus. In the MGB, a number of projections radiate to at least 6 (in cats and monkeys) anatomically and functionally delineable auditory cortical fields. On the basis of their frequency organization, two groups of cortical fields can be distinguished: i) a group with a strict cochleotopic organization (e.g., fields AAF, AI , PAF and VPAF in the cat) exhibiting a strong frequency gradient in one dimension and no frequency gradient in the orthogonal ('isofrequency ') dimensions, and ii) a group of cortical fields (e.g., field A II in the cat) that exhibit no or only a weak spatial frequency gradient. Cortical neurons that exhibit different types of binaural interaction are also spatially segregated in a number of cortical fields.

In general, the anatomical organization of the auditory system differs from that of other sensory systems in several important respects (8,9). They include:

1) Cortical neurons fed from the main-line auditory projection axis are 6th order neurons; input cortical neurons in the somatosensory and visual systems are 4th order neurons, i.e., there are two 'additional' levels of processing in the central auditory system.

2) There is an 'extra' dimension of representation of the auditory sensory epithelium. The predominant source of auditory nerve inputs is a single row of sensory cells (the inner hair cells) about 35 mm in length in humans. There is an orderly monotonic representation of resonant frequency along this sensory cell row. Within auditory nuclei, cochlear place = 'best' frequency is represented across an extended dimension, i.e., there is a point-to-line projection from the sensory epithelium to the cortex, rather than a point-to-point projection scheme, as in the visual and somatosensory systems.

3) Sound frequency is not represented in a simple way across the sensory epithelium of the inner ear, or across the auditory nerve array. The nervous system representation of pitch must be *extracted* from spatiotemporal patterns of discharge of auditory ganglion cells, possibly by the operation of correlation-based mechanisms in central nuclei (10,11).

4) Unlike visual and somatosensory sensory surfaces, there is no representation of sound location by place on the sensory epithelium. Auditory space is encoded by neurons in special central nuclei (principally in the SOC) which respond selectively to interaural time and intensity differences and to spectral differences produced by the direction-dependent filter characteristics of the outer ear.

5) Encoding of both frequency and sound location is dependent upon relatively high frequency temporal information. The high temporal acuity of the auditory system's operations is without parallel in other mammalian sensory systems.

6) There is a massive anatomical convergence of information in the isofrequency dimension along the main-line projection axis. After nucleus-specific treatment of inputs relevant to the encoding and representation of pitch, loudness and sound location in 12-13 brain stem nuclei, derived nucleus-specific information is converged in a broadly overlapping anatomical projection to the ICC. The outputs of this nucleus project, in turn, in a massively convergent-divergent projection to thalamic MGB subdivisions. The outflow from these

nuclei is again massively convergent, within the isofrequency dimension, to target cortical fields. As a consequence of this organization, input from the same frequency-representation domains of source nuclei are combined and shuffled in a complex fashion, projecting from a number of highly specific information-extracting binaural and monaural brain stem nuclei, to a series of auditory cortical fields that are activated in parallel.

3 FUNCTIONAL ORGANIZATION OF THE AUDITORY SYSTEM

3.1 Frequency Selectivity

Historically, studies of the processing of the most fundamental parameters of acoustical signals --- sound frequencies and intensities --- contributed substantially to the identification of component nuclei and cortical fields of the auditory system in mammals. Major divisions of the auditory pathway topographically represent the auditory sensory epithelium. Processing of inputs is by cochlear location, or within frequency 'filters' (e.g., 2,7,9,12,13,14,15). Spatially, the pattern of frequencies across the main spectral gradient of 'tonotopic maps' is roughly logarithmically distributed, closely corresponding to the frequency-specific spatial location of the peak resonance along the cochlear partition. Thus, the representation has been described as being 'cochleotopic' as well as 'tonotopic'.

The spectral filter characteristics of the first-order auditory neurons in the spiral ganglion of the cochlea reflect the mechanical resolving power of the inner ear. The filter properties of auditory neurons are most often expressed in terms of frequency threshold tuning curves (FTCs). The lowest intensity that produces an increment of the spontaneous discharge activity of a neuron is determined for each frequency; the response threshold vs frequency function is the FTC. For stimulus levels up to about 60 dB above minimum response threshold for a neuron, the FTC of an auditory nerve fiber is characteristically V-shaped (16) (see Figure 2). At higher intensities, lower frequencies are disproportionately effective for exciting auditory nerve fibers. At highest operational levels, all or nearly all frequencies below the most effective frequency can drive the auditory nerve at its maximum (saturated) action potential discharge rate. Thus, the sharpness of tuning is substantially degraded at high sound levels.

The FTC is commonly characterized by two simple measures that correspond to the center frequency and the sharpness of a filter curve: i) the frequency that requires the lowest intensity to evoke spikes is the 'characteristic frequency' (CF) and ii) a quality factor (Q 10 dB) defined as the CF divided by the bandwidth of the FTC measured 10 dB above the minimum threshold, i.e.,10 dB above the threshold value for the CF. At the level of the auditory nerve, a fairly strict relationship between CF and Q-10dB is observed. The quality factor or sharpness of filtering increases with increasing CF. (Note, again, that this description is valid only for sound intensities below 50-60dB SPL.) In the cat, for example, Q-10dB values increase from 1 to 2 for CFs around 0.5 kHz to about 6 to 10 for CFs around 10 kHz (Evans, 1975). In other words, the relative frequency resolution increases with rising characteristic frequency.

In the mainline ascending auditory pathway, all nuclei are cochleotopically organized. In

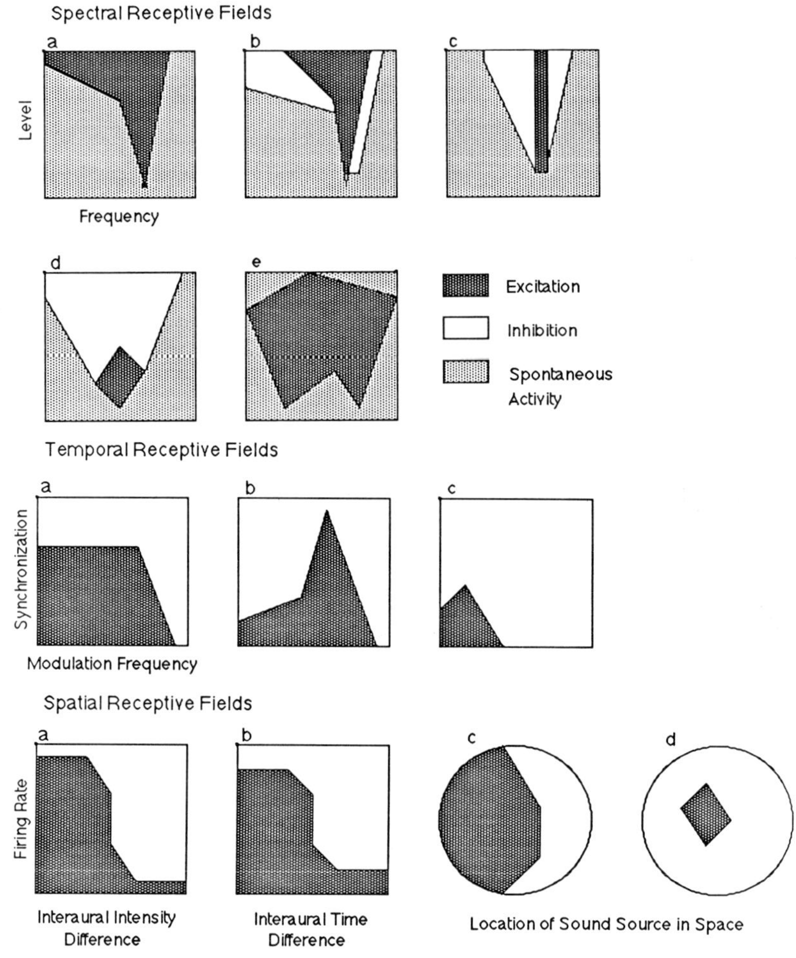

Figure 2: Elementary auditory receptive fields obtained with simple stimuli. ***Spectral receptive fields*** are schematically represented by frequency tuning areas (FTAs) or "response areas" obtained for pure tones presented at various frequencies and levels. Five typical FTAs with different combinations of excitatory and inhibitory activation are illustrated. Different nuclei of the auditory system contain neurons with different types of FTAs. Auditory nerve fibers , for example, exclusively have spectral receptive fields of type (a) with variable degrees of spontaneous activity. FTAs of type (b) can be found in all other nuclei beyond the auditory nerve. Type (c) is common within the inferior colliculus, medial geniculate body and auditory cortex. Type (d) is commonly found in the dorsal cochlear nucleus and in some fields of the auditory cortex. Broadly tuned FTAs (e) can be found, among other nuclei, in the inferior colliculus and non-tonopic fields in auditory cortex. ***Temporal receptive fields*** can be represented by temporal modulation transfer functions. A measure of the ability of a neuron to phase-lock (synchronize) its activity to the envelope of an amplitude modulated signal is plotted as function of the modulation frequency. The three examples of temporal receptive fields are typical for the auditory nerve (a), cochlear nucleus and inferior colliculus (b), and auditory cortex (c), respectively.***Spatial receptive fields*** of neurons can be represented through the effect of binaural intensity and/or time differences on response magnitudes (a and b). Free-field stimuli contain the manifold acoustical cues that can signal sound localization. Sound source locations that produce a significant response can be plotted on a two- dimensional representation of space coordinates (circle in (c) and (d)). Examples of resulting spatial best areas or receptive fields for a cortical neuron (c) and for a neuron in the superior colliculus (d) of the cat are shown.

three-dimensional structures, neurons with similar CFs are arranged into 'iso-frequency sheets' or 'bands' of neurons with a restricted range of CFs that extend across auditory nuclei from edge to edge (7). In the cortex, the iso-frequency representation is strip-like across one dimension of a cortical field (8,15). The main frequency gradient of the cochleotopic map in each nucleus is oriented orthogonal to the direction of iso-frequency neuronal sheets or bands.

There is abundant evidence that the representation of CFs appear to be highly ordered throughout most of the auditory nuclei on the main projection axis. Less information is available regarding the shape of the frequency or spectral receptive fields. However, it is clear that the shapes of the tuning curves at higher levels commonly deviate substantially from the shapes of FTCs in the auditory nerve (Figure 2). Extreme shapes include multi-peaked FTCs, U-shaped tuning curves with a roughly equal spectral bandwidth at all suprathreshold frequencies, broadly tuned FTCs responding strongly at near-threshold levels over an octave or greater frequency range, and non-monotonic FTCs that are responsive over a limited range of frequency only over a restricted domain of intensity. In the inferior colliculus and higher stations, U-shaped tuning curves become more common (3,7,18,19). The range of Q-10dB values encountered for a given CF is far greater at higher system levels than for auditory nerve fibers. Although systematic single neuron studies are still unavailable, there is some evidence from the inferior colliculus and the auditory cortex that the sharpness of tuning curves varies systematically across the isofrequency dimension of a cochleotopic map (20,21). In other words, the sharpness of the frequency receptive field or the spectral selectivity may constitute in itself a basic represented parameter of topographic neuronal maps.

In addition to the excitatory tuning curve, stimulus parameters with inhibitory effects on the neurons have to be included in any description of spectral receptive fields. There is significant mutual suppression of auditory nerve responses when tonal stimuli within an octave or so of frequency of each other are presented simultaneously. Inhibitory sidebands generated by centrally-originating inhibitory processes are a common feature of neuron populations in many central nuclei. Perceptually, the loudness of sounds are integrated when they fall within the spectral distance of a so-called "critical band" (22); at greater frequency separations, sounds are processed more independently. This perceptual quality varies little as the sound intensity is increased. Similarly, spectral interaction distances have been estimated in studies in central auditory nuclei; intensity-independent spectral integration at the midbrain level is of a scale comparable to perceptual estimates. Note, again, that the spatial spread of responses evoked by a simple tonal stimulus at the level of the auditory nerve is restricted only at relatively low stimulus levels; at higher levels (including normal operational levels, for example, in speech reception) any given tonal stimulus is commonly represented with a saturated (maximal-rate) action potential discharge over a broad sector of the auditory nerve array (23,24). By contrast, U-shaped tuning curves and intensity-independent windows of integration of spectral components are common features at higher neural levels (3,19). The neural processes by which this strikingly modified frequency filtering is achieved are not fully known; they almost certainly involve side-band inhibitory processes (3).

In summary, it can be stated that the 'location' of the spectral receptive field of most auditory neurons is characterized by its CF. The width or 'shape' of the spectral receptive field is ---- to a first approximation ---

characterized by its Q-10dB. In the auditory nerve, the CF and Q-10dB are closely related; this simple relationship breaks down at higher levels of the auditory system. Tuning curve shapes are substantially different at central nervous system as compared with auditory nerve levels. Intensity-independent processing of spectral information is a product of central processing that almost certainly involves inhibitory mechanisms operating at central levels. Cochlear location and its acoustic correlate, sound frequency, are represented in an topographic fashion at all levels in the main line auditory system.

3.2 Temporal Selectivity

In addition to the pure spectral content of a signal, the temporal patterning of complex sounds constitutes another important information-bearing signal parameter. Studies of the processing of temporal fluctuations of the envelope in frequency-analysis channels of cochleotopically organized subdivisions of the auditory system provide further insights into the neural representation of complex signals (e.g., 21, 25, 26, 27, 28, 29, 30, 31).

The responses of auditory neurons can replicate or 'follow ' in detail the time course of a *carrier* frequency within its spectral receptive field up to frequencies of 4-5 kHz in squirrel monkeys and cats. This action potential 'phase locking' of auditory nerve fibers and of neurons in lowest-level brain stem nuclei has been widely studied (16,23).

There have been fewer systematic examinations of the 'temporal receptive field' of neurons responding to *envelope* variations. These temporal fluctuations represent an important parameter continuum that has to be taken into account for a complete description of neuronal response selectivity in the time domain. The concept of "temporal modulation transfer functions" (tMTF) has been widely employed for this purpose (26,27,28,31,32,33,34,35; see Figure 2).

There is a wide range of these modulations of signal amplitude or envelope changes across different frequency channels in many species-specific vocalizations and in speech. Important features of complex communication signals contain envelope periodicities over a range extending from a few events per second to up to several hundred Hertz. The envelope or modulation frequencies for normal speech are in the range of 100 to 400 Hz. Other 'envelope' signals or periodicities utilized, for example, in musical perception, contain frequencies of up to approximately 1200 Hz, which correspond to the existence range of residue or periodicity pitch found in human psychophysics (e.g., 33,36).

A number of investigations have revealed some aspects of the temporal resolution of neurons at different levels of the auditory system. In general, responses of auditory nerve fibers show a high temporal resolution, i.e. the responses follow amplitude modulations over a wide frequency range (29,30,35). Lower temporal resolution was found in the cochlear nucleus (27,28) and, generally, even lower ability to follow amplitude modulation was seen in the inferior colliculus (21,25,26,37) and the auditory cortex (32,38,39,40). These findings can be summarized in part by stating that there is evidence supporting the hypothesis that overall temporal resolution decreases in successive stations along the ascending auditory pathway.

The temporal response characteristics of auditory nerve fibers correspond quite well to theoretical constraints imposed by the spectral filter properties of the nerve fibers. For all nerve fibers, the upper cut-off frequency for the representation of modulation frequencies is proportional to the bandwidth of the spectral tuning curves (35). In higher levels, this correspondence becomes weaker. That is, the temporal filter characteristics become increasingly more independent of the bandwidth of the spectral filter. For example, in the central nucleus of the inferior colliculus of the cat , a wide range of cut-off frequencies for the representation of modulation frequencies can be seen for a given CF and a given Q-10dB (21,37). The cut-off frequencies are, for the most part, lower than predicted by the spectral bandwidth. The high cut-off frequencies of the encountered range of modulation frequencies roughly corresponds to the limits imposed by the bandwidth of auditory nerve fibers.

An even more extreme reduction of the temporal response range is found in the auditory cortex. The majority of the upper cut-off frequencies for the representation of modulation frequencies is below 20 to 30 Hz, i.e. the temporal resolution is almost always a magnitude smaller than the spectral bandwidth of cortical neurons would accommodate (32).

In general, three levels of organization of temporal information can be distinguished in the auditory system: 1) The average frequency range of temporal variations represented with precision in the time patterns of neural reponses of the auditory nerve is higher than in the inferior colliculus, and the encoded temporal range of responses in the inferior colliculus is, on the average, substantially higher than in any of the cortical fields. Thus, at more peripheral auditory stations, relatively fast temporal variations in a signal are represented with high resolution by neuronal events with corresponding timing; only relatively slower or infrequent temporal variations are represented in the timing of neuronal activity at the cortical level.

2) There is a systematic topographic distribution (a 'map') of neurons with respect to their temporal response characteristics across the isofrequency dimension of the central nucleus of the inferior colliculus of the cat (21). Thus, the inferior colliculus contains highly organized topographic representations of acoustic signals based on *both* spectral and temporal signal aspects.

3) Different auditory cortical fields respond to slightly different, although overlapping, frequency ranges of temporal variations (32). Thus, there is evidence for a degree of some cortical segregation of temporal response features of neurons. The differences in the temporal resolution between cortical fields is most likely a consequence of task-intrinsic properties optimal for the field-specific representation or processing of acoustical information (3,9,18).

3.3 Spatial Selectivity

The perception of the spatial location of a sound source is mediated by neurons that receive input from both ears. The majority of neurons beyond the cochlear nucleus complex respond to binaural stimulation, i.e. to stimulation of both ears. The response characteristics of binaurally sensitive cells have been studied by using three major approaches to describe their properties. The simplest classification of binaural interaction is based on whether simultaneous stimulation of both ears results in i) summation or ii) suppression of the response activity evoked by stimulation of the most responsive ear alone. In the primary auditory cortex, neurons that are classified as binaurally summating or suppressive are spatially segregated. Elongated columnar aggregates or bands of neurons that show binaural summation alternate with aggregates or bands that contain neurons showing binaural suppression (41,42). Similar spatial segregations of these binaural classes can be seen in other cortical fields as well as in more peripheral stations like the inferior colliculus and the medial geniculate body (43,44,45,46).

A more detailed and functionally more relevant classification of binaurally sensitive neurons is based on their response to intensity differences and/or delay differences between the stimuli at the two ears, the two most important factors that determine the perception of the spatial location of a sound source (47,48). In birds, it has been demonstrated (49) that in subcortical stations these two different sound localization parameters can be independently processed before they are combined to produce a 'compound' spatial receptive field. In mammals, a similarly strict segregation of pathways that process temporal or intensity cues has not been seen. However, since the summation cells are often sensitive to interaural time differences and the suppression cells are often sensitive to interaural intensity differences, a certain spatial segregation of the neural representation of these two parameters seems to be maintained throughout the mainline ascending mammalian auditory pathway.

The sensitivity of neurons to sound-source location can be tested by employing free-field stimulation. With this method, fairly restricted spatial receptive fields have been identified in birds (50). Topographic representations of auditory space have been demonstrated in a midbrain structure of the barn owl homologous to a subdivision of the mammalian inferior colliculus (50), and in the superior colliculus of mammals (51,52). However, in the mainline ascending auditory pathway of mammals no topographic representation of auditory space has been demonstrated to date. Nevertheless, ablation experiments showed that the primary field (AI) of the auditory cortex is necessary and sufficient for the localization along an azimuthal sound source continuum (53). The strict cochleotopic organization of AI suggests that any space code superimposed on the frequency map is likely to be represented within the isofrequency dimension of the map.

4 CONCLUSION

The analytical description of the response of auditory neurons to complex stimuli requires a stimulus-independent and complete characterization of the spectro-temporal receptive field. Although some theoretical and experimental attempts have been made to derive such a description (5,6,54), most experimental characterizations of auditory receptive fields are still incomplete: Each of the spectral, temporal and spatial response correlates mentioned above is often determined in isolation without adequate knowledge of other response properties and their mutual interactions. Such a limited description of response properties by a few elementary features can be informative. For example, the study of topographic distributions of certain response properties has been successful, even if the stimulus property under consideration represents only one dimension of the relevant stimulus ensemble.

A more complete description of the signal processing and representation can be achieved by considering 'compound 'receptive fields derived from the linear combination of response characteristics for different simple signal aspects. Although such a characterization is limited to the types of simple signals employed in the assessment of the responses, the relationships among different response properties add an important component to the functional description .

Beyond the auditory nerve, the mutual influence of different signal aspects on the derived description of the neuronal activity has not as yet been very widely investigated. Consequently, descriptions of and predictions based on neuronal responses to isolated aspects of stimuli can provide only a limited picture of the signal processing when considering responses to signals that contain more complex combinations of signal features. The application of techniques that allow a more complete assessment of the 'spectro-temporal receptive field' of auditory neurons will be applied more widely in the future and will provide a better understanding of signal processing and representation in the auditory system.

5 ACKNOWLEDGEMENT

The authors were supported by NIH Grant NS-10414, the Coleman Fund and Hearing Research Inc..

6 REFERENCES

(1) Bullock,T.H. (ed.) Recognition of complex acoustic signals. Dahlem-Konferenzen, Berlin, 1977.

(2) J.E. Rose, R. Galambos, J.R. Hughes, In: Neural Mechanisms of the Auditory and Vestibular System (G.C. Rasmussen, W.F. Windle, eds.) Thomas, Springfield 1960, pp. 116-136.

(3) N. Suga, In: Dynamic Aspects of Neocortical Function, (G.M. Edelman, W.E. Gall and W.M. Cowan, eds.) John Wiley&Sons, New York 1984, pp. 315-373.

(4) J.W. Horst, E. Javel, G.R. Farley, In :Peripheral Auditory Mechanisms (J.B. Allen, J.L. Hall, A. Hubbard, S.T Neely, A. Tuleis, eds.) Springer, New York 1985, pp. 298-305.

(5) A. M.H.J. Aertsen, P.I.M. Johannesma, Biol. Cybernetics 38 (1980) 223-234.

(6) J.J. Eggermont, P.I.M. Johannesma, A.M.H.J Aertsen, Quart.Rev. Biophys. 16 (1983) 341-414.

(7) L.M. Aitkin, D.R.F. Irvine, W.R. Webster, In: Handbook of Physiology Vol.III (J.M. Brooklart, V.B.

Mountcastle, I. Darian-Smith, eds.) Waverly, Baltimore 1984, pp. 675-737.

(8) M.M. Merzenich, W. M. Jenkins and J.C. Middlebrooks, In: Dynamic Aspects of Neocortical Function, (G.M. Edelman, W.E. Gall and W.M. Cowan, eds.), John Wiley&Sons, New York 1984, pp. 397-424.

(9) M.M. Merzenich, J. Kaas , In: Progress in Psychobiology and Physiological Psychology, (J.M. Sprague and A.N. Epstein, eds.), Academic, New York, 1980, pp. 1-42.

(10) G. Langner, Exp. Brain Res. 52 (1983) 333-355.

(11) G. Loeb, M. White, M.M. Merzenich, Biol. Cybern. 47 (1983) 149-163.

(12) C.N. Woolsey, E.M. Walzl, Amer.J. Med. Sci. 207 (1944) 685-686.

(13) M.M. Merzenich, M.D. Reid, Brain Res. 77 (1974) 397-415.

(14) R.A. Reale, T.J. Imig, J. Comp. Neurol. 192 (1980) 265-291.

(15) MM. Merzenich, P.L. Knight, and G.L Roth, J Neurophysiol. 38 (1976) 231-249.

(16) N.Y.S. Kiang, In: Handbook of Physiology Vol.III (J.M. Brooklart, V.B. Mountcastle, I. Darian-Smith, eds.) Waverly, Baltimore 1984, pp. 639-674.

(17) E.F. Evans, In: Handbook of Sensory Physiology, Vol. 5/2 (W.D. Keidel, W.D. Neff, eds.) Springer, Berlin, 1975, pp. 1-108.

(18) N. Suga, In: Cortical Sensory Organization, Vol.3: Multiple Auditory Areas,(C.N. Woolsey, ed.) Humana Press, Clifton , New Jersey, 1982, pp. 157-218.

(19) G. Ehret, M.M. Merzenich, Science 227 (1985) 1245-1247.

(20) C.E. Schreiner, M. Cynader, J. Neurophysiol. 51 (1984) 1284-1305.

(21) C.E. Schreiner, G. Langner , Soc. Neurosc. Abstr. 10 (1984) 34.5.

(22) E. Zwicker, G. Flottrop, S.S. Stevens, J. Acoust. Soc. Am. 29 (1957) 548-557.

(23) D.O. Kim, C.E. Molnar, J. Neurophysiol. 42 (1979) 16-30.

(24) M.B. Sachs, E. Young, J. Acoust. Soc. Am. 66 (1979) 470-479.

(25) P.G. Nelson, S.D. Erulkar, and J.S. Bryan, J. Neurophysiol. 29 (1966) 834-860.

(26) I.A. Vartanyan, Neurosci. Trans. 10 (1969) 17-26.

(27) A.R. Møller, Exp. Neurol. 45 (1974)104-107.

(28) A.R. Møller, Exp. Brain Res. 25 (1976) 307-321.

(29) E. Javel, J. Acoust. Soc. Am. 68 (1980) 133-146.

(30) R. Smith, M. Brachman, Hearing Res. 2 (1980) 123-133.

(31) G. Langner, Exp. Brain Res. 44 (1981) 450-454.

(32) C.E. Schreiner, J.V. Urbas, Hearing Res. 21 (1986) 227-241.

(33) N.F. Viemeister, J. Acoust. Soc. Am. 66 (1979) 1364-1380.

(34) M.R. Schroeder, Acustica 49 (1981) 179-182.

(35) A.R. Palmer, Arch Otorhinolaryngol. 236 (1982) 197-202.

(36) J.F Schouten, In: Frequency Analysis and Periodicity Detection (R. Plomp, G.F. Smoorenburg, eds.) Sijthoff, Leiden 1970, pp. 41-54.

(37) A. Rees, A.R. Møller, Hearing Res. 10 (1983) 301-330.

(38) I.C. Whitfield, E.F. Evans, J. Neurophysiol. 28 (1965) 655-672.

(39) E.F. Evans, In: The Neurosciences Third Study Program (F.O. Schmitt, F.G. Worden, eds.) MIT Press, Cambridge 1974, pp. 131-145.

(40) O.D. Creutzfeldt, F.-C. Hellweg, C. Schreiner, Exp. Brain Res. 39 (1980) 87-104.

(41) T.J. Imig, H.O Adrian, Brain Res. 138 (1977) 241-257.

(42) J.C. Middlebrooks, R.W. Dykes, M.M. Merzenich, Brain Res. 181 (1980) 31-48.

(43) G.L. Roth, L.M. Aitkin, R.A. Anderson, M.M. Merzenich, J. Comp. Neurol. 182 (1978) 661-680.

(44) M.N. Semple, L.M. Aitkin, J. Neurophysiol. 42 (1979), 1626-1639.

(45) J.C. Middlebrooks, J.M. Zook, J. Neurosci. 3 (1983) 203-224.

(46) J.J Wenstrup, L.S. Ross, G.D. Pollak, J. Neurosci. 6 (1986) 962-973.

(47) J.E. Rose, N.B. Gross, C.D. Geisler, J.E. Hind, J. Neurophysiol. 29 (1966) 288-314.

(48) T.C.T. Yin, S. Kuwada, In: Dynamic Aspects of Neocortical Functions. (G.M. Edelman, W.E. Gall, W.M. Cowan, eds.) Wiley, New York 1984, pp. 263-313.

(49) T. Takahashi, A. Moiseff, M. Konishi, J. Neurosci. 4 (1984) 1781-1786.

(50) E.I. Knudsen, M. Konishi, Science 202 (1978) 778-780.

(51) A.J. King, A.R. Palmer, J. Physiol. 342 (1983) 361-381.

(52) J.C. Middlebrooks, E.I. Knudsen, J. Neurosci. 4 (1984) 2621-2634.

(53) W.M. Jenkins, M.M. Merzenich, J. Neurophysiol. 52 (1984) 819-847.

(54) W.J.M. Epping, J.J. Eggermont, Hearing Res. 24 (1986) 37-72.

Pattern Recognition of Spectral Information in Hearing

Egbert de Boer

SUMMARY

The perception of pitch of a multi-component sound stimulus can be seen as a collective phenomenon. The pitch is mainly extracted from the set of frequencies of the lowest-order components. These are just the components that are nearly completely resolved as a result of peripheral auditory filtering. It is known that the neural representation of a well-isolated component is less clear in terms of firing rate than in terms of synchrony with the stimulus. In this paper a mechanism is proposed that can serve to tie together the information contained in the firing patterns of all auditory nerve fibres that are stimulated by one component. The operation is based on the synchrony property but the synchrony between groups of nerve fibres is considered instead of the synchrony with respect to the stimulus signal. It is argued why good use must be made of the full complement of available nerve fibres: the inherent stochasticity of nerve fibre firings sets a limit to attainable discrimination.

1. INTRODUCTION

Many properties of the auditory system can well be understood from the signal transformations that occur in the cochlea (the inner ear). The acoustical stimulus has components of different frequencies and these are projected at different locations along the length of the basilar membrane (BM). This 'frequency-to-place transformation' provides the basis for the property that stimulus components of different frequencies appear to operate different "channels" of perception, at least when the frequencies are not too close together. The projection, however, is not infinitely sharp and the actual situation is more complicated, there is considerable overlap between the excitation patterns evoked by components of a multi-component stimulus. Despite this property, the determination of the pitch is a process involving the stimulus componenets as if they are isolated perfectly. It is the purpose of this paper to discuss the many factors involved in this process, with the aim to arrive at possible and feasible neural mechanisms that could assist in the job of component separation.

2. THE 'RESIDUE': SPECTRAL PATTERN AND REPETITION FREQUENCY

Many sounds in nature, and especially the vowels in speech, have periodic or nearly periodic waveforms. The pitch of such signals usually corresponds to the frequency of repetition. Fig. 1-A shows the waveform of a most stylized form of a periodic signal: a periodic pulse. The period is T_0 and the repetition frequency f_0. Fig. 1-B shows the corresponding spectrum, it shows a set of components with frequencies f_0, $2f_0$, $3f_0$, and so on. The meaning of the spectrum is that sinusoidal waveforms with these frequencies and amplitudes when added (with the appropriate phases) just produce the waveform of Fig. 1-A. The components in this case are known as harmonics and the spectrum is a harmonic one.

If the fundamental component in the spectrum, the one with frequency f_0, is not present in the sound, the pitch still corresponds to f_0. This phenomenon is known as the 'case of the missing fundamental'. Apparently, there are two subjective components of pitch, one evoked by the fundamental component of the sound, and another one, evoked by the combination of all the other components. The latter pitch component was separated further by Schouten (1), the pitch evoked by the large set of the highest components was called by him the 'residue pitch'. Later the word 'residue' has also been used to denote the common sensation produced by a limited set of components.

It has taken much trouble and a long time (literally, more than a century!) to find out which is the principal cue used by the auditory system for the determination of pitch. See (2) for a review. There are two candidates for this cue: the period T_0 of the waveform and the greatest common divisor of the set of component frequencies. In further studies directed at this problem special stimuli are used that contain only a few components. See Fig. 1-C for the waveform and Fig. 1-D for the spectrum of such a signal.

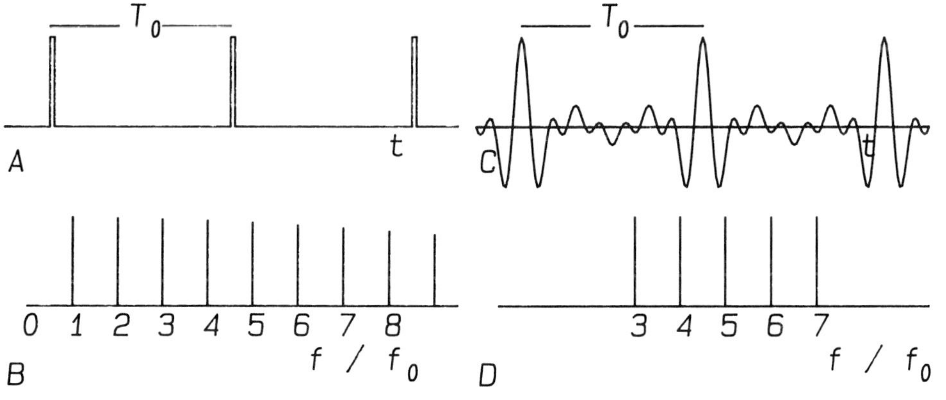

Fig. 1 Waveforms (A,C) and spectra (B,D) of periodic signals.

To understand more fully the fundamental problem it is fruitful to consider what happens to these stimuli as a result of the filtering process in the cochlea. Consider first a stimulus consisting of a few components of low order such as the one shown by Figs. 1-C and 1-D. The components are relatively wide apart from one another and each component can, in principle, be isolated in the cochlea. At the appropriate places along the BM one would observe nearly sinusoidal vibrations with frequencies that are integral multiples of f_0. The frequency f_0 would not be among them. Yet the pitch of the sound corresponds to f_0. It is clear that the pitch can only be determined by an analysis applied to the set of isolated components. The waveform of Fig 1-C would appear nowhere in the system and its period cannot serve as the principal cue.

The situation is entirely different for the case where the components are of high order. Take, for instance, a signal consisting of the 14th, 15th and 16th harmonics. These components are so close together in frequency that they cannot be isolated by the cochlea. Consequently, the components are processed together and, since together they produce a waveform with a pronounced periodicity, it seems reasonable to assume that the pitch in this case is determined from the length T_0 of the period. A characteristic feature of stimuli consisting of high-order harmonics is that their pitch is often ambiguous or multi-valued.

The question is, then, which type of cue is the most important one in the perception of multi-component periodic sounds such as occur in speech, for example. The answer came from experiments performed by Ritsma (3) and it was somewhat surprising: the residue pitch produced by a few low-order harmonics turned out to be dominant over the residue pitch produced by the (many) remaining high-order harmonics. We are therefore led to conclude that the residue pitch is far and above all determined by a process applied to the spectrum, i.e., the set of frequencies of isolated components. The dominant region comprises particularly the components of order 3 through 6.

In the field of physiology it is often stated that the period of a periodic waveform is the cue for pitch. This is true only inasfar as it applies to a signal consisting of many high-order harmonics. For a signal containing both low- and high-order components it is simply not true. The pitch of such a signal is solely determined on the basis of the spectrum, more precisely, on the set of frequencies of the lowest components. That the waveform's period corresponds to the pitch is just coincidental. This makes it, of course, extremely difficult to search for the 'physiological substrate' for pitch.

Let us come back to the pitch of a signal comprising only a few components. The notion of pitch being associated with the spectrum became much reinforced by the experiments of Houtsma and Goldstein (4) who mainly worked with stimuli containing only two components. On the basis of these results Goldstein (5) formulated a most ingenious theory in which, on just a few basic assumptions, nearly all psychophysical results in pitch perception could be explained. Since we need elements of this theory further on, its structure must be briefly outlined. The stimulus is subjected to peripheral auditory filtering. The components are more or less separated and, if this is successful, their frequencies are estimated. A principal point is that these estimates are essentially inaccurate. Hence the 'central pitch processor' is fed with a set of 'frequencies' that are more or less equal to integral multiples of a certain frequency f_0. It is the task of this central pitch processor to determine the frequency f_0 of which the harmonics agree best with the set of estimated component frequencies. Note that determination of the rank number of each component is included in this task, initially the rank numbers are assumed to be unknown. That the central pitch processor may commit errors explains the ambiguities of pitch that are inherent in stimuli with a small number of high-order harmonics.

Components that are separated by peripheral filtering take part in the process, components that are not separated are simply left out. The central point of Goldstein's theory lies in the finite accuracy with which the frequencies of the separated components can be derived. This accuracy is definitely less than that for a sinusoidal stimulus of the same frequency in isolation. Goldstein showed the difference to amount to a factor of 5 to 7.

Given this structure of the process of pitch determination we may ask how the steps are accomplished physiologically. The next section provides the necessary background for a more quantitative evaluation of this question.

3. PHYSIOLOGY OF THE AUDITORY PERIPHERY

Let us start with the mechanics of the cochlea. When the stapes (the stirrup ossicle) begins to vibrate, a wave starts to travel along the length of the basilar membrane (BM). When the vibration continues as a sinusoidal signal, a stationary pattern develops: waves continue to travel along the length of the BM in the direction from stapes to helicotrema. In their course the waves reach a peak, after that point is passed the wave's amplitude diminishes rapidly to zero. The location of the peak depends on the frequency, and it is in this way that different components are directed at different places. Notice that in the BM region of the peak the travelling wave has a different phase at different locations. These phase differences encompass more than one period.

Most measurements on the vibrations of the BM have been done with the Mössbauer method in which a tiny radioactive source is placed on the BM and the wavelength of the emitted radiation is monitored via a wavelength-sensitive radiation detector. With this method extremely small BM velocities can be detected, in the range of velocities that correspond to sounds of weak to moderate sound levels (20-60 dB SPL). To measure frequency selectivity sinusoidal stimuli are used of which intensity and frequency are varied in such a way that at the locus of observation (where the source is placed) a constant BM velocity is maintained. We will call a so-obtained curve of intensity versus frequency a 'tuning curve'. An example of a measured mechanical tuning curve, taken from data presented by Sellick et al. (6) is shown in Fig. 2. The data are obtained from an anaesthetized guinea pig. Two mechanical tuning curves are shown, one for constant velocity and one for constant amplitude. In both the resonance peak appears in the form of a deep dip. Similar mechanical tuning curves have been published by Robles et al. (7). They also measured the phase and from their data we can infer that in the region of greatest sensitivity (up to 30 dB above the deepest point of the tuning curve) the phase varies over approximately two periods (4 radians). These results give a good account of the properties of auditory filtering. Note that the data are taken at the lowest possible levels of stimulation to avoid complications from nonlinearity.

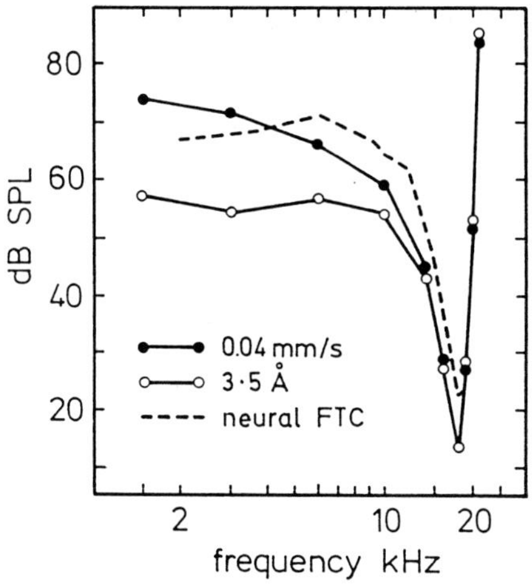

Fig. 2. Mechanical and neural tuning curves. See text.

Responses of fibres of the auditory nerve come in the form of trains of irregularly spaced action potentials ('spikes'). A neural tuning curve can be measured while monitoring the rate of firing of a fibre and keeping it at a constant level (e.g., 20 per cent above the level of the spontaneous rate of firing). One example of such a neural tuning curve is included in Fig. 2, it is measured for a nerve fibre that happens to have its frequency of greatest sensitivity (the so-called characteristic frequency) at about the same frequency as the mechanical tuning curve shown. The neural tuning curve resembles the mechanical tuning curve very much. Apparently, mechanical selectivity is preserved at the level of responses of auditory nerve fibers. No further sharpening occurs as a result of interactions in the cochlea.

When, with constant frequency, the intensity of a tone is increased, the average rate of firing of a fibre increases at first but beyond a certain intensity level that rate remains constant: the fibre shows saturation. Saturation occurs at an unexpectedly low level of stimulation, from 40 to 60 dB SPL. When a tone is presented at a level above 60 dB SPL, a fairly large range of nerve fibres is excited (notice the width of the tuning curves) and because of the saturation there is no sharp maximum in the rate of firing over the BM. Hence in terms of the rate of firing there seems to be a loss of frequency selectivity.

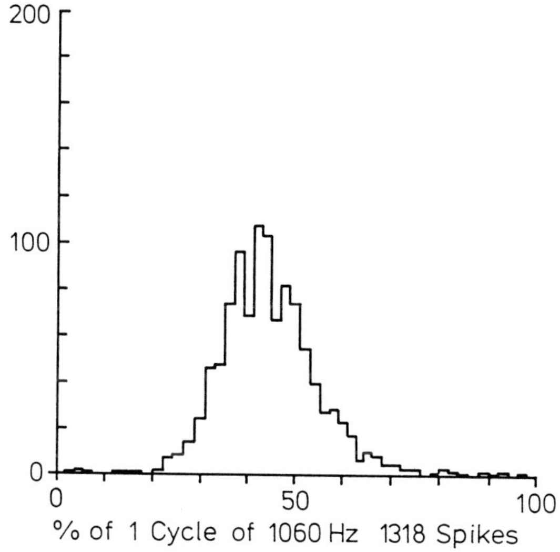

Fig. 3. Post-Zero-crossing Time Histogram (PZTH). Unpublished data.

Auditory nerve fibres have a tendency to synchronize their firings with the waveform of the stimulus. This property is detectable for frequencies of stimulation up to approx. 5 kHz and is a direct consequence of the physical and physiological processes involved in the initiation of nerve spikes. We will postulate that the separation of components of a multi-component stimulus by higher centres of the auditory pathway is partly based on temporal information provided by the nerve fibres. Note well that this postulate involving temporal analysis applies only to components that are more or less isolated by filtering. It should certainly not apply to the detection of the waveform's period.

Fig. 3 shows a time histogram of a nerve fibre's firings measured with respect to the phase of the stimulus waveform. The stimulus was a sinusoidal signal in this case. The zero point of time has been taken at the positive-going zero crossing of the electrical voltage across the earphone which delivers the stimulus to the experimental animal. The histogram is known as a PZTH (Post-Zero-crossing-Time Histogram). From the figure it is clear that the fibre does not always fire at a fixed phase of the stimulus. The instants of firing are distributed mainly over one half-period.

The situation can be understood as follows. There is an Effective Stimulus (to be called further on: ES) acting at the location where the nerve fibre under study originates. This signal is rectified and the probability that the fibre will fire is proportional to the (positive) excursions of the resulting waveform. The essential point is, now, that with increasing stimulus intensity the PZTH does not loose its sharpness. The average rate of firing may undergo saturation but the time histogram retains the property that the firings occur only in one half-cycle of the ES. If anything, the PZTH may even get sharper.

It should be noted well that a relatively smooth time histogram like the one shown by Fig. 3 can only be compiled from a large number of firings. The average rate of firing, for a moderately loud tone, is of the order of 50 - 100 spikes per second. The maximal firing rate is several hundred spikes per second. In any event, this means that, on the average, there is less than one spike per period. One fibre's firings thus do not contribute much to the information about synchrony. Only an ensemble of fibres can produce enough synchrony information so as to be of any use in the isolation of one component. This point brings us to the question of how the information content of the acoustical stimulus is handled by the auditory system and how much information would be needed for the job we have in mind. In order to provide the proper background, the next section treats detection and decision problems in a very general sense.

4. SIGNAL DETECTION THEORY

Questions about information content and sensory perception are usually reduced to questions about the smallest amount of information that can be recognized by the system or the observer. In the context of a psychophysical experiment this is translated by trying to find out whether or not a subject can discriminate between two stimuli that differ by a small amount in one physical parameter. Since the boundary between 'yes' and 'no' discrimination is not sharp but gradual (the same stimulus pair does not always lead to the same answer), detection can only be described within the framework of probability theory. This is, in fact, done by treating each detection as a statistical test between alternatives.

A psychophysical test is often carried out in the form of a two-alternative forced-choice (2-AFC) test. Let us take the case of intensity discrimination in hearing. Two sound stimuli are presented in succession, one of them being more intense. The order of presentation is random. The subject is asked to indicate which stimulus sounds as the louder one. Perfect discrimination leads to 100 per cent correct identification; if the subject cannot discriminate the signals at all, he will (by guessing) achieve an average score of 50 per cent. We could define a 'threshold' somewhere between these extremes.

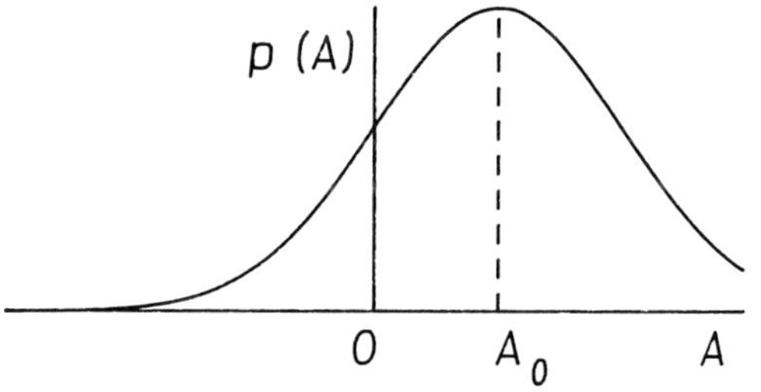

Fig. 4. Probability distribution of decision variable (A).

An abstract description of the decision process can be given with the help of Fig. 4. Consider the case where the first stimulus is the more intense one. The subject tries to estimate the loudness of the first burst and of the second one; let us assume that he is capable of subtracting these two estimates. The figure shows a hypothetical probability distribution of the difference, here designated by the variable A. Since the first burst is assumed to be more intense, the expectation value of A will be positive, A_0. Each trial is assumed to be equivalent to taking a sample of A from this distribution. If A turns out to be positive, the subject makes a correct decision; if not, he is wrong. The average probability of a correct decision is the area under the curve of Fig. 4 to the right of the $A = 0$ axis. It is easily seen that the conclusion remains true when trials in which the second stimulus is the more intense one, are included.

Clearly, it is the value A_0 relative to the standard deviation σ of the distribution that determines the probability of correct detection. The 'threshold' of discrimination can (arbitrarily) be chosen at the point where A_0 equals σ. When the distribution is assumed to be gaussian, that threshold corresponds to a correct-detection probability of 0.84.

This abstract description can also be used for the neural network that is involved in the task. Consider, then, the network dealing with the spikes that are delivered by the set of auditory nerve fibres. Take for the variable A the total number of spikes registered during one presentation of the stimulus minus the number registered during an equivalent period with no stimulus. This A is a random variable since the spike trains are irregular. Consequently, in some way the performance of the auditory system must be limited by the stochasticity of the spikes at the periphery. It has turned out that the limit due to stochasticity is quite close to the discrimination limit of a human listener, see for monaural listening Siebert (8), for binaural listening Colburn (9) and for pitch perception Srulowicz and Goldstein, (10). As a matter of fact,listeners are only a factor of 2 worse than an ideal observer.

In an extremely simplified way we can become convinced that these stochastic limitations lead to a theoretical discrimination limit of the right order of magnitude. Optimal intensity discrimination of a tone is obtained with observation intervals of approx. 200 msec. Assume that about 300 nerve fibres are involved and that each has an average rate of 50 spikes per second. In the observation interval 3000 spikes would be counted on the average and this number is subject to statistical variations with a standard deviation of $\sqrt{3000}\approx55$ spikes. Let us assume, further, that the average firing rate is proportional to the stimulus amplitude. The relative standard deviation of approx. 0.018 would then correspond to a just noticeable level difference of 0.16 dB. This result is of the correct order of magnitude; measured intensity discrimination limits range from 0.3 to over 2 dB depending upon the experimental circumstances and the subjects.

Because of the already cited saturation of the firing rate the actual situation is much more complicated. Variations of the intensity will be followed only by those fibres that are located on the fringes of the excitation pattern on the BM where fibres are not saturated. Calculation of the theoretical capability of a neural system was successfully completed for the first time by Siebert (8). The result bears the message: stochasticity of nerve firings imposes a limit on the ability of intensity discrimination which is quite close to the degree of discrimination that is psychophysically measurable.

Now we turn to the case of multi-component stimuli, and, in particular, to those that give rise to the dominant residue pitch. When the components are very weak, they give rise to well-isolated peaked excitation patterns on the BM. This is no longer true for signals of the order of 60 dB SPL (per component) or more. In the context of the present paper we hypothesize that the analysis of the stimulus components is done with the help of synchrony. Of course, synchrony should be understood in the sense it has been described above: the firings of nerve fibres form a random process, there are only few firings per period per fibre and these are distributed over half a period. The main lesson to learn from the foregoing is that we must understand synchrony in an over-all sense: information from all available nerve fibres should be processed to obtain sufficiently accurate results.

5. SYNCHRONY AND COHERENCE - MORE DETAIL

In this section we take a closer look at the coherence between nerve fibres as a result of their synchrony to a common stimulus component. In Fig. 3 we see the distribution of firings with respect to the phase of the actual stimulus signal. We now construct a hypothetical signal from the latter by shifting it in phase until the best agreement with the PZTH is obtained. The resulting signal is the Effective Stimulus (ES), and we envisage the fibre under study as a device having a firing probability which is proportional to the ES whenever the latter is positive. In this case the ES is sinusoidal; in a more general case the ES is the acoustical stimulus filtered by the cochlear filter.

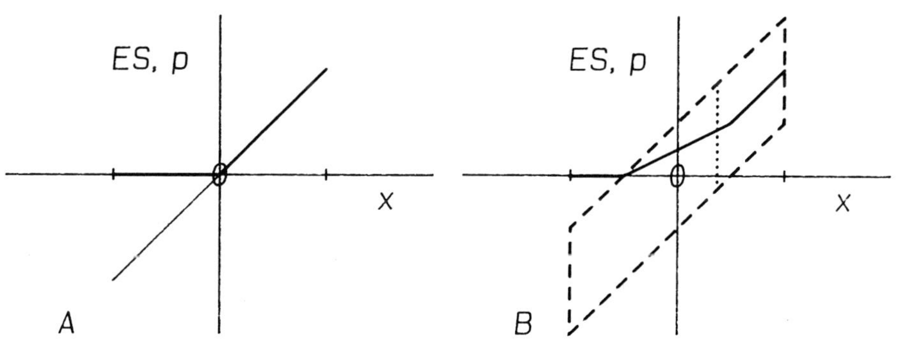

Fig. 5. Firing probability against instantaneous value x of desired component.
A: Single-component stimulus. B: Illustrating effect of a disturbing signal.

Instead of plotting firing probability against time we are going to plot it against the instantaneous value x(t) of the ES signal. For the case of a single stimulus component the result is simply the graph of a one-way rectifier since the nerve fibre can only fire when the ES is positive. See Fig. 5-A where the firing probability p is plotted against x(t), abbreviated x. The marks on the abscissa indicate the maximal excursions of x(t). The thin line shows the ES, the thick line p. The graph would not be so simple had we chosen a signal of a different phase as the reference: it would show part of an ellipse (the Lissajous figure) and it would be bi-valued over part of its course.

Let us next consider the case where the fibre is stimulated by the sum of two signals, the original one x(t) (to be called the 'desired' signal) and an extraneous one, possibly harmonically related, possibly not. And let us continue to use the instantaneous value x of the desired signal as the abscissa of the figure. The actual ES signal will be the sum of the latter signal and another one which we will designate as the 'disturbing signal'. Fig. 5-B shows by way of the dashed lines the maximal excursions of the total ES signal. The actual excursions fill the area so outlined. The firing probability will follow the total ES as long as it remains positive.

Consider now one point of the abscissa, i.e., one phase of the desired signal. It is indicated by the vertical dotted line in the figure. Over the course of many cycles, the value of the total ES for the same x will vary along this line. If all values are equally likely, the average firing probability for that value of x will correspond to the centre point of the part of the dotted line above the x-axis. If the disturbing signal is sinusoidal, the average firing probability will be slightly larger. Over the course of all possible x-values the average firing probability will approximately follow the thick continuous line in the figure.

What we actually have done here, is to consider distributions of signals, conditional upon one value of the phase of the desired component. This technique is fully analogous to the one introduced by Rajbman (11) in his development of the 'conditional regression function' and is equally helpful in obtaining insight here as it is in the theory of nonlinear networks and their identification.

From Figs. 5-A and 5-B we see the difference in synchrony brought about by the disturbing signal. Over the domain of positive x-values the average firing probability will increase with x just as in the first case but over the domain of negative x-values it will no longer be zero. Translated back into terms of the PZTH of Fig. 3: there will be firings in the half-cycle where formerly there were none. Such a conclusion could have been reached by a more qualitative reasoning (see also 12) but the present method via the conditional ES signal has the advantage of providing a good starting point for quantitative elaborations.

From the discussion so far, it is clear what happens when the disturbing signal becomes stronger. A certain degree of synchrony with the desired signal will still be detectable but, as a result of saturation, it will be smaller (synchrony suppression, cf. 12). What is more important: there will be many firings occurring in the 'wrong' half-cycle. This is, then, the main feature in synchrony that betrays the presence of a disturbing component in the ES. Note that we describe synchrony here solely in terms of a hypothetical x(t) signal; this is an artifice, of course, since the system does not have access to anything like this signal. It is fairly easy to see that the firings of one group of fibres can serve as the time reference for another group, provided these fibres are solely or mainly excited by one stimulus component. A system operating on this basis cannot only 'unite' fibres, it can also de-select fibres that are predominantly excited by other components.

6. STEPS TOWARD A SYNTHESIS

Consider a certain limited region of the BM, and the nerve fibres issuing from it. That region has a characteristic frequency (CF) and by way of filtering it acts to isolate stimulus components in a small range of frequencies around that CF. Let us assume for a moment that it is one component of a multi-component stimulus which excites the region and that its frequency is equal to the CF. Then we can think of all nerve fibres carrying information about that component as converging upon one coincidence station and exciting it. For simplicity we assume that all these fibres carry their information in the same phase, delay lines are taking care of any phase differences. We will call such a coincidence station a 'node'. In the ideal case where one component of the proper frequency is present, the firings at the node's output will all occur in one half of the period of the stimulus. When the stimulus is still consisting of one component but its frequency differs appreciably from the CF, the firings will no longer be so neatly distributed because the delay lines do no longer compensate correctly for the phase shifts. We imagine the circuit to be dimensioned so that the collective cut-off point due to mistuning occurs for fairly large frequency deviations. To make things practical, let us fix this point at relative deviations of 20 per cent.

The output of a node should be processed in such a way that it is preserved whenever the pertinent region is excited by a single component of the acoustical stimulus and blocked when the excitation involves more than one component. To achieve this goal we introduce a 'detection circuit', this takes the form of a coincidence station C as shown in Fig. 6, upper part. It is fed by two (output) signals coming from the same node, one the normal output signal ("O"), the other being shifted in time so as to be in opposite phase (" ") for a frequency equal to the CF. Normally, when one component of this frequency excites the node, the two outputs will be mutually exclusive, there will be no coincidences and C will not be activated. When, however, the output of the node includes firings occurring in the 'wrong' half-cycle, there will be an overlap between the "O" and "π" signals. In that case C will pass a signal. The lower part of Fig. 6 shows an inhibitory station I that, when activated, blocks the transfer from the node. It is tacitly assumed that the inhibitory influence lasts for more than one stimulus period. The threshold of inhibition cannot be independent of the threshold for detuning, of course. As a practical compromise, we take both thresholds as being approximately 20 per cent.

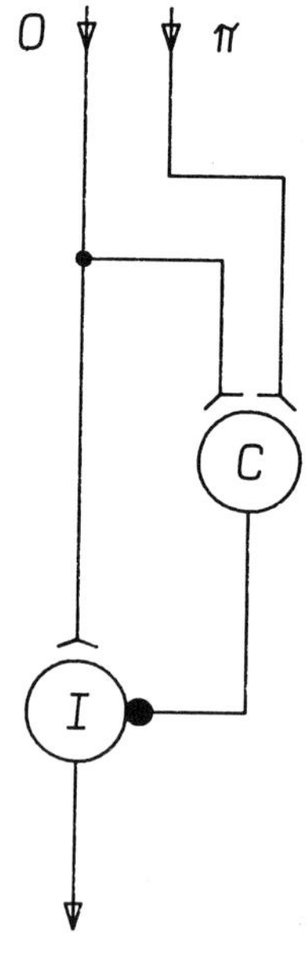

Fig. 6. Elements of neural selection circuit. See text.

In some respects the simple circuit of Fig. 6 shows the desired behaviour. For instance, a single stimulus component with the CF as its frequency is passed. When a component of that frequency is (severely) contaminated by another one, the output will be blocked. Such a situation is unlikely to occur, however, since the location in question is highly resonant for frequencies near the CF.

We will next discuss simple modifications of the circuit. In the first place it is necessary to provide the circuit with a phase reference taken from a different location. Consider, then, the same circuit with the "π" input taken from another node, for instance, one situated farther apicalward along the length of the BM. In this case the delay line making up the "π" signal can be shorter or absent. For the case where only one component makes up the ES's at both locations, the situation is not much changed. The reason is that the phase shifts at both locations are approximately linear functions of frequency for small frequency deviations and thus have a nearly constant group delay difference. The circuit works well when the "π" input is excited by one and the "O" input by two components. However, when the "π" input is excited by two and the "O" input by one component the "O" channel is blocked where it should remain open. This consideration leads to the next variation: the "π" input should only be active if it is carrying information about one component. That is, the "π" input should not be taken directly from a node but via that node's own inhibitory station.

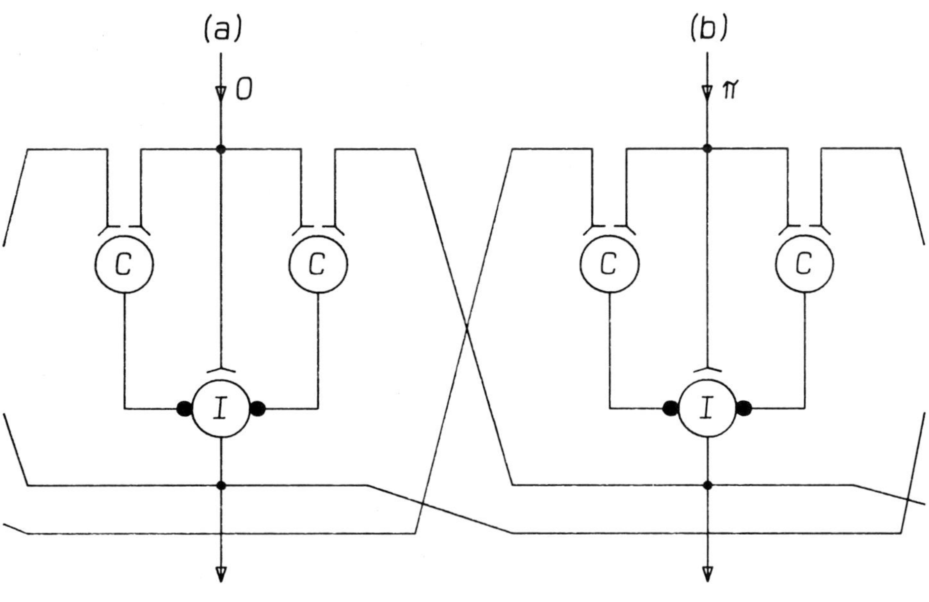

Fig. 7. More complete selection circuit.

The result is the structure drawn in Fig. 7. Both nodes (a and b) are controlled by an inhibitory station (I) that is fed by coincidence stations (C). One input of each C is taken directly from the same node, the other one from the controlled output of another node. Only one of these connections is shown in the figure, from the controlled output of node (b) to node (a). For simplicity node (b) is chosen such that its phase for the CF of node (a) is "π". We may note that the connections cannot be symmetrical because of the dispersion occurring in the cochlea. In this way the nodes are effectively decoupled so that there is no danger of unstability of the entire circuit.

The two nodes of Fig. 7 cooperate to process information about components with frequencies in the range between the CF's of the nodes. In a way the node pair 'demarcates' a region of the BM and shields it against 'invasions' from one (in this case the left) side. We now realize why node (b) should have a counterpart C controlled by a node farther to the left: this is to shield against 'invasions' from the other (the right) side. The argument applies, of course, to all nodes. In Fig. 7 the two C's operate on one I; it is well conceivable that there are two of such I's connected to each node. Anyway, the circuit shows one method to group together the output signals that are stimulated coherently (i.e., by one component). Many variations are possible, too many to discuss them here.

Let us now try to review the over-all operation of the circuit. A multi-component stimulus is presented and we consider the case where the components are more or less separated by peripheral filtering. We assume that there are no real 'gaps' of neural activity between the excitation patterns of the components. One node where the CF corresponds to the frequency of one of the components has a nearly sinusoidal ES and consequently the firings will be neatly confined to one-half period. The locations "$-\pi$" and "π" from which input comes to that node's coincidence circuits, are within the region where this particular component dominates (remember that the total BM stretch excited by a strong component encompasses a phase range of plus and minus 2π radians). As a consequence, that node's output is passed by its inhibitory station and the information radiates to both sides to inputs of other coincidence circuits. If such a site is predominantly excited by the same component, that signal is passed by the appropriate inhibitory station I. In other cases the output will be blocked. In this way gaps of neural activity are created at both sides of the region excited by one component. Simultaneously, the information extracted from one component is channeled into a group of neural pathways thus facilitating the determination of that component's frequency.

7. DISCUSSION - OTHER ASPECTS - CONCLUSIONS

In the preceding section a possible neural circuit is developed that can serve to isolate components of a multi-component acoustical stimulus. At this point it is good to summarize the main reasons why we think a circuit with these properties should be present:

1. The pitch of a harmonic sound is dominated by the residue pitch evoked by a few of the lowest-order harmonics. These components are almost completely resolved by the peripheral auditory filter, hence the process of pitch extraction should be a process applied to the spectrum. The best theory explaining that process, that of Goldstein (5), describes pitch extraction as a two-stage process, 'analysis' followed by 'synthesis'. The first step involves the determination (not error-free) of the component's frequencies and the second the determination of the best-fitting greatest common divisor of these frequencies.

2. Although components may well be resolved by peripheral filtering, their representation in the form of nerve-fibre firing rates is far from ideal. In particular, the rate pattern is severely distorted for higher sound levels. In terms of firing instants the response of a nerve fibre - or of a group of fibres - is more true-to-type: it reflects the fibre's ES quite well. Apart from synchrony suppression there is little distortion. The proposed circuit can serve to characterize a particular component by way of its timing pattern and it has the innate property that its operation is nearly independent of stimulus level.

The circuit proposed can serve to isolate components but we cannot specify, as yet, how the frequencies of the components are determined. Whatever the mechanism, the accuracy of frequency determination is affected by the inherent stochasticity of nerve fibre firings. It is evident that this accuracy will be less for an isolated component of a multi-component stimulus than for a single pure tone: there are simply fewer nerve fibres involved.

About the action of the central pitch processor we can say even less. It should work according to a pattern detection mechanism that compares the set of extracted frequencies with a template of harmonically related frequencies and seeks to determine a fundamental frequency f giving the best fit. It might, for a moment, be thought that this mechanism would work on partial coincidences of the temporal information passed on by our nodes. This is unlikely, though, because in that case the contribution from the lower-order components would by no means be as accurase as that extracted from the set of high-order components. This is in conflict with the principle of dominance.

A few words are in order about the possible location of the circuit proposed in the preceding section. Two nuclei of the auditory pathway emerge as candidates: the cochlear nucleus (CN) and the medial superior olive (MSO). Since the circuit needs accurate time information the CN would be more likely at first sight: there is less synaptic delay and thus less extra variability. At the MSO the first binaural processing of signals takes place and we know from psychophysical experiments that lateralization thresholds correspond to interaural time differences of the order of 10 to 30 μsec. Hence the information present at the MSO is extremely accurate with respect to time and this makes the MSO an equally likely candidate. Moreover, binaural processing necessitates the presence of time delay circuitry and it seems likely that this is also present in the MSO.

At the conclusion of our discussion we pose the question: would it be possible to verify experimentally elements of the mechanisms proposed? The provisional answer is: yes. The output of a circuit like the one shown in Fig. 6 is typically conditional upon the presence or absence of competing components. For once, this property could be tested by using well-chosen two-component stimuli. On a more advanced level, the coincidences necessary to operate the inhibitory station are conditional upon certain instantaneous values of the desired component (see Fig. 5-A and 5-B). This feature is typical of the conditional regression function (Rajbman, 11) and it has been shown that specific analytic methods can be based on this concept (13, 14, 15). There is no reason why a similar development would not be possible for the case at hand.

REFERENCES

(1) J.F. Schouten, Proc. Kon. Acad. Wetensch. (Neth.) 43 (1940) 356-365 and 991-999.

(2) E. de Boer, In Handbook of Sensory Physiology, Vol. V, Part 3, Springer, Berlin, 1976, pp. 480-583.

(3) R.J. Ritsma, J. Acoust. Soc. Am. 42 (1967) 191-198.

(4) A.J.M. Houtsma, J.L. Goldstein, J. Acoust. Soc. Am. 51 (1972) 520-529.

(5) J.L. Goldstein, J. Acoust. Soc. Am. 54 (1973) 1496-1516.

(6) P.M. Sellick, R. Patuzzi, B.M. Johnstone, J. Acoust. Soc. Am. 72 (1982) 131-141.

(7) L. Robles, M.A. Ruggero, N.C. Rich, In Peripheral Auditory Mechanisms, Springer, 1986, pp. 121-128.

(8) W.M. Siebert, In Recognizing Patterns, MIT Press, Cambridge (USA), 1968, pp. 104-132.

(9) H.S. Colburn, J. Acoust. Soc. Am. 54 (1973) 1458-1470 and 61 (1977) 525-533.

(10) P. Srulowicz, J.L. Goldstein, J. Acoust. Soc. Am. 73 (1983) 1266-1276.

(11) N.S. Rajbman, Chto takoe identificatsija? (What is identification?) NAUKA, Moskow, 1970, pp. 25 ff.

(12) D.D. Greenwood, In Auditory Frequency Selectivity, Proc. 1986 Cambridge workshop (to be published).

(13) H.C. Lammers, E. de Boer, Proc. IEEE 67 (1979) 432-434.

(14) E. de Boer, Proc. IEEE 67 (1979) 317-318.

(15) J.G. Cornelis, Proc. IEEE 68 (1980) 931.

Pattern Recognition of Spectral Information in Hearing

Discussion
edited by *Peter I. M. Johannesma*

Several of the attendants were not well acquainted with the investigations presented by de Boer. As a consequence a number of questions were directed toward a better understanding of pitch perception. The central issue of the discussion turned out to be the neural coding of sound. Two type of questions arose:

1. What aspects of sound are coded into neural activity?
2. Which aspects of the neural activity carry the information concerning the acoustic environment?

Concerning the first question it was observed that encoding of the full spectral information, not only amplitude but also phase, would imply a complete neural encoding of sound. In this situation the temporal form of the acoustic stimulus is also present in the neural activity. As a consequence, there would be no fundamental difference between neural encoding of spectral or of temporal features of sound. Only when the coding is incomplete can the dichotomy of spectral versus temporal arise. Since it is evident that the activity of the auditory nerve is related to spectral as well as temporal features and that moreover the neural representation is incomplete it could well be that loss of information has to be defined in an integrated spectro-temporal description of sound.

The second question was concentrated upon the stimulus induced activity of the auditory nerve fibres. What features of the activity are informationally significant? Firing rate or firing pattern? Is it the number of events during an interval of time (10 ms?) or the precise timing of the action potentials? Experimental data on correlation function of acoustic stimulus and neural events indicate a rather precise time relation. Central use of this information cannot be derived from measurements in the peripheral auditory system. Psychophysical observations are difficult to interprete in neural terms. Recording of activity in many fibres synchronously could form a bridge between neural models and psychophysical observations.

A tentative conclusion of the discussion is the necessity to formulate models for the auditory system and to carry out psychophysical and neurophysiological investigations which concentrate on the coherent aspects of both the acoustic stimulus and the neural activity.

Information Flow and State in Cortical Neural Networks: Interpreting Multi-Neuron Experiments

by *George L. Gerstein*

ABSTRACT. We review the experimental and analytic methodology available for simultaneous and separable extracellular recordings of multiple neuronal spike trains, particularly in the context of the experimental structure that is imposed by various non-stationarities. Inferences drawn from such work in a number of laboratories working on diverse preparations are remarkably consistent: the cooperative phenomena in a neural assembly are dynamic on several different time scales. We examine the possible interpretations and significance of these observations.

1. RECORDING METHODOLOGY.

The objective is to make parallel extracellular recordings of the individual firing activity of a number of neurons, together with the occurrence of events like stimulus, response, reward, etc. Although early work of this sort has been done with anesthetized preparations, it is obviously desirable do such experiments with unanesthetized, behaving animals. Only then, and with appropriate choice of target tissue, can we be reasonably certain that we are studying a neuronal assembly that is actively engaged in processing some fraction of the information involved in the observed behavioral task. Finally, it is important to know the absolute, or at least the relative positions of the neurons under observation.

Various electrode and associated technologies have been developed, ranging from simple bundles of thin wires to complex integrated circuit structures. A fully satisfactory device has yet to appear. Reviews of electrode and associated problems are in Gerstein 1983 and Krüger 1983.

If a given electrode is in the vicinity of several (two to four) active neurons, the various geometric factors will likely cause the activity of each individual neuron to have a different, unique waveform in the recording. It is obviously efficient to sort out such a compound record, both because of the increase in observed number of neurons and because here we are studying spatially contiguous neurons. Review and discussion of the technical means to such sorting of recorded waveforms is given by Gerstein 1983 and Schmidt 1984a and 1984b. One of the yet unsatisfactory aspects of such waveform separation is the inability to deal rapidly with the compound waveform that occurs if two of the neurons fire in close (overlapping) temporal sequence. This situation is potentially among the most interesting, since it may indicate close functional connectivity between these observed neurons.

2.TEMPORAL DESIGN OF EXPERIMENTS.

Three salient properties of the nervous system are a) the apparently stochastic nature of its activity, b) the prevalence of spontaneous activity, and c) the non-stationarity. Attempts to deal with these properties during any measurement influence both the experimental design and the subsequent analytic process. In particular the stochastic and spontaneous properties make it extremely difficult to make measurements that are "instantaneous" in the sense that they define some property of the nervous system at a particular moment in running time. Instead we are forced to average our measurements over longer periods of observation. Unfortunately, the longer the averaging period, the more we gloss over the possible effects of the non-stationarities.

For example, the stochastic and spontaneous activity aspects are best accomodated by an experimental design that repeats a given stimulus many times, allowing an averaging process over these repetitions, as in the familiar PST histogram. Non-stationarities of a short time scale that are on the average locked in time to (modulated by) the stimulus are revealed by the analysis, while the non-stationarities on a longer time scale are averaged out and their effect eliminated.

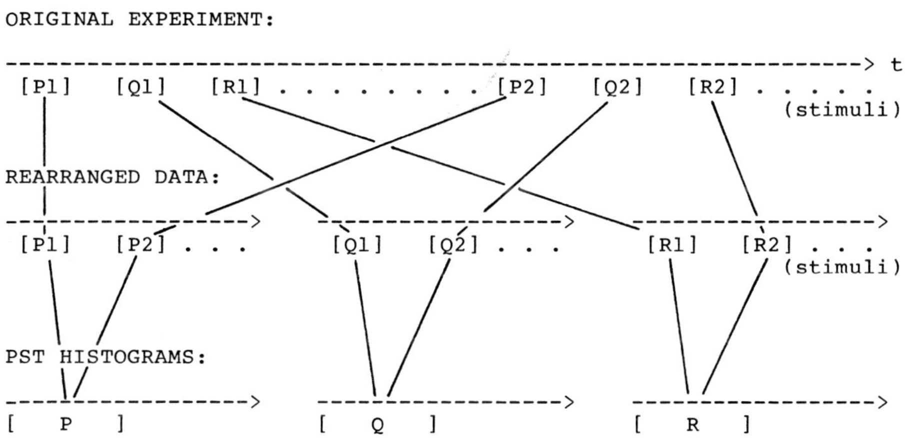

Figure 1. Temporal structure of experiments and analyses.

If our assignment is to compare measurements related to two different values of stimulus (P and Q for example), the appropriate experimental design is ALTERNATE (or pseudo-random order) presentations of the two stimuli. Presenting a long sequence of stimulus P followed by a long sequence of stimulus Q clearly results in measurements that are influenced by non-stationarities of a medium time scale; the alternating stimulus presentation averages out such effects. Conceptually, for purposes of analysis, we may consider "cutting" the data into P-pieces and Q-pieces; each type of piece is then "glued" together into a long strip, and treated by normal methods. (Note that there are time discontinuities at the "glued" boundaries; some questions and analyses must take these into account.)

In the above we have considered non-stationarities of a short time scale, particularly if time locked to the stimulus, and of a medium time scale that is involved in comparing for example two stimulus conditions. Clearly there are also non-stationarities of a much longer time scale involved in processes like habituation or learning; these would perhaps affect measurements made hours apart or on successive days. These are beyond the scope of our present concerns.

We summarize in Figure 1 the type of experimental design and rearrangement of the data pieces suggested here.

3. ANALYTIC METHODOLOGY.

The problem is to detect and quantify significant temporal relationships among the activities of the several recorded neurons, possibly also with regard to stimulus or behavioral events. Most of the available calculations to date are in some way related to the concept of cross-correlation between point processes. An ample review of many of the appropriate calculations, together with excellent references to the original literature are found in Glaser and Ruchkin 1976. Some subsequent developments for three neuron calculations are in Abeles 1982.

Figure 2. arranges the various types of calculation according to a) the short, medium, or long time scale of the phenomena that are emphasized, and b) the multiplicity of neurons (spike trains) involved. Several of the calculations are newer than the review (Glaser and Ruchkin 1976); these will be presented in somewhat more detail below together with physiological results that bear on properties of neuronal assemblies.

	Within P	Compare P to Q	Long Term
Single neuron	Peri-Stimulus-Time histogram	Inter-Spike-Interval, Joint ISI, Serial and Auto-Correlation	Segmentation of data, followed by all in left columns
Pairs	Joint PST Scatter diagram	Cross-Correlation	!
Triples	3D version of Joint PST Scatter diagram	"snowflake"	!
Many	??	Gravity Condensation, Pattern Repetition	V

Figure 2. Tabulation of analytic methods.

3.1 Revisiting the Joint PST Scatter Diagram.

This is an ancient measurement for two spike trains (Gerstein and Perkel 1969, 1972) that allows the detailed, stimulus locked time structure of the cooperativity between two neurons to be assessed. The scatter diagram is the joint firing probability of the two neurons relative to the stimulus mark event, and averaged over many presentations of the stimulus. In such a diagram, increased dot densities that appear in a band parallel to either axis represent stimulus locked individual activity of the two neurons. A diagonal band of increased dot density reflects the cooperative firing of the two neurons.

In general the intensity of such a diagonal band (i.e. of the measured cooperativity) will vary along its length, indicating that the stimulus is modulating the intensity of the cooperative firing. However, note that this cooperative firing is indicative of TWO underlying processes: 1) the intrinsic "effective connectivity" between the two neurons and 2) their (stimulus modulated) firing rates. If we want to get at the intrinsic "effective connectivity" alone in order to see whether it is a constant or a stimulus modulated quantity, we must make a correction for the direct effects of the modulated firing rates.

In the old work on the Joint PST scatter diagram, we tried to accomplish this correction by comparing the scatter diagram of the actual and "shifted" data, i.e. by using a standard "shift" predictor (see Gerstein and Perkel 1972). This required subtracting relatively sparse matrices, an unsatisfactory exercise at best. In a recent reexamination of the problem we have realized that a much better comparison function from the statistical viewpoint is the product of the two PST histograms, and that a final correction should normalize (divide) the difference at each X,Y value by the product of the standard deviations at the corresponding points in the two PST histograms. This procedure results in a "cross-correlation" surface (Habib and Sen 1985) whose values all lie in the range +/- 1.0. A more mathematical description of the several possible normalization and correction procedures for Joint PST scatter diagrams is in preparation; the word-description above should suffice for the present purpose.

In Figures 3 and 4 we show application of the Joint PST scatter diagram and its normalization procedure to data taken from two pairs of neurons that were recorded in cochlear nucleus of an anesthetized rat. There are two panels within each Figure, the left corresponding to the "raw" JPST scatter diagram, while the right panel corresponds to the normalized result. Within each panel, the scatter diagram is shown on the left. The corresponding individual PST histograms appear below and at left of the scatter diagram. In the right portion of each panel are the two integrations of the scatter diagram. The histogram along the major diagonal represents the (smoothed) counts in the specified diagonal region of the scatter diagram. The histogram at the extreme top-right of each panel represents sums along each diagonal file of bins in the scatter diagram (each sum adjusted for the number of contributing bins). This is essentially the correlogram of the two spike trains (Gerstein and Perkel 1972) in the "raw" count panel, but is corrected for time varying firing rates in the normalized panel.

We concentrate our attention on the smoothed diagonal histogram of the normalized (right) panel in each Figure. These histograms represent the stimulus related variations (if any) of the intrinsic "effective connectivity" between the two observed neurons. Full correction has been made for the effects of the modulated firing rates. We see that the diagonal histogram in Figure 3 has elevations corresponding to the ON and OFF responses to the tone burst, but that these are fairly similar. In contrast, in Figure 4, the elevation during the ON response is much greater than during the OFF response. The contrast between these Figures shows that the "effective connectivity" within each pair of neurons is differently modulated by the stimulus, weakly in one case, strongly in the other. Note that the modulation time scale is in the 10's of milliseconds.

In this section, we have presented evidence that the "effective connectivity" between two neurons may in some cases be a rapidly varying function of time. We defer discussion of the interpretation and significance of this observation till later.

3.2 Gravity calculations.

The various correlation measurements that are available for the analysis of data from simultaneously recorded neurons have been applied to a variety of neuronal preparations with satisfactory results. It is possible to make inference about the type of "effective connection" between the members of each pair of observed neurons; the description is in terms of a) direct neuronal interaction (perhaps synaptic), b) shared input from unobserved sources, and c) stimulus related coordination. Analogous methodologies and possibilities of inference exist for three neurons considered at a time. (Again, see the review by Glaser and Ruchkin 1976, and Perkel, Gerstein, Moore 1967b.)

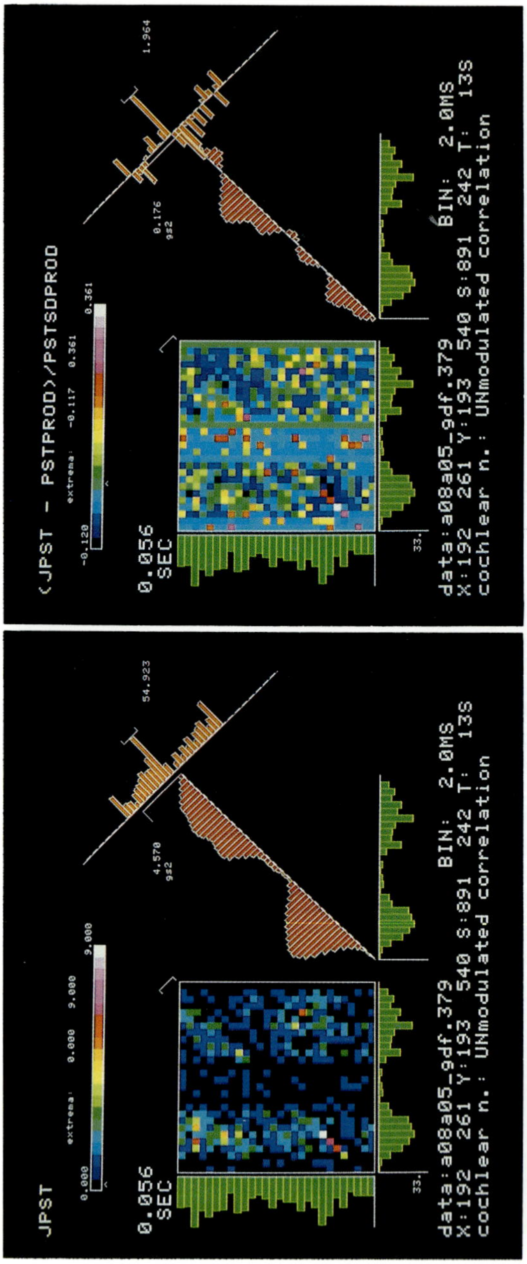

FIGURE 3.

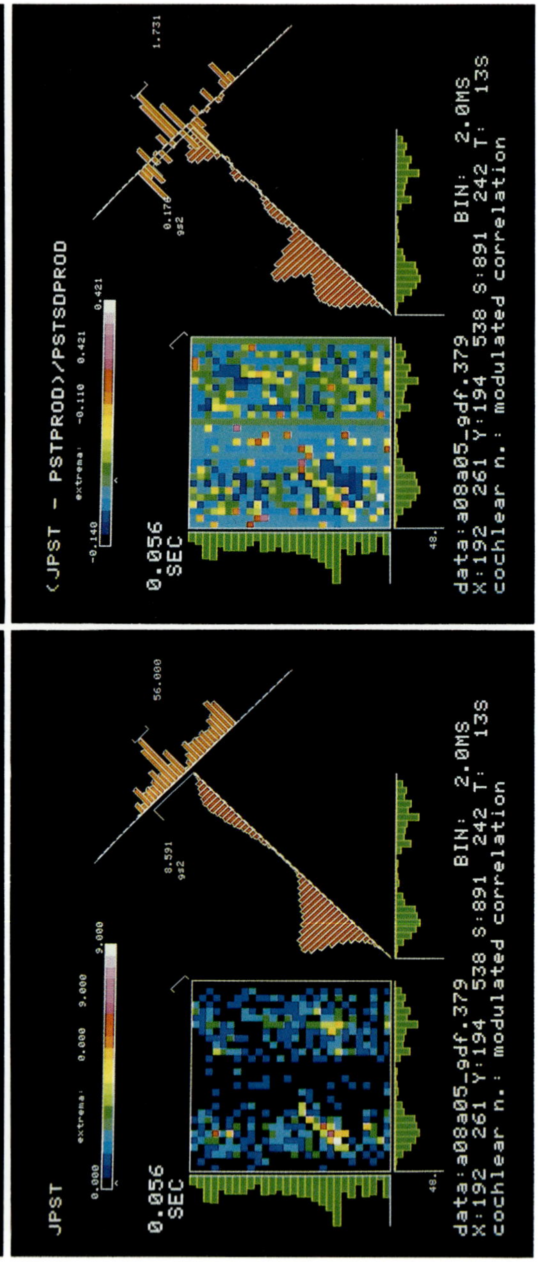

FIGURE 4.

FIGURE 3. (Facing page TOP) Weak modulation of "effective connectivity" by the stimulus. Left: "raw" Joint PST scatter diagram and integrals; Right: "normalized" (correlation surface) and integrals. See text for description of integrals. Time and count scales are indicated. These two neurons were recorded from cochlear nucleus of an anesthetized rat. (Data from Gochin et al. 1986.)

FIGURE 4. (Facing page BOTTOM) Strong modulation of "effective connectivity" by the stimulus. Same conventions as Fig. 3. The same neuron is represented along the abscissa in Figs. 3 and 4; a different (simultaneously recorded) neuron is along ordinate in each Figure. (Data from Gochin et al. 1986.)

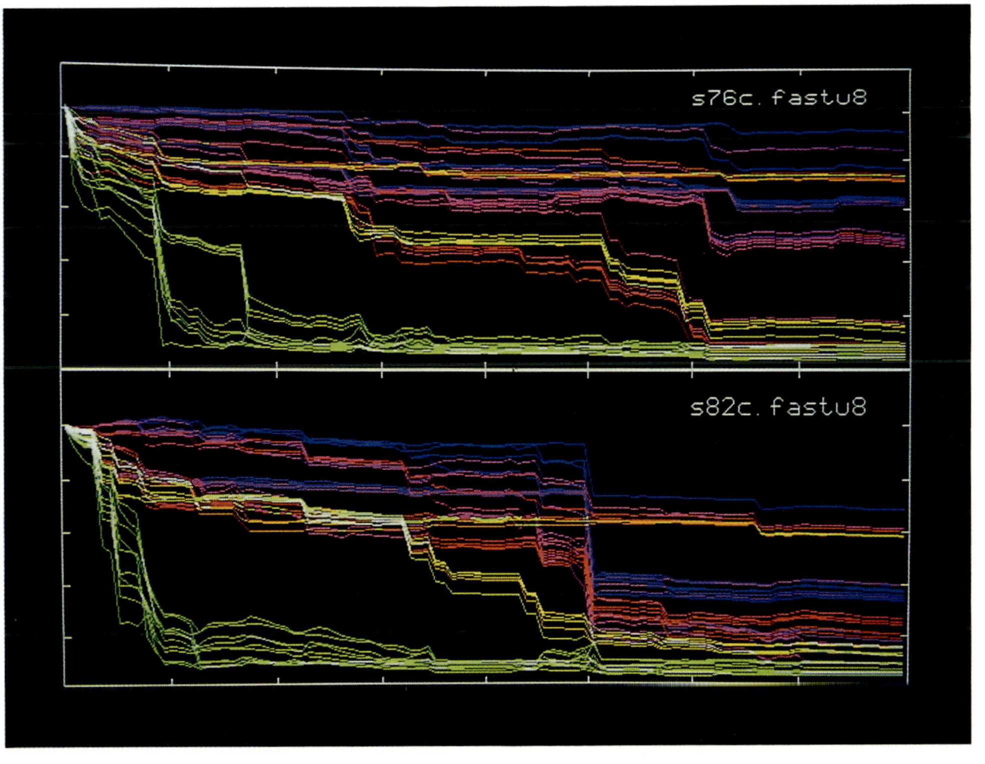

FIGURE 5. Pair distances during gravity condensations for data from 10 neurons in auditory cortex of a cat during two different stimulus conditions. (Top: tone C preceded by tones A,B; Bottom: tone C preceded by tones B,A; in both cases analysis is made ONLY during the presentation of tone C.) The color scale represents the 45 particle pairs on an arbitrary linear scale from deep blue through reds and to light green. The distribution and identity of pair distances is different in the two stimulus conditions. Abscissa: time, 8 sec total span; Ordinate: distance. Initial inter-particle distances are all set to 100 units. (Data from Espinosa and Gerstein 1984, 1985, 1986)

When we try to study properties of neuronal assemblies with these calculations, there is a purely combinatorial difficulty. Dealing with many simultaneously recorded spike trains as pairs or triplets produces an extreme proliferation of calculation time, and far worse, of inspection time by the experimenter. A few minutes of recording from 10 neurons studied with 10 stimulus conditions results in thousands of correlograms and histograms. Clearly, it is desirable to invent calculations that analyze the entire observed group of neuronal activities as a single entity, rather than as a multiplicity of pairs. We have developed the "Gravity" calculation for this purpose during the past several years (Gerstein et. al. 1985a, Gerstein and Aertsen 1985, Aertsen et. al. 1986, Aertsen et al. 1987), and are now increasingly applying it to ongoing multi-neuron experimentation.

The basic idea transforms a problem in recognition of temporal structure among the several recorded spike trains into a problem in recognizing time varying spatial structure. Each of the N original neurons is represented by a particle in an N-space; initially all particles are equidistant, i.e. at the corners of a hyper-cube. Each particle is given a "charge" (analogous to gravitational mass or electric charge) which is a low pass version of the corresponding spike train: the "charge" is incremented by a fixed amount at the moment when a spike occurs, and subsequently decays with an appropriate time constant. It is usually appropriate to subtract the mean "charge" , thus obtaining an working "charge" that has zero mean. Now we set up dynamic rules so that each two particles interact through a force (vector) that is directed along the line which joins them and whose magnitude is proportional to their instantaneous "charge" product. It has turned out useful to have the force independent of inter-particle distance, unlike in similar physical systems. Finally we stipulate a high viscosity, so that velocity, rather than acceleration is proportional to the force.

As amply demonstrated in the above references, the gravity transformation produces a mechanical system whose behavior is extremely sensitive to cooperative aspects of the neuronal firing. In particular, particles representing those neurons which have correlated spike trains will tend to approach each other. If we are observing several sets of separately interacting neurons, the corresponding gravity "model" will demonstrate condensation into several clusters of particles in the N-space. Thus the condensation process produces a spatial representation of the "logical" Venn diagram that describes the neuronal interactions. We have described various modifications of the basic gravity transformation that extend the ability to distinguish pre- and post-synaptic aspects of cooperative firing. Other modifications allow removal of the direct effects of stimulus modulated firing rates, so that we may extract a dynamic description of the intrinsic "effective connectivity" among the observed neurons. The significant improvement inherent in this class of calculation is that the entire observed neuronal population is analyzed simultaneously rather than as a collection of neuron pairs, and that functional neuronal assemblies are automatically identified by the clustering process.

The output of the gravity transformation can be examined in two convenient forms: a) interparticle (pair) distances versus time, and b) a geometric (or other appropriate type of) projection of particle position from the N-space to a two-space as a function of time. The latter output is best observed as a movie film, where each frame represents a successive instant of time. However, note that it is generally inappropriate to make inferences about clustering of the particles in the N-space from such projections; only in certain situations can the projection plane (or method) be so chosen that the actual condensation in N-space is well represented. In general, one should use direct cluster detection methods in the N-space.

For the purposes of the present paper, we demonstrate the application of the gravity transformation to multi-neuron data recorded from auditory cortex of a lightly anesthetized cat. A repeating series of various auditory stimuli were presented at appropriate intervals; we will be interested in contrasting the configuration of the observed neuronal population during two of these stimuli. This means, in terms of the description above of experimental time structures, that we have cut the data into many short segments corresponding to each presentation of each stimulus type and have then "glued" together all segments of one type. The time discontinuities at the joints are essentially irrelevant for the calculations of the gravity transformation.

In Figure 5 we demonstrate that the collapse of pair distances for these data varies in time structure as stimulus is changed. Note that we have compensated for the direct stimulus effects, so that these different condensations are reflections of different inherent "effective connectivity" and DO NOT simply represent the different apparent cooperativity caused by different stimulus responses.

In Figure 6 we present the same data in the form of selected frames from movies of the condensation process in an appropriate projection. The differences in arrangement at these selected moments is obvious, and again reflects differences in the inherent "effective connectivity" which could be extracted from the approach velocities of the various particles. We caution the reader again that the actual clustering process of particles in the N-space is not necessarily properly represented by such a projection. The figure does, however demonstrate that the condensation is different for the two different stimulus situations.

The astute reader may note that no distances in Figure 5 increased beyond the starting values; this does not happen for data taken during spontaneous conditions. This is probably an indication that we are not

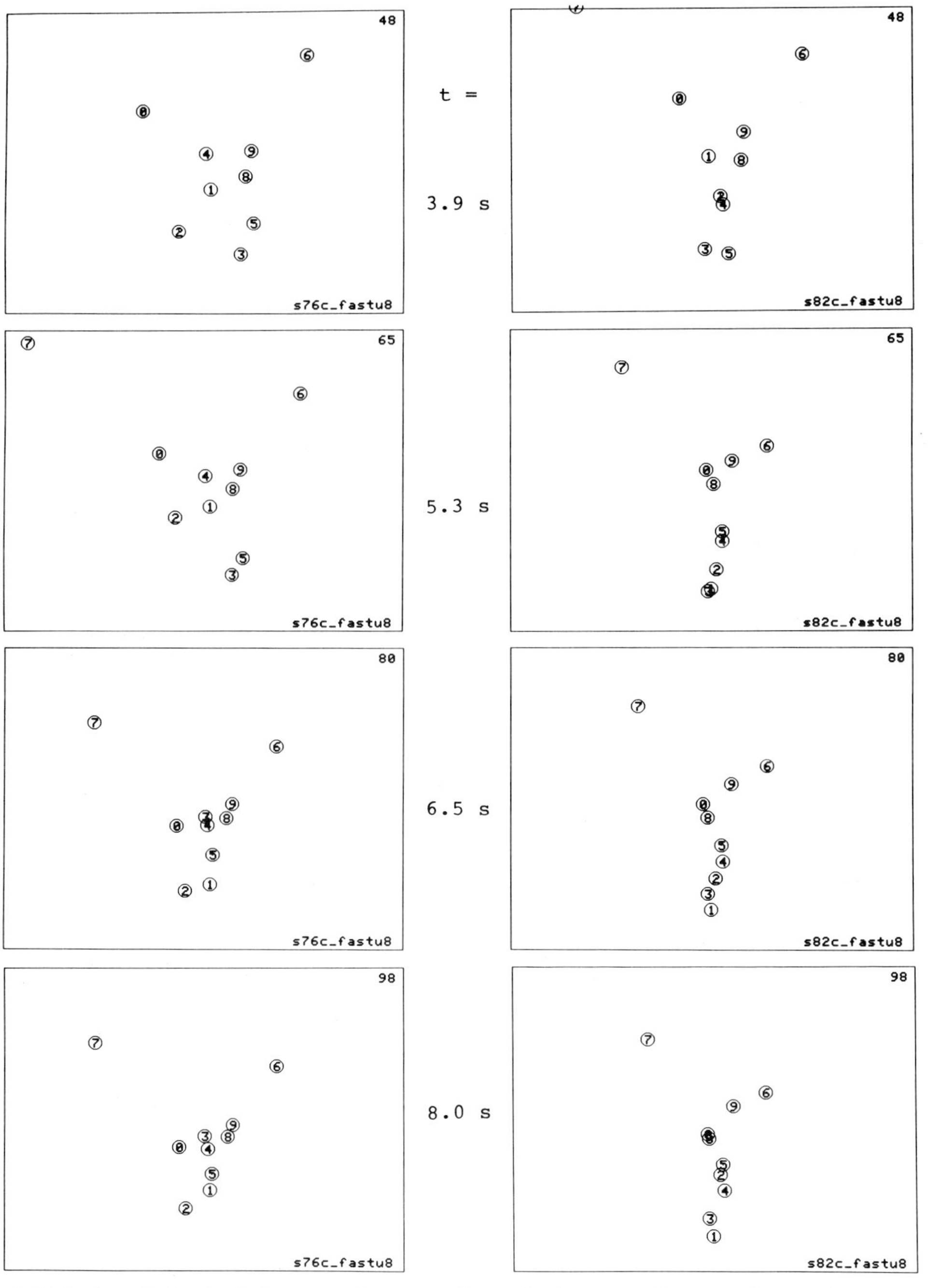

FIGURE 6. Selected frames from movies of the gravity condensations for the same data used in Fig. 5. (Left column: tone C preceded by tones A, B; Right column: tone C preceded by tones B, A; in both cases the analysis is made ONLY during the presentation of tone C.) This projection shows considerable differences of particle arrangement in the two stimulus conditions. See text.

making quite good enough a correction for the stimulus induced firing. In analogy to the work described above about the Joint PST scatter diagram, it is necessary to also incorporate the variances of the individual PST histograms.

We have demonstrated in this section that the structure of firing cooperativity among a number of simultaneously observed neurons depends, on the average, on the current stimulus conditions. In fact, since the particular data used here were gathered during presentation of stimulus sequences, the firing cooperativity may depend in addition on the immediately preceding stimulus conditions. (We were contrasting average cooperativity during presentation of stimulus C preceded either by stimuli A,B or by stimuli B,A.) Such differences in firing cooperativity may be interpreted as evidence for dynamic changes of organization in neuronal assemblies on a medium time scale. Note that these changes are on a much longer time scale than the dynamics demonstrated above by means of the Joint PST scatter diagram. Although the results demonstrated here were calculated with the gravity transformation, similar results have been obtained (more painfully) with cross-correlation of pairs of spike trains.

3.3 Distinctive Repeating patterns.

In all the above we have been considering measures of assembly organization that are *averaged* over time or over a repeating stimulus situation. This situation is forced on us because we are observing too small a portion of the nervous system to allow the ensemble averages that presumably are actually used in normal information processing by the brain. Could we get away from the averages and somehow identify distinctive "events" among our recorded data that would correspond to distinctive "states" of the observed portion of nervous system?

One possible approach to this problem is to seek detailed firing patterns among the observed neurons that repeat more than would be expected by chance. We can make a musical analogy here. If we write out the firing activity of the observed neurons in ordinary musical notation, where each neuron is assigned a different note (pitch), we are looking for repeating musical phrases (chords, arpeggios etc.). There is no preconception of how such a phrase should look, except that we may set an arbitrary upper limit to its duration, perhaps 0.5 sec. If we find such repeating phrases, and can identify those which are occuring significantly above chance levels, we may have an indicator of particular repeating (organizational) states of the observed neuronal population. The underlying assumption here is that a repeating phrase represents a repeating sequence of activation among the observed neurons; this would occur upon repetitions of a particular informational process.

We had earlier investigated repeating events in the firings of a single neuron (Dayhoff and Gerstein 1983a, 1983b), and discovered that repetitions exceeding the .01 confidence level occur in about half of the data we examined. Since these data came both from crayfish claw neurons and cat visual cortex, we concluded that such pattern repetition was a relatively general property of an active nervous system. In recent work (Abeles and Gerstein 1987) we have extended the investigation to parallel multi-neuron recording.

Additional theoretical and experimental work is needed on this project. Nevertheless, even the early results show that multi-neuron pattern repetitions above chance level DO occur often in cortical recordings, even though at this stage we do not understand the circumstances and (internal or external) conditions that are associated with them. We note that several groups are now actively pursuing the pattern problem, (Perkel 1985, Frostig 1985, Landolt 1985, Johannesma 1987) so that we may expect some interesting progress in dealing with such distinctive brain events.

4. SUMMARY OF RESULTS FROM MULTI-NEURON OBSERVATIONS.

We present these results in the form of a numbered list in order to facilitate the subsequent discussion of the possible interpretations.

A. Results from MIDDLE time scale "averages".

1. Many neuron pairs which show "raw" correlation of firing also show residual correlation of firing after making all corrections for firing rate modulations by the stimulus.

2. The "raw" and corrected correlation structure among the firings of a group of observed neurons each differ with presentation of different stimuli.

   Here we compare the cooperativity during (repeated) presentation of stimulus P as contrasted with the (repeated) presentation of stimulus Q; "averages" are over total P time and over total Q time. Corrections, when made, compensate for local firing rate modulations by each stimulus.

3. The "raw" and corrected correlation structure among the firings of a group of observed neurons may each differ even for presentation of the SAME stimulus if the immediately preceding stimulus conditions are different.

   Note that the measurement of cooperativity is made ONLY during the time windows where that SAME stimulus is being presented, NOT during the times at which the differing preceding stimuli are presented. Here stimuli P and Q are the same, but have different preceding sequences of other stimuli; averages are over total P time and over total Q time.

4. In comparing cooperativity across a whole family of stimulus parameters, we may find that "tuning curves" determined from cross-correlation measurements of a spike train PAIR may be much narrower than "tuning curves" determined from PST histograms of the individual spike trains.

This observation may be made with BOTH "raw" and stimulus-corrected cross-correlations.

B. <u>Results from SHORT time scale "averages".</u>

1. The "raw" and corrected stimulus-time locked correlation between the firings of two neurons may each be modulated by the stimulus. The time scale of such modulation is comparable to that seen for firing rate modulations in the PST histogram.

This observation is made with the Joint PST scatter diagram and with the "correlation surface", its normalization that fully corrects for stimulus induced firing rate modulations.

C. <u>Results on "distinctive events".</u>

1. Particular patterns of firing can repeat far more than would be expected by chance. Such patterns can be found in the firing of a single neuron, or across the firing of several neurons. Circumstances which favor such repeated pattern production are not known.

D. <u>Experimental sources for above results.</u>

Most of the cited observations come from more than one laboratory group, using diverse preparations, neural systems, recording technologies, and analytic programs. Thus we may take most of these results as robust observables about multi-neuron properties. In detail: A.1 and A.2 are reported from Freiburg, Jerusalem, Marburg, Nijmegen, Philadelphia, and Tübingen (Aertsen et al. 1987, Frostig et al. 1983, Eckhorn et al. 1987, Eggermont et al. 1983, Gerstein et al. 1985b); A.3 and A.4 are reported from Philadelphia (Bloom et al. 1984, 1986, Espinosa et al. 1984, 1985, 1986); B.1 comes from Freiburg, Philadelphia, and Tübingen (Gochin et al. 1986 and in preparation); C.1 comes from Los Angeles, Jerusalem, Philadelphia and Waterloo (Frostig 1985, Abeles and Gerstein 1987, Landolt 1985).

5. INTERPRETATION OF MULTI-NEURON OBSERVATIONS.

A primary distinction that must be made before interpreting the observed cooperativity phenomena is whether we consider the "raw" cooperativity, or whether we make correction for stimulus related firing rate modulations. The "raw" cooperativity is the quantity that is actually available to the neurons in the brain during the processing of information; it can be considered as a potential code, or a carrier of information within and among neuronal assemblies. In contrast, the "stimulus corrected" cooperativity is purely a construct of the experimenter. It is never directly available to the neurons of the brain. However, from it, the experimenter may make useful and interesting inference about underlying neuronal organization. We have expressed such organizational properties above as "effective connectivity". We stress again that this is NOT the anatomical connectivity, although of course it is related to it, and often represents the subset of anatomical connectivity that is currently active.

How then should we interpret the observed stimulus-related changes of correlation among the firings of a group of neurons (A.1, A.2, A.3, A.4, and B.1)?

Consider first the changes in "raw" cooperativity. These could be viewed as the consequence of different waves of activity passing through a mass of connected neurons. A convenient (although spatially inaccurate) analogy would be ripples in a lake. Different stimuli would correspond to stones thrown in at different locations; the ripples passing several selected points of observation would correspond to different sequences of correlated firing. From this viewpoint, we may simply be seeing the propagation of different units of information through a large distributed network.

Now consider the changes in corrected cooperativity. These suggest that the "effective connectivity" which defines the assembly configuration changes with the stimulus. The time scales for such changes are both medium and short in the terminology we have used above. There are known physiological mechanisms like presynaptic inhibition which are capable of modulating synaptic efficacy on the appropriate time scale. Nevertheless, a context related, rapid modulation of synaptic efficacies in neuronal assemblies seems somewhat unpalatable.

There is a possible alternative interpretation of the changes in the corrected cooperativity. Return to the analogy of ripples in the lake. The corrected correlations and the inferred "effective connectivity" can be interpreted simply as the actual path that the disturbance took through the neuronal mass. The stimulus dependence of corrected correlation (A.2, A.3, and B.1 above) would, with this interpretation, signify different paths through the neuronal mass for each different stimulus disturbance. If we assume that different stimuli enter a large network at different points (or in a different spatial or temporal way) then the paths taken by each corresponding disturbance may differ without requiring context dependent changes of "effective connectivity". Note that we are here making a distinction between the disturbance (which presumably codes information) and the path along which it propagates.

Choice between these several interpretations is presently impossible, and will require additional development of both theory and experiment.

6. OTHER ASSOCIATED PROBLEMS IN UNDERSTANDING ASSEMBLIES.

A. The issue of local versus distributed assemblies.

Consider a model that consists of local assemblies inside a block of tissue. As one passes stimuli into the tissue, one might find that each one of these assemblies responds only to one possible stimulus, and one would only find local activity patterns associated with that one stimulus. In the alternate model of a completely distributed network each neuron might take part in the representation of many different stimuli, and activity patterns would be widespread.

We clearly need additional models and experiments to distinguish between these alternatives.

B. The significance of pattern events.

When we observe these distinctive repetitive patterns: Who reads them? How do we find out who reads them? What information do they carry?

C. The organization of networks.

When we have networks that project onto each other: How do we find out whether the high level organization is in the form of cascade processes (feedforward) or of feedback processes? Are there models and experiments that are critical to the difference between the two?

It seems clear from all the above that there is a great need for new "granular" (rather than "gaseous") and dynamic theory for describing neuronal assemblies. At the same time there should be great efforts to design new, theory critical multi-neuron experiments.

7. ACKNOWLEDGEMENTS

This paper was written while on a Sabbatical stay at the Max Planck Institute for Biological Cybernetics in Tübingen, West Germany; I am grateful to Professor Valentin Braitenberg for this work opportunity. I thank Dr. Ad Aertsen for discussions and for critical reading of the manuscript. Volker Staiger contributed essential photographic skills. Various portions of the work were supported by NIH (NS-05606), the System Development Foundation (SDF 0013), and by the Office of Naval Research (ONR N 00014-83-K-0387).

8. BIBLIOGRAPHY

Abeles, M., (1982), Local Cortical Circuits, Springer, Berlin.

Abeles, M. and G. Gerstein, (1987), "Detecting spatio-temporal firing patterns among simultaneously recorded single neurons", in preparation.

Aertsen, A., G. Gerstein, P. Johannesma, (1986), "From neuron to assembly: neuronal organization and stimulus representation", In: Brain Theory, A. Aertsen and G. Palm (Eds.), Springer, Berlin.

Aertsen, A., T. Bonhoeffer, J. Krüger, (1987), "Coherent activity in neuronal populations: Analysis and interpretation", In: Physics of Cognitive Processes, E. Caianiello (Ed.), World Scientific Publishing, Singapore. (In press)

Bloom, M. and G. Gerstein, (1984,1986), "Dynamic changes in cortical connectivity during sound localization", Soc. Neurosci. Abstr. 10: 245 and 12: 1274.

Eckhorn, R. and H. Reitboeck, (1987), "Multi-neuron observations on cat visual cortex", private communication.

Eggermont, J., W. Epping, A. Aertsen, (1983), "Stimulus dependent neural correlations in the auditory midbrain of the grassfrog (Rana temporaria L.)", Biological Cybernetics 47: 103-117.

Espinosa, I. and G. Gerstein, (1984,1985,1986), "Correlation among spike trains in cat´s auditory cortex during presentation of three-tone sequences", Soc. Neurosci. Abstr. 10: 245 and 11: 249 and 12: 1274.

Frostig, R., Y. Gottlieb, E. Vaadia and M. Abeles, (1983), "The effects of stimuli on the activity and functional connectivity of local neuronal groups in the cat auditory cortex", Brain Research 272: 211-221.

Frostig, R., Z. Frostig, R. Frysinger and V. Schechtman, (1985), "Multineuron analysis reveals complex patterns of interactions among neurons", Soc. Neurosci. Abstr. 11: 1020.

Gerstein, G. and D. Perkel, (1969), "Simultaneously recorded trains of action potentials: analysis and functional interpretation", Science 164: 828-830.

Gerstein, G. and D. Perkel, (1972), "Mutual temporal relationships among neuronal spike trains", Biophysical J. 12: 453-473.

Gerstein, G., M. Bloom, I. Espinosa, S. Evanczuk and M. Turner, (1983), "Design of a laboratory for multi-neuron studies", IEEE Trans. on Sustems, Man and Cybernetics, SMC-13: 668-676.

Gerstein, G., D. Perkel and J. Dayhoff, (1985a), "Cooperative firing activity in simultaneously recorded populations of neurons: Detection and measurement", J. Neuroscience 5: 881-889.

Gerstein, G., A. Aertsen, M. Bloom, I. Espinosa, S. Evanczuk, and M. Turner, (1985b), "Multi-neuron experiments: observation of state in neural nets", In: Synergetics, H. Haken (Ed.), Springer, Berlin.

Gerstein, G., and A. Aertsen, (1985), "Representation of cooperative firing activity among simultaneously recorded neurons", J. Neurophysiology 54: 1513-1528.

Glaser, E. and D. Ruchkin, (1976), Principles of Neurobiological Signal Analysis, Academic Press, New York.

Gochin, P., J. Kaltenbach, G. Gerstein, (1986), "Stimulus linked temporal changes in interactions of neurons in rat dorsal cochlear nucleus", Soc. Neurosci. Abstr. 12: 1269.

Habib, M. and P. Sen (1985), "Non-stationary stochastic point-process models in nuerophysiology with applications to learning", In: BIOSTATISTICS: Statistics in Biomedical, Public Health and Environmental Sciences, P. Sen (Ed.), Elsevier, North Holland.

Johannesma, P., (1987), "Patterns and Surprise", private communication and in preparation.

Krüger, J., (1983), "Simultaneous individual recordings from many cerebral neurons: techniques and results", Rev. Physiol. Biochem. Pharmacol. 98: 177-233.

Landolt, J., S. Reinis, D. Weiss, (1985), "Identification of local neuronal circuits in the visual cortex of the cat", Soc. Neurosci. Abstr. 11: 1010.

Perkel, D., G. Gerstein and G. Moore, (1967), "Neuronal spike trains and stochastic point processes. II. Simultaneous spike trains", Biophysical J. 7: 419-440.

Perkel, D. (1985), "Temporal patterns in a population of impulse trains", private communication of working notes.

Schmidt, E., (1984), "Instruments for sorting neuroelectric data", J. Neurosci. Methods 12: 1-24.

Schmidt, E., (1984), "Computer separation of multi-unit neuroelectric data", J. Neurosci. Methods 12: 95-111.

Information Flow and State in Cortical Neural Networks: Interpreting Multi-Neuron Experiments

Discussion
edited by *Ad Aertsen*

Gravitational Clustering

REITBOECK: You mentioned the observation of stimulus-dependent coupling between neurons; we made this observation too. There is another point I would like to bring up. Gravitational clustering is one of the very few novel concepts in multi-unit analysis. What makes me feel uncomfortable about the method, however, is that there is an inherent source of error at the basic concept: When there are three or more signal sources, the correlation between any two of them can make other sources appear to be correlated to the pair (or cluster), although they may not be correlated at all (when two points are moving toward each other in Euclidean space, there is, in general, also a relative movement with respect to other points). Can you give an estimate on the magnitude of this error?

AERTSEN: Consider the case of three neurons. At the outset of the analysis the positions of the three corresponding particles form an equilateral triangle. If now two particles are approaching, while the third one remains fixed, this implies that the distance between any of these two and the third one will also decrease, to a possible minimum of 87 percent (being half the square root of 3) of the original distance. This decrease, however, does not reflect a genuine coherence of firing. This systematic effect should be added to the more obvious lower limit for effects to be significant, set by the diffusion type influence of the stochastic nature of neuronal firing. For various reasons therefore, small decreases in particle distances should not be overinterpreted. The results that Gerstein presented, however, both from his data and from our analysis of Krueger's data, show clustering which clearly is very much stronger than this threshold for significance, which I estimate to be in the order of 10 to 20 percent of the original interpair distance.

BIENENSTOCK: Would you agree that the gravitational method is mainly a tool for analysis?

GERSTEIN: Yes.

BIENENSTOCK: In this case, why do you go to an N-space and then project that to a 2-space? Why don't you just start and do all this in the plane? You could combine the gravitation approach with the annealing technique, and then you would probably see the same: some particles will attract and some will not. What you are interested in after all is a simple behavior, a behavior you understand and which can be looked at only in the 2-space.

GERSTEIN: I disagree strongly. What we see in the 2-space is only a very poor approximation of what goes on in the N-space: consider clustering into various configurations in N-space and what these would look like after projection to a plane, depending on the projection chosen. (the choice of projection plane is very critical). On the other hand, a statement like "particles condense into a line in the N-space" is a very significant one. I do not think that you could work in the 2-space with this kind of problem; it seems to me that the initial conditions, the starting configuration of the particles will be prohibitive: within the plane the starting positions of N particles (N>3) cannot be equidistant, and this clearly is an essential requirement.

PELLIONISZ: Continuing on Bienenstock's remark, I think that the gravitational method is not just a tool. You can go far beyond that if you consider the geometry of the N-space, in fact it could be just the way the system works. The first question would be: what is the metric? This can be provided by the eigenvectors, which in turn may show up in a stabilizing pattern in the multi-neuron activity. From these experimentally determined eigenvectors you could construct the metric in the N-space and then you can construct your projections in accordance.

Stimulus-Dependence of Neuronal Correlation

AIPLE: In your presentation you seem to favor the interpretation of a network with changing effective connectivity. In our experiments we saw a most clear example of correlations changing with stimulation, namely when we compared left eye stimulation to right eye stimulation. One then has different forms of excitation in the two cases, from this we favor the view that we excite the same neural network with constant synaptic efficacies, but that the excitation simply takes different paths, and therefore the correlations look different; just like your 'ripples in the lake'.

GERSTEIN: You throw stones in at different places, right. As I said, that certainly is a possible interpretation of this phenomenon. The phenomenon itself, I think, we need not argue about, it certainly is real: it is repeated in many different places, analysed from many different systems. The change of correlation with stimulus across a network is real. There are at least two different ways to interpret this. We should talk about how to design experiments that would differentiate between these possibilities.

KRUEGER: To inspect ten thousands of correlograms (I actually did that) need not be so hopeless as you suggested. In fact, studying correlograms so far is the only way we can learn about poly-synaptic connections. Neuroanatomy can only teach us about mono-synaptic connections. For instance, anatomy cannot tell us whether one mono-synaptic link is weaker or stronger than a thousand bi-synaptic links. The answer to this question, in general, is not immediately clear. In cross-correlograms we get all these different kinds of interactions in all causal directions together, anatomy only gets the mono-synaptic interactions. From cross-correlograms we have to disentangle all these different interactions. Cross-correlograms are not an endpoint, but we want to know, for instance, to what they in turn correlate.

For example, how is direction selectivity actually generated? A specific question might be: do non-directional neurons inhibit directional neurons? If you separate all observed neurons into directional and non-directional ones, you end up with four possible combinations of these into pairs of 'reference' neuron and 'target' neuron: d-d, d-nd, nd-d and nd-nd. One can then lump together all the pairs of a particular combination, then the correlograms of these pairs (there may be hundreds of them) can be averaged. Now you have to look at only a few averaged correlograms. We have results from this procedure. It is a method, and it is sensible to do this.

GERSTEIN: I do not want to overstress this point of the thousands of correlograms. Indeed you have thousands of correlograms, but there are many ways of handling that problem, other than either by condensing them along some question - as you have done, using, by the way, a procedure whose validity remains to be discussed - or by just staring at them. One possibility which comes to mind is the expert system. If you can formulate a rule - you gave an example of a possible question or rule: non-directional neurons inhibit directional neurons - you could take a database of tens of thousands of correlograms. You could then ask that question and the computer would do it for you. In terms of data processing there are many interesting possibilities which we simply have not explored yet.

VON SEELEN: Is it good to analyse a system without knowing anything about it? Let us go back to the days of Galilei in experimental physics. Would it not be better to make a model, estimate its parameters and investigate the predictive value of this model, rather than to rely on an expert system? Is that, in fact, not a step back?

GERSTEIN: I do not think it is a step back, I rather see it as eliminating a lot of unpleasant work. When you look at thousands of correlograms, i.e. data which represent relationships, you must look at them along certain rules. The way that we humans seem to do this is that we look for possible rearrangements of these data which will form some orderly pattern.

Local versus Distributed Assemblies

SHAW: I want to mention Mountcastle's organization scheme as a third possible model for generating stimulus-dependence of neuronal correlation: interacting minicolumns with fixed connectivity, that generate a multitude repertoire of complex spatio-temporal patterns.

Another aspect is that your gravity method seems extremely well suited for local analysis within a minicolumn, rather than for a global analysis: within a minicolumn with a hundred of neurons, ten electrodes give a sampling ratio of roughly 10 percent; in a macrocolumn with ten thousand neurons that ratio is rather 0.1 percent.

GERSTEIN: What would you call local? For instance, Krueger's electrodes cover a span of about half a square millimeter, and within that region one has several thousands of neurons; then it indeed becomes a small sample. So you are probably right that, in view of what we know about columns, everything we talked about today should be restricted to one column.

FREEMAN: My question concerns the nature of the model you have for the substrate of the activity. Essentially you have here data which have a certain covariant structure, described primarily in terms of the time variable, less so in terms of the space variable, although I expect that will come too. Looking at structure, do you have some idea about the basal state from which this structure is emerging, the null state? One such model would be a stochastic process with a certain probability distribution, say Gaussian, against which you measure the emergence of structure. My first question would be: using that as a model, what would be the fraction of the total variance that these covariances, resulting in clustering in the gravitational scheme actually represent? My hunch is that very weak attractors can give rise to the form of convergence that you clearly demonstrate: something in the order of 1 to 5 percent of the total variance may be responsible for that, in which case you are left with a large amount of unexplained variance (noise).

An alternative model would be that it is not noise in the classical sense, but chaos, in which case it has a deterministic structure, reflected in some form of reduced dimensionality which is not apparent in any ordinary statistical test. It looks like noise, but it is not noise. One would have to use procedures to calculate this reduced dimensionality (Hausdorff dimension, Lyapunov exponent) and investigate whether that is responsible for this gravitational clustering.

KRUEGER: For our data from visual cortex - recordings over a region of half a millimeter wide, within one layer, at positions separated by 160 microns, i.e. relatively far from each other - a typical value would be that 1 out of 100 to 1000 spikes of two neurons are synchronized. That can still be seen in the cross-correlogram, if you record long enough. I think that 5 percent is too high a number, that would be a very strong correlation.

GERSTEIN: I agree, the numbers are closer to 0.1 or 1 percent, rather than above that. Mind you, we talk here about the synchronized fraction of spikes from two neurons; the synchronization itself can be brought about by various mechanisms, such as direct connection or shared input. In each particular realization the synchronized fraction will have its own relation to the underlying connectivities.
The condensations we measure with gravity are considerably above noise level. The variations in a set of simulated, non-interacting neurons cause changes in distance that look like diffusion, and we can measure an effective 'diffusion constant'. The condensations I showed you for physiological data are clearly much stronger than that diffusion.

Feedforward versus Feedback Systems

JOHANNESMA: You discussed the finding that correlations may be stimulus dependent. There is a sort of mirror image to this: the single unit receptive field may depend on stimulus or stimulus context. From theoretical considerations one can show that, if a system of neurons satisfies certain conditions, the receptive field should be stimulus-invariant, and correlation functions are essentially independent of the stimulus and can be used to make the relation with the connectivity. If these conditions are not fulfilled, all of this is no longer the case. The conditions are quite simple: either the system is linear (which it probably is not) or it contains no effective recurrent connections, i.e. it is a cascade system. Under these conditions the system can be analysed, possibly with a lot of work. From the correlations the connections can be calculated. From here one might turn the argument around: would it be possible in the experiment to manipulate the feedback connections in the system, for example temporarily block them by pharmacological or behavioral manipulations, and thereby study the simpler cascade system, and then gradually increase the feedback and return to the more complex, natural system?

GERSTEIN: It sounds fascinating, but I have no real suggestions that would implement this idea into an experimental paradigm for eliminating recurrent connections.

Design of Multi-Neuron Experiments

PELLIONISZ: Gravity is a very attractive way of looking at the activity of N neurons acting together as an N-vector. But: this vector is a vector in a space. Now one could assume, as you did, that this space is a Cartesian space, but you need not do that, in fact one should not do it like that. There is really a frame there, and we should investigate the geometry of this N-space. How do we know that this N-vector is a covariant vector or a contravariant vector? We have to find that out. You could design an experiment with N neurons (with N larger than three but probably less than one hundred, say) in a system where we know something about the frame, for instance in the sensori-motor system. In that case the frames are known anatomically. Then we would know whether this N-vector is covariant or contravariant, and we would know something about the geometry of the N-space. If you then apply your clustering procedure and you observe the N particles collapse into a point in this N-space, you have really put together experimentation, interpretation and theory, and you could have a geometrical brain theory, which would be an incredible advance.

GRZYWACZ: If you want to test the hypothesis that assemblies are changing by a (conditioning) stimulus, you need another stimulus as test stimulus. How difficult would it be to stimulate electrically from one of your recording electrodes and investigate the assemblies under those conditions? In that case you would test whether the assembly is actually changing, or whether only 'ripples' have passed through.

GERSTEIN: As far as I know micro-stimulation has not been considered in this context. It might be a good idea, although it may also be very difficult and it gets you into another combinatorial problem: which electrode would you use for stimulation, what should the stimulus look like, and which effects would you expect. On the other hand, some of the experiments I showed do involve a combination of conditioning and test stimulus. In the tone sequences we were testing during a particular tone stimulus, which was preceded by different possible combinations of other tone stimuli. That is of the nature of a set-up stimulus, followed by a testing stimulus, not a simultaneous testing stimulus.

Acknowledgement

This report was made on the basis of audio recordings of the discussion made by Ulrich Leinhos. I wish to thank George Gerstein for critically reading this text, and Shirley Wuerth for checking the English.

Discourse on

Current Issues in Multi-Neuron Studies

edited by *Ad Aertsen*

Multi-Unit Analysis and Brain Theory

AERTSEN: By way of introduction to this discussion on current issues in multi-neuron studies, I will try to summarize what I learned from George Gerstein's presentation on this theme and from the discussion that followed it. Obviously this summary contains the necessary amount of personal bias!

Considering the 'state of the art' in multi-unit work I want to distinguish three different aspects: experiment, data analysis, and theory. Concerning the experiment, it is fair to say that over the past decade it has become a feasible enterprise, at least in a growing number of laboratories (5 to 10, say), to record simultaneously and separably the activity from some 5 to 30 neurons at a time. This became possible through improvements in both multi-electrode technology and spike waveform separation techniques. This development resulted in a rapidly increasing number of multi-unit recordings from different parts of the central nervous system; from the nature of these experiments it follows that the experimenter is faced with rather massive pieces of data to be analysed. For the analysis of multi-unit spike trains different statistical tools have become available: e.g. cross-correlogram, joint-pst, 'snowflake', 'gravitational clustering'; essentially these methods are all based on cross-correlation procedures introduced some twenty years ago. Although this might be taken to imply that analysis methodology has come of age, there are still quite fundamental problems that have not been treated in a satisfactory way: for instance the treatment of spike trains which show (stimulus induced) nonstationarities and/or deviations from Poisson, the separation of stimulus effects and neuronal effects and the related issue of proper normalization of correlation results. Twenty years after the introduction of cross- correlation of pair recordings, the methodology of true multi-unit spike train analysis remains a relatively underdeveloped area.

Notwithstanding these deficiencies, the results from multi-unit experiments on different preparations, recorded in different laboratories, and analysed with different statistical techniques, convincingly demonstrated that:

1. in the central nervous system there is neuronal correlation (coordination, cooperativity, coherence) which clearly goes beyond purely stimulus induced effects (modulation of single unit firing rates), and

2) this neuronal correlation in itself is dependent on the stimulus, the immediate stimulus past or the stimulus context in general.

All this is strongly suggestive, although not strictly proof of an organization in terms of neuronal assemblies, moreover an organization which is not fixed but dynamic, constantly being modulated by external events. Fine! So what is next? By this time, if not earlier, the multi-unit experimenter has to start looking at what theory has to offer. Without guidance from theory there is no obvious way to proceed. There is, however, a slight problem: most of theory in brain science is expressed in terms which are wildly beyond the terms used in experiments; this makes it prohibitively difficult to reconcile the two. What seems to be needed for this purpose is an 'intermediate' type theory, rather than a global one: a theory which is expressed in terms of variables (state variables) which are or can be expressed in measurable quantities, so that it relates to the experimental data and can be confronted with them. This type of theory is necessary to suggest what to look for in multi-neuron recordings, so that one can truly study the function of neuronal interaction, rather than discuss the function of the study of neuronal interaction. What seems to be needed is a theory which addresses neuronal problems, rather than one which solves travelling salesman problems; however interesting the latter may be for its own sake, that does not make it a neuronal theory.

There is another reason why one needs theory in the field of multi-neuron experiments, and it directly relates to the experiments themselves: theory should be used to suggest better multi-unit experiments. Most multi-neuron experiments as I perceive them in fact are single-neuron experiments, where one happens to have more than one electrode in the tissue. Questions that are posed and, accordingly, stimuli that are presented are directly transposed from successful single unit experiments. This, however, is not necessarily an adequate strategy if one wishes to investigate neuronal interaction. The single unit type experiment and stimulus quite naturally induces a certain emphasis in the analysis of multi-unit data: a very common approach is to try and explain single unit characteristics, such as receptive field properties, from observed multi-unit correlations. The analysis of functional connectivity is used here in an attempt to understand, retrospectively, the known single unit behavior (e.g. can directional preference of a neuron be explained from the wiring of more elementary elements?). Without questioning the validity and usefulness of such an approach, from the type of goal setting, I would characterize it as a 'regressive' approach, in contrast to a more 'progressive' approach, which takes neuronal interaction as a starting point and tries to go beyond the single neuron paradigm, directly adressing the question of neuronal organization and the way it functions dynamically. Also for this reason we need an 'intermediate' type theory, truly connected to experimental variables; a theory which is capable of suggesting the proper multi-unit experiments and adequate ways to analyse the data in view of alternative hypotheses.

In the above I sketched a few arguments why the field of multi-unit studies would need theory, at the same time pointing to some requirements or boundary conditions such theory would have to meet from the experimenter's point of view. I rather stressed the point of the theory being at an 'intermediate' level: a relatively modest representation in terms of variables which in principle could correspond to measurable quantities, rather than a highly formalized and stylized structure where questions regarding a possible relation with down to earth issues such as experiments are not considered a sign of good taste. It is my impression that a reasonable strategy to develop such a theory is an interactive one, i.e. through a critical discussion on issues as they naturally arise in the multi-unit field, issues such as those which were discussed in connection to Gerstein's presentation on the subject this morning.

GERSTEIN: Let me make the following suggestion. Suppose you make not a grand global theory about one hundred million neurons in some infinitely complex net, but instead make a rather granular, relatively local theory: one hundred-neuron clumps which interact. One could play with such elements on various size scales and using various rules of interaction. Then one could, in simulation, do multi-neuron experiments on such a system, and try to see whether one can distinguish between different connection rules, different parametric arrangements and the like. If so, then you could do such a multi-neuron experiment on the real nervous system and test whether it is composed of similar building blocks with similar type of interaction. As far as I know, this approach of having a granular theory on which you do a multi-neuron 'experiment' has never been tried, and I think it would be very worthwile.

PELLIONISZ: I think one should discriminate between a theory and a model; in fact, you have data, you have models and you have theory, and clearly you need all of them. What I would suggest is that you already have adopted a theory by saying that, having recorded from N neurons, this represents a point in an N-space. This is a germ of a theory. What if we look at this point in the N-space in an experimental setting, where this point is connected by some kind of identifiable reference frame to an identifiable physical object. Do you think it would be interesting for you to think of some kind of sensori-motor setting for this approach that you have?

GERSTEIN: How can I say anything but an unqualified 'yes and go to it'? But let me suggest something more modest. The technology of the experiment lets us measure only a very small number of neurons, somewhere between ten and one hundred. Any theory that you build may require a much larger sample in order to be tested.

PELLIONISZ: Not really. Suppose that you work with the eye. The eye has six muscles; suppose that you measure from six motor neurons, each belonging to a different eye muscle. Then you would not only have a vector in 6-space, in addition you see the eye moving. So far one has only looked at one pair of muscles at a time. The six neuron recording would be a multi-dimensional extension of this. As an experimental paradigm one could consider oblique saccadic eye movements, and at the same time monitor the wanderings of the vector in the 6-space. There may be, in fact there is, a functional geometry in this 6-space: this geometry governs the movement of the eye. If there is this geometry, then there is a metric in that 6-space, and if there is a metric, then we know there are some special directions for that metric: the eigenvectors. These can be calculated.*

GERSTEIN: But how does this relate to any possible brain model in which you have a population of elements, connected according to some rules, which is basically what a brain model has to be at some level?

Correlation and Connectivity

TOULOUSE: I have a much more elementary question. I did not know about these results on the change of cross-correlation with stimulus. What would be the interpretation so far, is it a change of the synaptic efficacies or is it interpreted as due to the history of the firing patterns?

GERSTEIN: The experimental fact that we see is a change in the correlation structure of the multi-neuron firing. As I said before, I think there are two possible types of interpretation and I know of no possible test to differentiate between the two. One such interpretation would be to suggest that in the function of the brain, as each stimulus comes in, it somehow changes the configuration of the system that is accepting it. This could be accomplished by, for

* Note added in proof:
A result of this discussion is: *Pellionisz, A.* (1987), Vistas from Tensor Network Theory, in: *Computer Simulation in Brain Science* (R. Cotterill, ed.) Cambridge University Press.

instance, presynaptic inhibition. Changes of correlation are thus interpreted as a time variation in the effective synaptic connectivity. The other possibility I would call 'ripples in a lake', i.e. you throw in a stone, the stimulus, and some kind of 'disturbance' travels through the system, which is a distributed, large system that you sample at some points. If now the disturbance comes from one direction, you observe one possible sequence of events, and if it comes from another direction, another sequence is the result. This would introduce similar phenomena, i.e. changes in correlation structure depending on stimulus circumstances. I should stress that, when I am talking about correlation structure, we have normalized this in all the ways we have so far been able to, i.e. we are looking at the pure, residual correlation effect that is left after correction for everything that you control, such as stimulus induced rate variations etc.

TOULOUSE: Could this issue of two possible interpretations be resolved by studying tissue cultures?

AERTSEN: In Tuebingen we are setting up precisely this type of experiment: to record with multi-electrodes from tissue cultures and try to address the issue which keeps coming up: the relation between functional connectivity (correlation structure) and structural connectivity (synaptic connectivity). The possibility of having fast changing synapses has been raised for theoretical reasons, e.g. by Von der Malsburg. We did some simulations on a large network with fixed 'synaptic' connectivity, from which we made a 'multi-unit recording' while presenting different stimuli to it. From correlation analysis of the 'recorded' spike trains we inferred the functional connectivity, and found that under those conditions one may very well observe a changing functional connectivity despite a fixed structural connectivity. This obviously does not disprove the possibility of changing synapses, but it certainly leaves room for an alternative explanation.

AMARI: If you record from a large network, and you measure the activity of only a small number of neurons, say four to five, then one should expect that what cross-correlation analysis actually measures of the connectivity greatly depends on the number of inputs from different neurons that is required to make a single neuron fire. If one pulse from a neuron is sufficient to make another one fire, then cross-correlation is able to capture the connectivity between the two, but if on the contrary the neuron requires a large number of incoming pulses, then only a small fraction of them will be measured, and consequently the correlation will be very weak. Do you know anything about the number of inputs that is required in order to make a neuron fire?

AERTSEN: It clearly depends on the area one is recording from, but I would say that in cortex you rather need a large number of inputs in order to make a neuron fire.

BRAITENBERG: I would like to present some anatomy which seems appropriate at this point. I was once again struck by what George Gerstein presented: how nicely, almost too nicely, it matches what you simply read off the structure of the cerebral cortex regarding the probability of two pyramidal cells being connected. I only talk about pyramidal cells, because you mainly see excitatory connections, and these are either between pyramidal cells or from the input on to the pyramidal cells.
One is struck, looking at the probability of connection between two pyramidal cells, even when they are close together, at a distance of say 100 microns, that it turns out, by various ways of calculation, to be very low: by far the highest probability is that they are not connected at all (roughly 0.9), that they are connected with one synapse has a probability of almost 0.1, while for two or three synapses it is vanishingly low. Thus most of them are not connected, and if they are, the connection is very weak. Why weak? Because one synapse is one out of five thousand in the mouse cortex (maybe ten

thousand in larger mammals); and although we do not really know what the threshold is, there are some indirect indications that it is in the order of one hundred, at least. Therefore you expect, when one tries to find connections 'accidentally', by physiological means, that the influence of one neuron onto another is absent or very weak.

You may find more in the input, though. In the axonal feltwork present in a certain place in the cortex, one particular pyramidal cell produces in the order of 10^{-5} of the axons there present, however, one arbor of a thalamo-cortical afferent produces in the order of 10^{-3}, which is very much stronger. Therefore, if you find strong effects, you would expect them to be connected with the input. So there you have the same suspicion: when you have common input, it is mainly direct thalamo-cortical input and therefore is very strong.

When else can you expect strong effects? Since the connections between the pyramidal cells are so weak, you can only find strong effects in that network when many pyramidal cells are active in a certain orderly way, so as to converge onto one. This, of course, leads to the idea of cell assemblies. Since in our opinion these cell assemblies are strongly connected with the input patterns, this again agrees very much with the experimental fact that the statistically visible effects of one pyramidal cell upon the other are very specifically dependent on the input, as revealed by their analysis of Krueger's data. So, on the whole it fits almost too nicely.

JOHANNESMA: There is a common aspect in the presentations of Merzenich, Gerstein and Koenderink. Merzenich talked about the receptive field, as if it existed. After we inquired, he said it does exist, but it may be context-dependent. We found similar things: receptive fields do exist but, depending on the type of neuron, they may depend on the stimulus; therefore it is a limited concept, it is not completely clear and well defined. Correlation functions were presented by Gerstein, and he stated that under certain conditions he can derive a connectivity from them. But he also made clear that you have to be very careful here, that only after observation of many examples you can induce, rather than deduce, certain structures from these correlations. He also mentioned unique patterns of activity, which may happen only a few times, but still significantly more often than you would expect; something special, although it occurs only a few times. Koenderink talked about correlation functions, and he in fact derived from them the structure of the sensorium, related to the resolution of the sensory field.

I want to present some simple but hard results, which relate to this. They have to do with the difference between networks with and those without feedback. These two classes of networks are not essentially different when the system is linear: in that case a simple mathematical transformation will suffice to reformulate one into the other. If the system is nonlinear, however, these two are essentially different, and the situation is more complex. The case without feedback I call 'reactive', since it only reacts to the stimulus. Such a system belongs to the class of Volterra systems, the class that Wiener theory is about. That is a simple system, it can only show one type of behavior: if you push it, it may behave in a complicated way, but asymptotically it just relaxes to a single, stable state. If the brain were such a system, then the computation of all pair correlations of different orders between the elements would solve the problem of analysing the connectivity. If for such a reactive system you calculate the correlation of the activity of one element with the stimulus, then you find a function which is related to the receptive field. Given enough computer power, the problem in principle is soluble.

The second case is completely different: we now have nonlinear feedback, this I call a 'creative' system. Now one should make a clear distinction between 'syntactics' and 'semantics', between 'signifier' and 'what is signified'. For a simple nervous system with a sensorium and a motorium, there is a relation between each neuron and its sensorium and its motorium; this relation is the 'meaning' or semantic aspect of the neuron. The syntactic aspect is the way activities of different neurons are related. Correlation functions can no longer be used: a correlation function is some sort of average, and an average can only be used as long as it exists. For these systems, however, such an average does not exist, which has to do with convergence of certain expressions. I can give you many examples of systems in this class for which no average exists, and the more you average, the worse are the results (distributions with long tails). For instance, in the case of a chaotic system, averaging will give completely misleading results. For 'creative' systems we reach the limit beyond which averaging and correlation fuctions no longer make sense. This implies that we have to investigate the data as they are, and not averages of them.

If the system you are studying belongs to what I called the 'creative' class, there is no general procedure for finding the structure of it. So you have to rely on the individual intelligence, the perceptual abilities of the observer. If we view the neural activity as a model of the outside world, this suggests that we should look at neural activity in terms that relate to this outside world: in as far as we have a receptive field - which may be context-dependent - use this to express the activity of a neuron, by means of the receptive fields, in terms of the outside world: look at what is signified, rather than at the signifier. The evaluation of this translation of activity clearly should incorporate the behavioral context of the animal.

GERSTEIN: Let me come back to the 'unique event': a certain pattern of activity which repeats in the data more frequently, with some degree of tolerance, than one would expect by chance. That is an example of unique events, not in an average sense, which we can play with. When we did this with spike trains from one neuron, this kind of event indeed occured repeatedly in a lot of the data we looked at: you find some 'word' which repeats much more than by chance. In the last year, together with Moshe Abeles, we found methods of doing the same thing, but now across N neurons, and indeed it looks preliminarily as if it happens there too: unique repeating events of multi-unit spike patterns, which look like a melody in musical notation. These are examples of unitary events, not associated with averaging. Such an event is presumably associated with passage of some kind of information every now and then, but not associated necessarily with the stimulus.

JOHANNESMA: You mentioned that these unique patterns can be compared to musical melodies. This relates to ideas we are pursuing (Aertsen: 'neurophone'): to represent multi-neuron activity by giving every neuron 'what belongs to it'. If they are auditory neurons, whenever one of them fires, you generate the sound which is associated with the auditory receptive field of that particular neuron. If they are visual neurons, whenever one of them fires you generate the image associated with its receptive field; this I would call a 'neuroscope'. We should make an instrument to listen to and to look at the nervous system, and this listening or looking should be in terms of the outer world, not in terms of neural activity itself.

AERTSEN: I have been listening to multi-neuron recordings in that sense, and my experience so far with the 'neurophone' is that indeed you hear things. My feeling is that such an approach may be useful to generate hypotheses, which then could be tested using statistical or other qantitative analyses of the data. This has an advantage above a brute force approach where you would have to test for any pattern you can possibly think of. If only for that very reason, an hypotheses-generator like the neurophone/neuroscope seems to be a useful device.

Spatial and Temporal Scales

SHAW: What could we hope to find from all this? I think we would like to find the code (or codes) used for information processing in the brain. That involves more than just populations of neurons firing at certain rates. The analysis George Gerstein gives shows just that: because it depends on the sequence, and since the same neurons are involved in several different stimuli, I think it is most likely a spatio-temporal code; it is not just the high frequency firing. The temporal codes so far have been completely neglected. For instance, in the higher association area you find that when you show a bar some twenty percent of the neurons are involved; you cannot explain that unless you take time as an integral part, otherwise you have no storage capacity, no information processing.

Another aspect: what is the appropriate scale one is looking for? It is clear that it is not the single neuron, one neuron in cortex cannot fire another one. So you have to look at the appropriate spatio-temporal scale. Mountcastle has made it very obvious in his scheme that the spatial scale is like the minicolumn, we see it is functional and we know its size. For the temporal scale, from a number of different experiments, from psychophysics to oscillations in auditory cortex, it all points to 25 to 100 milliseconds as the appropriate temporal scale. Using thirty electrodes over spaces that involve at least ten thousand neurons, you have a sampling problem. However, over the last thirty years people have been using and publishing pst-histograms, and believe there is some meaning in it. I believe there is too, and I also believe in sampling. If you go to coarser electrodes, or look at hash, you find that the neurons, on average are behaving quite similarly. That means that if you calculate a pst-histogram by averaging over thirty to fifty trials, you are sampling that local population. So when looking on a large scale, one should not calculate the cross-correlation, first one should do the pst-histogram. Then one might say, you have thrown everything away.

Well, you have not, all of the timing is in there. And the way you get the correlations is that you have a specific neural model, you put that back in and see how it propagates. That is what we are trying to do with Krueger's multi-electrode data.

The specific models that I have in mind have that spatio-temporal scaling and, unlike the Hopfield-type model which ignores timing, you can get an incredibly large repertoire of spatio-temporal patterns out of the macro-column: hundreds to thousands of patterns depending on the stimuli just in that small region. The current models have very simple connections. In addition, you can modify these connections with learning rules and enhance whatever you want to, but already with the simple network we get hundreds to thousands of patterns which propagate, and that does the correlations for you, the correlations of the whole network.

GERSTEIN: You have raised an interesting issue: is trial averaging followed by correlation equivalent to correlation, followed by trial averaging? I think there is a fundamental problem of order since we are dealing with a basically nonlinear system. You are examining correlations between the results of trial averaging. I question whether the inverse order would not bring you more.

SHAW: That may be so when you are looking at very close, local connectivity. But when looking over a more global scale, when measuring five neurons out of ten thousand, say, you will be lost in the noise.

AIPLE: You are in fact using a pst-average over, say, 50 trials and take that to be equivalent to a local ensemble average over 50 neurons. I assume you would, in fact, prefer an electrode which measures the combined activity of such a local group of neurons?

SHAW: Yes, for this purpose it does not matter if the recording is not a single-unit recording.

AIPLE: I think this discussion refers to different conceptions about what a neural assembly is, or rather, what the elements of a neural assembly are: single neurons or local groups of neurons with similar characteristics.

SHAW: I think the single neuron is not the appropriate unit to look at information processing. There is local processing, that is clearly important within the layers, and there is the more global processing, the result of which will be transmitted from 17 to 18, say. In Krueger's data, where electrodes were separated by a few hundred microns, you are clearly not within the same 'microcolumn'.

AERTSEN: My impression is that we are currently discussing an issue which can be settled experimentally. It relates directly to the difference between multi-electrode recordings with distant recording sites (like in Krueger's experiments) and close recording sites (recordings on one electrode, separated by spike waveform analysis). You are assuming that neurons within a local neighborhood have similar characteristics, and that assumption can be tested.

EPPING: We have recent data in Marburg, recorded on one micro-electrode, consisting of large spikes from a single neuron and background activity with small spikes from other neurons in that same region. The average over the small background spikes turned out to have stimulus-response characteristics which were completely different from the large single unit spikes. It is not yet clear what this means, but it suggests that from averaging over a local collection of neurons as the spatial unit another picture emerges compared to the single neuron as spatial unit.

KRUEGER: I feel somewhat uneasy about taking the trial average (pst) as representing the activity of several neurons. Take for example the giant cells of Meynert, there is only one of those at a particular location. These Meynert cells are surrounded by small cells with other characteristics. In layer VI it is highly likely you record from such a Meynert cell, as I did quite often. For these neurons it would not be reasonable to think that a trial average over thirty repetitions of the same stimulus is meaningful in the sense that it represents the activity of thirty neighboring Meynert cells, certainly these are not there.

BRAITENBERG: By virtue of the arguments I drew before, there is very little coupling, even neighborhoud coupling in the network of pyramidal cells, which are most of the cells. The coupling, the columnar structure comes from the fact that there are thalamo-cortical inputs with very dense ramifications that have about the size of one of those Mountcastle columns. They produce common input for large groups of cells, but within each of these groups there is of course a great deal of so-called spontaneous activity. I refuse to believe that there are one or ten thousand times more cells than the number of grains in the functional picture, which Shaw's approach seems to imply.

SHAW: But the vertical coupling is stronger than the horizontal coupling, to first order, right?

BRAITENBERG: Maybe, yes.

SHAW: So that gives you the strong kind of assembly thing. And then clearly there is horizontal coupling. The idea is to look at things on various scales. There is incredible coupling locally within a 100 microns, and there is a lot of organization, I agree.

BRAITENBERG: I have to make my point again. I can propose as an alternative idea, that this so-called incredible coupling only comes out when you are very close to the sensory input, because the strong connections, visible in the cortical anatomy, are given by the common input to a whole column of cells from the ramifications of one thalamo-cortical or otherwise input fiber. But this is only a very small part of the components. The vast majority of the components are pyramidal cells that are very weakly coupled. Now you may say: 'they are weakly coupled, but coupled they are, and it is always ten thousand or so which do the same things'.

SHAW: No, no. I mean that the group size is about one hundred neurons, not ten thousand. And then these groups are coupled.

HOFFMANN: If the connections are weak, as has to be assumed, then looking at correlations and strong coupling between neurons does not seem to be the most straightforward thing to do. It may be much more important to look at coding properties, which could be amongst neurons without them being strongly coupled. If you use a large array stimulus you would have to look at the coding of the stimulus covering the receptive fields of many neurons and that can be done very well by looking at these neurons simultaneously, instead of scanning the stimulus across the receptive field of a single neuron, which has been done. So it seems to be obvious to look at the joint coding of many neurons involved in a larger scale stimulus.

A second point I want to make is that an action potential normally is generated to travel a long distance. So looking at couplings between action potentials of neurons very close by does not seem to be very obvious. One should rather look at systems where the action potentials, after travelling for a long distance, converge onto a functional system and one very obvious suggestion for such a system, as done by Pellionisz, would be the eye. So there are two different aspects: first, looking at correlated activity, couplings, between neurons which are involved in joint aspect, for instance in movement; and the other aspect would be to look at the coding among populations of neurons which do not have to be coupled.

AERTSEN: This point of the synchrony as a code was also brought up in the presentations of Gerstein, Merzenich and Koenderink. Over the last years, I think, we have seen a gradual shift in the multi-unit field from the use of correlation functions to unravel connectivity ('wiring diagram' approach) to taking synchrony for what it is, covariation of activity as a possible code.

FREEMAN: In this context I want to report on some experiments I did last spring together with Bob van Dijk in Spekreyse's lab in Amsterdam on the primary visual cortex of monkey.

We recorded from the pial surface with an array of 35 surface electrodes (electrodes 0.5 mm apart) that were implanted by Fernando Lopez da Silva. We recorded from 16 of these electrodes, where the maximal window was about 3 by 4 cm. The monkey had been trained to respond to visual cues. We analysed the surface-eeg patterns from 16 of these electrodes with respect to the sequence of events: the animal was watching with fixed attention a flickering spot (diameter 3'), waiting for it to disappear; at this point, having a time window 400 to 600 ms later it had to pull a lever in order to get some juice. So the animal is waiting for the flicker to end, he sees it stop, there is an evoked potential (retrieved by averaging) at the end of the flickering between 100 and 200 ms, and then the period when the animal looks down at the lever which he is going to pull, and then he gets his reward. This sequence of events is related to the visual stimulus. By proper analysis of the eeg-activity we get clear retrieval of spatial information which relates to each of the events in the sequence: the animal waiting for the stimulus, he sees the stimulus, he looks for his u.c.s. and then he gets the u.c.s.

This information has the same basic form as the kind of information that we found in the olfactory system: oscillatory potentials in the range of 20 to 50 Hz, where the amplitude of the activity is the key bearer of the information, and where there is a commonality of the waveform of this event over the entire structure. Although the content of the information from one state to another obviously is different, since the spatial patterns are different, the density of information is uniform.

So in this case we are observing a very widespread pattern, over an area of some 3 by 4 cm, whereas in most of the multi-neuron experiments one concentrates on dimensions of less than 1 mm. I would suggest that what is significant in terms of the carrying of information, of the operations on information, in fact may be very widespread over the visual cortex. The introduction of information into the visual cortex is in the form of single axons, carrying information regarding various feature detectors. This information is put into the cortex in a variety of local domains, and somehow has to be integrated and combined with past experiences in something under a tenth of a second. The periods of commonality which relate to these external events are in the order of 75 ms, so that is the time course for which one can assume stationarity. So you cannot really make measurements across periods in excess of 100 to 200 ms, and expect to get any kind of coherence with respect to the behavioral variables which you recorded from the animal.

In a real sense then, there is a very severe time constraint, but the information in fact is not in the time domain, it is in the spatial domain. So I would suggest that the region of commonality of signal, of carrier if you like, the spatial domain where you can expect to find the behaviorally significant coordinations or correlations, vastly exceeds the range that you are looking at in terms of local columns. What in fact is important is that information which is shared over the whole of the cortex by the processes of global integration, and becomes accessible to large areas of cortex. Therefore the techniques that you are proposing to use should not be restricted to local windows of less than a millimeter in diameter, but in fact should cover a substantial region of cortex involving several square centimeters.

GERSTEIN: There may be a number of different functions that cortex indulges in. For instance one has pre-attentive discrimination of one kind or another, and I think that what you describe relates to that, i.e. where one looks at a whole field and can parse it almost immediately: it takes no attention. And then there is the other type of process, where you have to read the whole page to find the name, because it is just one name among one thousand. So you are probably quite right, and it is perhaps possible that this particular task was being done by the monkey in one of these pre-attentive modes, and that surely a very large patch of cortex is involved in a pre-attentive parser. On the other hand, you have the other kind of processes which have to do with detailed pattern vision, where one has to foveate, scan etc. These may be very different actual set-ups of the brain.

FREEMAN: I am convinced they are. In the olfactory sphere it would have to do with olfactory concentration, where an animal is trying to track a concentration gradient and he has to inspect over a sequence of steps and has to remember what he did. Just as when you foveate, you have to remember what the positions of your eyes and your head were. It is a combining of somatosensory, of proprioceptive information with the sensory information. My point is that, whichever of these processes is going on, the activity is global, not local.

SHAW: There are clearly a number of scales. There is the single neuron, there is the microcolumn and there is global activity. Are there intermediate patterns that come together? We know there are time scales of milliseconds, there are reaction times of 300 ms, but there are also oscillations still in between. It is wrong to say these other things are not important, we do not know that yet. We know there is structure at these intermediate scales, and what they mean has to be examined; in fact all of them have to be examined, I do not think it is an either-or.

AERTSEN: But before we start doing all that, to me it seems appropriate to note that we are all finite in time, resources and similar commodities. It is important though, that implicit to all these proposals are models or theories. So before doing the experiment, one should try to make this theory explicit, maybe even formulate an alternative, so one has something to test against, instead of obtaining more data which then has to be confronted with more other data.

Acknowledgement

This report was made on the basis of audio recordings of the discussion made by Ulrich Leinhos and on notes made by Juergen Krueger. I wish to thank Valentin Braitenberg and George Gerstein for critically reading this text, and Shirley Wuerth for saving the English language from too devastating attacks.

Review:

Principles of Cortical Network Organization

by *Christoph von der Malsburg* and *Wolf Singer*

SUMMARY

The word "organization" can mean two things - the structure of a system, or the process by which the structure is brought into existence. We want to concentrate here on the process. Even then there are two interpretations of the word "organization". It could refer to the realization of a preexisting plan, in which case one could speak of "assembly". We here will make use of the word "organization" exclusively as referring to the genesis of structures through cooperative mechanisms, as optimal equilibria of the participating forces. (To make the distinction, the term "self-organization" may also be used.)

One often speaks of some structural trait of an organism as being "genetically programmed". This seems to imply the genes contained a blueprint describing the organism in full detail. However, all the stages of brain organization (not just evolution) involve, more or less strongly, an element of self-organization, of creativity. It has often been emphasized that the genes cannot contain, in any naive sense, the full information necessary to describe the brain. Cerebral cortex alone contains on the order of 10^{14} synapses. Forgetting considerations of genome size,* one can hardly imagine how ontogeny could select the

* To specify one out of 10^{10} cells one needs about 33 bits. For the specification of 10^{14} synapses one therefore needs a trifle more than 33 10^{14} bits. Vertebrate phylogeny has taken less than 10^9 years. Counting a year for each generation, and assuming an accumulation of a few bits of genetic information in each generation, genetic information is only a few billion bits.

correct wiring diagram out of all alternatives if these were of equal likelyhood. Besides, judging from the variability of the vertebrate brain structure, the precision of the ontogenetic process is not sufficient to specify individual connections.

The conclusion one has to draw is that ontogeny makes use of self-organization: of general rules, of principles of error correction. Above all, ontogenesis can produce only structures with a high degree of regularity, e.g. homogeneity, repetitivity or continuity. This conclusion is of course fundamental to the venture of doing research on cortical structure: any success in it has to be based on regularities. One cannot understand the function of cortex without knowing its structure, and one cannot know the structure of cortex without knowing the principles of its self-organization.

1. *Abstract Scheme of Organization*

There are well-studied paradigms of pattern formation, especially in physics, in physical chemistry and in astronomy — convection, crystallization (or, more generally, phase transitions), reaction-diffusion systems (the emergence of spatial chemical patterns, e.g., in the Zhabotinski-Belusov reaction), star and galaxy formation. We will attempt to give here a general description of the mechanism of organization. It has to be verbal since, anyway, there is no canonical mathematical description of organization yet.

Organization takes place in systems consisting of a large number of microscopic elements (atoms in a liquid or crystal, small sub-volumes of liquid in convection, in a reaction-diffusion system or in an evolving star system, or, in the application interesting us here, synapses in a nervous network). Initially, the system is in a relatively undifferentiated state (atoms are in random motion, all parts of the liquid are in the same state of motion or of chemical composition). Then, some small deviations from that state begin to form (e.g., some convective motion sets in in a fluid). The first such deviations may arise spontaneously as the result of some random forces. To stress the random nature of small deviations they are called fluctuations.

To make things more concrete, let us take as an example convective pattern formation - the Bénard phenomenon. A liquid is heated in a flat vessel from below. If the temperature gradient is below a certain threshold, heat is conducted from the lower to the upper surface without bulk movement of the liquid. If, however, the temperature gradient is above that threshold, the warmer, lighter, liquid near the bottom rises and colder liquid from the top flows down. Under carefully controlled conditions this flow pattern is very regular and has the form of hexagons or of rolls.

There are the following three very general aspects to organization:

1) Fluctuations self-amplify. This self-amplification is analogous to the reproduction in Darwinian evolution. In the Bénard system, fluctuations are created by thermal motion. If the movement in a small column of liquid is upward, more warm liquid is drawn in from the bottom, the column becomes less dense and is accelerated upward. Downward movement accelerates analogously.

2) Limitation of some resource leads to competition among fluctuations and to the selection of the most vigorously growing (the "fittest") at the expense of others. In the Bénard system, upward movement in one place requires downward movement in other places. The columns with least density will win and rise.

3) Fluctuations cooperate. The presence of a fluctuation can enhance the fitness of some of the others, in spite of the overall competition in the field. (In many systems the "fitness" of a fluctuation is identical with the degree of cooperativity with other fluctuations.) The liquid near a column of rising liquid is dragged up by viscosity.

The identification of these three aspects in a concrete system is sometimes ambigous. In the Bénard system competition in terms of upward movement might also be called cooperation between upward movement in one place and downward movement in another. Whole coherent patterns of movement, again, compete as far as there is local contradiction between them: liquid cannot move up and down in the same place.

A fundamental and very important observation on organizing systems is the fact that global order can arise from local interactions. The intermolecular forces acting in a body of liquid are of extremely short range, yet the patterns of convective movement they give rise to may be coherent and ordered on an arbitrary scale. This fact will be of extreme importance to the brain in which local interactions between neighboring cellular elements are to create states of global order, ultimately leading to coherent behaviour.

The stage for the organization of a pattern is set by the nature of the participating forces and by initial and boundary conditions. In our example, the Bénard system, there are hydrodynamic forces, there is gravity, thermal conduction and expansion. Boundary conditions are set by temperatures at the upper and lower boundary and by the form of the vessel. In the nervous system, the stage for ontogenesis is ultimately set by the genetically determined rules for the interaction of cells, cellular processes and signals, and by the environment, inside and outside of the organism, that the nervous system finds itself in. The important and exciting difference between the nervous system and most other systems studied so far is the fact that neighborhood relationships in the nervous system are not necessarily shaped by its embedding in three-dimensional space. Rather, its interactions are shaped by its connectivity, connected cells being "neighbors" although they may be located at different ends of the brain. This gives rise to genuinly new phenomena. Some of the ordered structures within the nervous system may not "look" ordered to our eye, which relies essentially on spatial continuity. In most of the concrete cases considered below, however, ordinary space still plays a dominating role.

An organizing system may contain a symmetry such that several equivalent organized patterns could arise. These compete with each other during organization. In the Bénard system, if set up in a circular pan, any organized pattern could be rotated around the center of the pan by an arbitrary angle to obtain an equivalent pattern. One of these has to be spontaneously selected during pattern formation: Symmetry is spontaneously broken. This usually is a very slow process. However, the boundary or initial conditions may be slightly deformed to destroy the original symmetry such that one organized pattern is favored. Organization is then much quicker. In general, systems react very sensitively to symmetry breaking influences.

2. *Neural network organization in general*

Two types of variables are relevant to network organization: signals and interconnections. Signals are action potentials propagated along axons. Connections are characterized by a weight variable, which measures the size of the effect a nervous impulse arriving at a nerve terminal exerts on the postsynaptic element. Correspondingly, organization takes place on two levels, activity and connectivity.

We are here interested mainly in network organization. A network reorganizes by creating a signal ensemble and by reacting to this ensemble plastically. The signal ensemble created by a network has certain statistical properties. The existence of excitatory connections between two cells increases their probability to fire simultaneously, the existence inhibitory connections decreases that probability. Synaptic strengths react to those statistical properties of signals. In order to obtain reorganization of a network (instead of stabilization of it) the feed-back loop between changes in synaptic strengths and changes in signal statistics has to be positive, so that coherent deviations from an undifferentiated state can self-amplify. The process is constrained by the requirement that modifications in a synaptic connection should be based on signals which are locally available. These are the presynaptic signal, the postsynaptic signal, and possibly modulatory signals which are broadcast by central structures. The postsynaptic signal could be a local dendritic signal or the outgoing axonal signal.

These requirements, locality and self-reinforcement, suffice to specify the mechanism of synaptic plasticity in excitatory synapses: A strong synapse leads to coincidences of pre- and postsynaptic signals, and in turn the synapse is increased in strength by such coincidences (the transmission delay is neglected in this argument). Donald Hebb (1) gave this formulation:

> *"When an axon of a cell A is near enough to excite cell B and repeatedly or persistently takes part in firing it, some growth process or metabolic change takes place in one or both cells such that A's efficiencey, as one of the cells firing B, is increased".*

One refers to this rule as to "Hebbian plasticity" (although the idea is much older). The corresponding rule for inhibitory synapses would have a synapse strengthened which was successful in inhibiting the postsynaptic element. This rule, however, would lead to a difficulty. A spike in a cell is a relatively rare event. Most inhibitory signals, therefore, hit silent cells and are in this sense "successful". Thus the rule would require too many inhibitory synapses. (A way out could be to require activity in the postsynaptic cell shortly before the arrival of the inhibitory signal.) It is probably due to this difficulty that most authors consider inhibition as a rigid service system not taking part itself in network reorganization.

Hebb's rule corresponds to the self-reproduction of the general scheme of organization. In order to stabilize the system, some competition for a limited "resource" has to be introduced. Most likely, there is a mechanism of isostasy, by which each cell keeps the temporal average of its activity (taken over the span of some hours) constant. As a consequence the increase in strength in some synapses has to be compensated by a decrease in others. Only the more successful synapses can grow, the less successful ones have to get weaker (and, eventually, to disappear). For technical reasons, some models discuss a simpler competition rule for synapses, in which the sum of the synaptic weights of all synapses converging on a cell is kept constant. This rule leads to functional deficits and is probably not realistic. Synaptic plasticity, constrained by competition, implements the characteristics 2 and 3 of organizing systems, as listed above.

A single synapse on its own is not efficient in producing favorable events. It has to cooperate with other pathways. Those pathways cooperate with each other which couple the same pair of cells. These pathways may be indirect, involving intermediate cells, and they may also be constituted by casual chains in the external world.

The rules of cooperation and competition act on a local scale. The phenomenon of self-organization is the emergence of globally ordered states, as discussed with the example of global convection patterns in the Bénard phenomenon. These states are ordered in the sense of optimal mutual consistency of all local rules with each other. The fact that the external world takes part in the game leads to adaptation of the nervous system to it.

If the rules for synaptic growth introduced so far constituted the only kind of control over synaptic growth the nervous system would perhaps produce ordered connection patterns but it would not necessarily develop to optimal biological utility. For this central control is necessary. The state of the whole system is evaluated by central structures (which have evolved according to the Darwinian mechanism) in terms of its degree of biological desirability. If the state just reached proves to be useful, a gating signal is sent to all of the brain or to an appropriate part of it to authorize synaptic plasticity, which then stabilizes the state just reached or contributes to the likelihood of it happening again. If the state proves to be undesirable, plasticity could be inhibited by the central signal.

Central control as the only criterion for growth or decay of synapses is not sufficient: Let us assume our nervous system evaluated the usefulness of its state once a second. It then would create 3 10^9 bits of information in our lifetime, for that is the maximum number of seconds given to us. This certainly is not sufficient to regulate the strengths of all of the 10^{14} synapses of our cerebral cortex. On the other hand, this information is certainly sufficient to select among the much fewer ordered connectivity patterns made possible by the rules of local co-operation and competition.

After this brief introduction of principles we now enter a more detailed discussion of a few paradigmatic cases of network organization. We stress the application to the visual system, which happens to be our specialty. Retinotopy, ocularity and retinotopy have been intensively studied both theoretically and experimentally over several years or decades. Associative Memory and the perceptron have been, for a long time, the corner stones of brain theory and have inspired and motivated many experimental studies.

3. *Establishment of Topological Fiber Projections.*

At some stage in the development of the vertebrate embryo the fibres of retinal gangion cells grow out through the eye stalk towards the brain and establish there retinotopic connections, neighboring cells in retina connecting to neighboring positions in the target structure. The most intensively studied case is that of retino-tectal connections in amphibia and fish. The interesting problem is, how do fibres find their correct target position? This problem has long been recognized as an important paradigm case for the organization of structured networks. There are many topological fiber projections within cortex, and there are similar structures which may be treated by analogy. Retinotopy is one of those rare biological model problems which have been experimentally addressed with such intensity that theoretical issues can be decided.

For a long time the puzzling aspect of the retinotopy system was a mixture of rigid genetic determination on one hand and plasticity on the other. The orientation of the projection, e.g., nasal retina projecting to caudal tectum, is genetically fixed. On the other hand, the magnification and the position of the map flexibly responds such as to map all of the existing retina to the existing tectum, both after experimental manipulation and in physiological conditions. If, for instance, growth in retina and tectum is disparate (retina in a concentric fashion, tectum from front to back), the already existing fiber projection shifts in an ordered fashion to attain systems matching. The picture is further complicated by results according to which fibers are able to follow tectal tissue which has been grafted to a new position within tectum (for a review of various experiments see (2)).

The puzzle of the retino-tectal system is solved by the following statements:

A. There is a mechanism to guide fibers to tectum.

B. There is a mechanism to position fiber terminals within tectum. This mechanism is responsible for genetical constraints on the mapping. The nature of the mechanism is not yet known, but there is the wide-spread conviction that it is based on genetically determined chemical marker gradients in retina and tectum.

C. There is a fiber-sorting mechanism which improves the precision of the mapping over that attained by mechanism B alone and which is activity dependent (3). This mechanism is able to account for the plasticity observed in the retino-tectal system.

In spite of many controversies going on in the retino-tectal community, the above statements have gained wide acceptance. The division of labor between mechanisms B and C and the nature of C have first been formulated in (4). (That paper, however, failed to recognize how precise mechanism B is by itself.) In the present context our interest concentrates on the fiber-sorting mechanism C since it deals with genuine network organization and can be generalized to other interesting cases.

Basic to the fiber-sorting mechanism is the fact that the spontaneous activity of neighboring retinal ganglion cells is correlated, due to excitatory connections in retina. These correlations carry complete information about neighborhood relationships within retina (without coding for retinal position directly!). They are made use of for fiber-sorting in the following way. Retinal fibers establish tentative contacts on tectum. Through these contacts they impress their activity patterns on tectal cells. Due to excitatory links within tectum, activity in neighboring tectal cells is correlated. Retino-tectal contacts undergo a selective growth process, according to which successful contacts grow in strength and less successful ones decay. Success is measured by the degree to which a contact is able to correlate pre- and postsynaptic signals with each other. This success, for a given fiber, is strongly influenced by the number and strength of other contacts connecting neighboring cells (or the same cell) in retina to neighboring cells (or the same cell) in tectum. Competition among contacts is introduced by the inability of a tectal cell to receive more than a certain number of contacts and by the inability of retinal fibers to support more than a given number of contacts.

According to the rules for selective stabilization or elimination of contacts just formulated the total pattern of retino-tectal contacts evolves and changes while always favoring local configurations of parallel fibers: fibers coming from neighboring positions in retina and going to neighboring positions in tectum. Under appropriate conditions the projection system reaches, and stabilizes, in a global optimum in which this situation prevails throughout and which accordingly constitutes a retinoptopic mapping. This has been shown in several sets of computer simulations and by mathematical analysis (5). There is, however, the danger that the system ends in a state of local optimum in which there are several part-maps of retina onto tectum which are not compatible with each other. This can best be avoided by starting the sorting process in a configuration which already contains some global order. This starting configuration also fixes the orientation of the final mapping. For these reasons it is necessary to postulate the positioning mechanism B.

The sorting mechanism has the systems matching property. If an actual fiber projection does not match all of retina to all of tectum, as after some surgical manipulation or in the physiological case of disparately growing retina and tectum, the sorting mechanism is able to coherently move fibers over the tectal surface until systems matching is achieved. In this way the positioning mechanism B is important for getting the process started, but the fiber sorting mechanism C dominates the system with its plasticity in the long run. If a tectal graft of sufficient size is transposed to another position in tectum in a regeneration situation the positioning mechanism establishes a new "initial" projection in which the corresponding part of the projection is displaced along with the graft. This creates a local optimum which cannot be repaired by the sorting mechanism, which is limited to coherent modifications of an existing map. The observation that plastic map transformations are limited to coherent, i.e., neighborhood-preserving, changes has also been made by Merzenich in the somatosensory system (personal communication).

4. *Discussion of Retinotopy*

Let us put the activity dependent retinotopy mechanism in the perspective of our abstract scheme of organization and of neural network organization and let us derive some more specific organizational aspects. The "fluctuations" of our abstract scheme are represented here by density deviations from an undifferentiated initial state. These self-amplify, by setting up retino-tectal signal correlations which in turn are the condition for the contacts growth. This rule directly corresponds to Hebbian plasticity. The example demonstrates the central importance of cooperation between fluctuations — in this case between "parallel" fibers.

One of the initially most surprising aspects of the fiber-sorting mechanism is its ability to create global order from local rules in interaction without having to hunt through the huge combinatorial space of all possible permuted projections. This ability is, however, not a matter of course. It has to be installed by proper control of structural parameters. The general rule is that the self-amplification of fluctuations has to be so weak that only globally ordered systems of cooperating fluctuations are just able to grow. The consequence of this is that growth is slow: the price to be paid for global order is the amount of time necessary for it to be established. The process is particularly time-consuming if there are several equivalent globally ordered systems of fluctuations which compete with each other, usually due to a symmetry. In the retino-tectal case this is realized by ordered projections of different orientation. It is therefore highly advantageous to break the symmetry by the establishment, through an independent positioning mechanism (B), of an already partially ordered initial mapping. (B has, of course, the same problem, but it may use mechanisms which have an inherently larger physical interaction length, e.g., reaction-diffusion.)

The stage for retino-tectal fiber-sorting is set by the pre-existing networks in retina and tectum, by mechanisms which bring fibers to tectum and establish an initial mapping (A and B), and by the genetically determined rules for the growth of contacts. The important aspect is that it needs rather little genetic information to set this stage and that the ensuing organization process creates a huge mass of information in the form of detailed cell-cell contacts. There is no violation of a "law of conservation of information" because by varying the genetic information one cannot create *any* connection pattern but only select from a narrow universe of regular patterns.

The two logical steps of retinotopical network organization are the creation of a signal ensemble which expresses the structure of the network and the plastic reaction of the network to the correlations in this signal ensemble. Very little activity organization is necessary to establish local correlations in retina and in tectum and between regions in retina and tectum which happen to be strongly coupled by the tentative contacts.

Various ways to interpret this process suggest various ways of applying the retino-tectal model to other situations. From the point of view of a retino-tectal contact the occurrence of correlations between signals on the fiber and on the tectal target cell indicates the presence of a causal connection. In this system this is due to loops of retinal—retino-tectal—tectal connections. In other cases there may be causal connections in the external world. When positively reacting to these correlations, our individual contact tends to "model" the indirect causal connection by a direct link.

From the point of view of a tectal cell responding to retinal afferents there appear recurring patterns, buried in the otherwise random signals on the afferents. When selectively strengthening those contacts which carry the patterns the tectal cell starts to turn into a selective detector for those patterns. In the retino-tectal system these patterns correspond to local clusters of retinal cells. When there are several separate pattern-carrying subsets of afferent fibers (here: groups of fibers coming from different loci in retina) our cell has to decide among them, due to the competitive aspect of contact growth. In this way tectal cells develop restricted receptive fields, which are shaped by retinal patterns of signal correlation and by competition among afferent fibers. Afferent patterns may also be created by interactions external to the nervous system. After eye opening, retinal activity patterns are strongly influenced by light patterns, which may, if there still is sufficient plasticity, influence the shape of receptive fields (see the section on orientation specificity).

In this process of development of pattern selectivity different tectal cells compete with each other for retinal patterns. (This competition is due to that between contacts made by one retinal fiber.) The system therefore tends to distribute representation of the afferent patterns evenly among the available tectal cells (such that, in this particular case, different retinal loci are represented by different tectal cells). In other contects, this important aspect of the mechanism is called "competitive learning".

Finally, different tectal cells which excite each other tend to respond at the same time and therefore tend to specialize to the same afferent pattern. This aspect of the mechanism is, of course, of paramount importance for establishing retinotopy. In other cases it serves to establish a continuous representation of patterns which is able to interpolate between those patterns which actually occured.

The last aspect of the retinotopy mechanism which we would like to discuss is that of homeomorphic graph matching. The initial network consists of the retinal and the tectal subnetworks (which, mathematically speaking, are graphs) plus a diffuse system of connections between them. The two subnetworks have similar structure: two-dimensional sheets of cells with local connections. In a loose sense they may be called homeomorphic (or isomorphic if they are of the same size). Now, the fiber-sorting mechanism is able to "detect" the homeomorphism between the graphs and to represent it by selectively connecting corresponding nodes in the two graphs. This process is called graph matching by mathematicians and may play an important role for cortical organization, by "discovering" structural relationships between different objects represented in the brain, or between an external object and a structure within the brain.

5. *Ocularity Domains*

In many vertebrates the visual fields of the two eyes overlap. In these animals one finds structures in the brain which receive overlapping retinotopic projections from the two eyes. There seems to be a general tendency for these projections to align such that corresponding points in the two retinae project to the same point centrally. (This alignment presumably is the result of a retinotopy mechanism of type B.) For the fiber-sorting mechanism C this situation presents a conflict. A cell in the target structure receives two sets of fibers from two retinal spots. Fibers within one set are correlated in their signals, but there are no correlations between signals of the two sets. The situation is analogous to one in which there is only one retina and a target cell has two receptive fields in distant parts of the retina. Consequently, the two sets of fibers compete with each other, and one of them is disconnected. Thus, target cells decide for one ocularity. Due to short-range excitatory connections within the target structure neighboring target cells are likely to decide for the same retina: the target territory is divided into ocularity domains.

What is the spatial scale of this segregation, and what is the likely shape of ocularity domains? In order to answer these questions it is important to discuss the different influences acting on fibers during the ocularity decision. As already mentioned, the local competition of fibers converging on each target cell carries over to neighboring target cells. The scale and shape of this *ocularity competition* is set by the local excitatory connections within the target sheet. If afferent fibers disconnect terminals from one position in cortex they have to strengthen connections elsewhere in order to conserve axonal arbor. This leads to a strengthening of the ocularity just disconnected, or to mutual *cooperation between ocularities*. The scale of ocularity cooperation is set by the lateral extent of the arborization of afferent fibers. Domains can only form if the range of ocularity cooperation (the extent of afferent axonal arbor) is larger than that of ocularity competition (controlled by excitation within the target sheet). The interaction between the two fiber types then looks qualitatively like the difference of two Gaussians. The growing domains have a scale comparable to the size of the central, negative, part of the interaction curve.

The form of ocularity domains is a considerable more complicated issue. Let us draw the analogy to the Bénard problem. There, the role of ocularity is taken by upward or downward convection of liquid. The interaction between the two types of movement has a shape very similar to ocularity interactions. Two kinds of ordered domains are possible in the Bénard system: "rolls" and "hexagons". Rolls would correspond to ocularity stripes, hexagons to a patchy distribution. The decision depends in a complicated way on the form of the interaction forces. It has been shown in simulations (6) that a system set up to produce two overlapping retinotopic mappings is able to generate ocularity domains in the form of stripes. Ocularity stripes have been observed in visual cortex (7), and in experimentally manipulated frog (8). The latter observation is particularly interesting since the overlap between two projections has been produced artificially in this experiment, making it highly unlikely that there is a specific genetic program for ocularity stripes. The formation of ocularity domains has been shown to be abolished by intraocular injection of TTX(9), supporting the activity dependence of the mechanism.

What determines the orientation of ocularity stripes within the target structure? If the relevant fiber interactions were isotropic in their spatial shape and range stripe orientation would have to be decided by chance. If, however, there is the slightest systematic anisotropy in those interactions, stripe orientation is determined by them. If, for instance, the local magnification factor of the retinotopic projection was larger along the x-coordinate than along the y-coordinate, the stripes most likely would be oriented parallel to the y-axis. It would be interesting to test this prediction experimentally.

The above considerations apply to the case in which there is an about equal balance between both ocularities. If the number of fibers from one eye outweighs that from the other, or if one eye is shut off from visual stimulation, the extent of the territory of the dominant eye in the target sheet is increased at the expense of the other eye. Domains are then reduced to islands of minority territory in a sea of the other ocularity. Experimental manipulation of the ocularity distribution has been a convenient means of investigating the central control of plasticity.

6. *Orientation Specificity*

Three problems are addressed here. What is the mechanism which gives neurons in primary visual cortex orientation sensitivity, what induces orientation sensitivity to be distributed continuously over the cortical surface, and how does this organization arise ontogenetically?

As to the mechanism of orientation sensitivity the simplest theoretical possibility would be an arrangement of afferent fibers which would give individual cells elongated receptive fields, or some other afferent organization appropriate for selective response to oriented edges or bars of light. Unfortunately it has not been possible in spite of extensive experimental efforts to demonstrate the existence of the corresponding receptive field organization. A second problem with this idea is that the orderly distribution or orientation sensitivity would be difficult to explain.

We know that the majority of fibers impinging on a cortical cell — especially on a cell outside layer IV — are of intracortical origin. Moreover, the response times of cortical cells to afferent stimulation are long enough for many rounds of exchange of intracortical excitation and inhibition. The response characteristics of cortical neurons are therefore with all likelihood determined largely by intracortical and not so much by afferent organization (10). This realization is the basis for the cooperative mechanism of orientation selectivity, which has been proposed by (11, 12, 13). According to the cooperative mechanism several groups of neurons compete for response to a visual stimulation. The groups inhibit each other and are integrated by excitatory connections within themselves. To each possible orientation corresponds one group (groups for neighboring orientations being partially identical). With the cooperative model individual neurons need very little orientation specificity in their direct afferents, indeed so little that it may be undetectable by itself. When, however, a cortical region is stimulated with an oriented retinal stimulus, the different groups of cells collectively add up the amount of afferent stimulation which they get and compete with each other in terms of this sum. The winning group may have started with only 1% more afferent excitation than its best competitor, but after repeated exchange of excitation and inhibition this edge rapidly improves and finally, after a few dozen of milliseconds, it succeeds in silencing its competitors and winning the response.

With the cooperative mechanism the problem of ontogenesis of orientation specificity is greatly reduced. It is no longer necessary to explain specific complicated connection patterns. What is left to establish is a reliable association between stimulus orientations and cortical activity patterns. For this it is only necessary to slightly modify synaptic strengths in the afferent fibers. Let us first assume that this process takes place in the presence of actual visual stimulation, i.e., after birth. The necessary synaptic modifications then are easily accounted for (11) on the basis of the principles discussed in connection with the ontogenesis of retinotopic projections. The only difference here is that activity patterns in retina and in the postsynaptic sheet are somewhat more complicated. Let us assume that intracortical interactions consist of short-range excitation and longer-range inhibition. Let us further make the very natural assumption that these interactions have a very slight but consistent anisotropy, caused, for instance, by slightly anisotropic growth of the cortical sheet. Patterns of activity in cortex will then locally look like parallel bands, oriented along one of the axes of anisotropy. Like with many natural phenomena — ripples of particles on the surface of a tea cup or of sand on the beach under the unisotropic influence of a current or wind — the degree of long-range order will be low, just as in the deoxuglucose pictures of (14). At different times, waves of different phase or position will be active, all having the same general orientation. All these waves patterns in a small cortical region form a family which can be labeled by one parameter, running form 0° to 180°.

Now take the position of a cortical cell. It receives afferent fibers from a local patch of retina (via the CGl). On these fibers there appear patterns of activity which correspond to bars and edges of light.

(The filter properties of retina and CGl suppress homogeneous surfaces and emphasize edges (11). Also these patterns form a family which can be labeled by the numbers from 0-180°. Correlated fibers cooperate to strengthen their synapses. Our cortical cell specializes to one of the patterns, i.e., to one of the possible orientations. This eventually would lead to receptive fields in the form of bars or edges of light. If, however, plasticity in afferent synapses is reduced in the relevant period, the receptive field is only deformed to a small degree.

Different cells which are active in the same cortical activity pattern undergo receptive field deformations which correspond to the same retinal activity pattern or orientation. A cortical group thus acquires a composite receptive field which gives it a sharp orientation specificity. Neighboring, overlapping groups — waves of different phase — specialize to neighboring orientations. Groups which inhibit each other specialize to different orientations. The system will soon fall into an equilibrium in which different retinal orientations map to different cortical patterns such that neighboring orientations map to neighboring patterns. With other words, when recording from cortical cells along an electrode track at right angle to the local activity waves one will find a regularly processing sequence of optimal orientations.

If it is true, as is suggested by experiments (15), that the association between oriented visual stimuli and cortical activity patterns is already fixed at birth, then the ontogenetic situation is somewhat less clear but there is no organizational difficulty. There could be waves of activity which are created spontaneously in the retinae of the embryo and which take on the role of oriented stimuli in the mechanism just outlined. Alternatively there could be a small minority of retinal cells which are connected such as to be orientation specific and which establish patterned connections to cortex already before eye opening (16).

How do orientation domains and ocularity domains interact with each other? It was assumed in our discussion so far that formation of domains (ocularity domains or activity patterns) started from a homogeneous, though noisy, initial distribution. Domain formation then takes the form of a race of different self-amplifying patterns of imbalance. If there is no disturbing force, the race is won by that pattern which has most internal cooperation and least competition. If, however, there are small but systematic forces able to influence the growing pattern these forces get their way and determine the form of the winning one. (In our analogy, an extremely shallow temperature pattern on the floor of the Bénard vessel determines the convection pattern.) Now assume ocularity domains had already formed and you observed the development of an activity pattern under the influence of afferent excitation which happened to be dominated slightly by one eye. Afferent excitation would then be consistently stronger in the regions of the corresponding ocularity domains. Consequently the developing activity pattern should be a replica of the ocular dominance pattern. The apparent independence of ocular dominance domains and of orientation domains creates the problem of explaining how the two processes can take place in an independent way. One possibility would be a fixation of the form of the activity patterns of orientation columns before segregation of ocularities. If ocularity develops slowly, this process is only influenced by the average of many cortical activity patterns. This average is likely to be homogeneous in its statistical properties and thus unlikely to disturb the precise form of ocularity domains. It would be interesting to learn from experiments whether this is indeed the sequence of events.

REFERENCES

(1) Hebb, D.O.: The Organization of Behaviour. New York: Wiley (1949).

(2) Fraser, S.E.: Cell interactions involved in neuronal patterning: An experimental and theoretical approach. In: Morelucar Bases of Neural Development, G.M. Edelman, W.E. Gall, and W.M. Cowan, Eds. Neurosciences Research Foundation (1985).

(3) Harris, W.A.: The effects of eliminating impulse activity on the development of the retino-tectal projection in salamanders. J. Comp. Neurol. 194,(1980), 303-317.

(4) Willshaw, D.J. and Malsburq, C.v.d.: How patterned neural connections can be set up by self-organization. Proc. R. Soc. London B194, (1976), 431-445.

(5) Häussler, A.F. and Malsburg, C.v.d.: Development of Retinotopic Projections -- An Analytical Treatment. J. Theor. Neurobiol. 2, (1983), 47-73.

(6) Malsburg, C.v.d.: Development of ocularity domains and growth behaviour of axon terminals. Biol. Cybernetics 32, (1979),49-62.

(7) LeVay,S., Hubel,D.H. and Wiesel,T.N.: The pattern of ocular dominance columns in macaque visual cortex revealed by a reduced silver stain. J.Comp.Neurol. 159, (1975), 559-576.

(8) Constantine-Paton,M. and Law,M.I.: Eye specific termination bands in tecta of three-eyed frogs. Science, 202, (1978),639-641.

(9) Meyer,R.L.: Tetrodotoxin blocks the formation of ocular dominance columns in goldfish. Science, 218, (1982), 589-591.

(10) Dinse,H.R.O., Best,J., Krüger,K., Mallot,H.P. and v.Seelen,W.: The receptive field organization of visual cortical neurons of area 17, 18, 19 and PMLS is time-variant due to non-separable spatiotemporal interactions. In preparation.
v.Seelen,W., Mallot,H.P., Krone,G. and Dinse,H.R.O.: On information processing in the cat's visual cortex. in: Brain Theory. Eds. Palm,G., Aertsen,A. Springer, 1986, pp.49-79.

(11) Malsburg,C.v.d.: Self-organization of orientation sensitive cells in the striate cortex. Kybernetik, 14, (1973), 8-100.

(12) Creutzfeldt,O.D., Kuhnt,U. and Benevento,L.A.: An intracellular analysis of visual cortical neurones to moving stimuli: Responses in a co-operative neuronal network. Exp.Brain Res. 21, (1974), 251-274.

(13) Finette,S., Harth,F. and Csermely,T.J.: Anisotropic connectivity and cooperative phenomena as a basis for orientation sensitivity in the visual cortex. Biol.Cybern. 30, (1978), 231-240.

(14) Löwel,S., Freeman,B. and Singer,W.: Topographic organization of the orientation column system in large flat-mounts of the cat visual cortex: A 2-deoxyglucose study. J.Comp.Neurol. 255, (1987), 401-415.

(15) Hubel,D.H. and Wiesel,T.N.: Ordered arrangement of orientation columns in monkeys lacking visual experience. J.Comp.Neurol. 158, (1974), 307-318.

(16) Malsburg,C.v.d. and Cowan,J.D.: Outline of a theory for the ontogenesis of iso-orientation domains in visual cortex. Biol. Cybern. 45, (1982), 49-56.

Generation of Spatial Order in the Developing Nervous System

Alfred Gierer

SUMMARY

In the course of brain development neurons acquire position-specific features. In the lateral dimension of sheeted structures (such as the cerebral cortex), mechanisms contributing to spatial order include internal generation of morphogenetic fields (presumably by reactions involving autocatalysis and lateral inhibition), induction by adjacent cell sheets, and secondary fields arising from margins. In the dimension across the sheet, spatial order is generated by cell movement: Cells of different types are produced in one layer and sort into another layer, guided by contact mediated cell interaction.

The establishment of connections makes use of the positional and directional cues encoded in the brain. A model is discussed for the formation of the retinal tectal map according to which few graded markers in retina and tectum specify mapping for a large number of fibers. The axonal growth cone is proposed to strongly enhance slight directional cues encoded in graded markers. Functional checks by synaptic activity may lead to sharpening of the map by stabilizing appropriate neighbourhood relations. Guiding gradients may but need not be stable during development; changes may lead to shifting connections until the final map is established.

Possible relations between generation and function of the brain are discussed. Shifting connections as well as stabilization of connections by electric activity may be involved in both. Of special interest is the formal relation between generation and function, in particular with respect to autocatalysis and lateral inhibition; it suggests general relationships between pattern formation and pattern recognition which may even be relevant for a theory of aesthetic experience.

1. PRINCIPLES OF SPATIAL ORGANIZATION IN BRAIN DEVELOPMENT

This article reviews some facts and notions on brain development and discusses possible relations to brain function. A large part of the brain in higher organisms consists of sheets made up of a specific sequence of cell and fiber layers. Within the sheets different functional areas with discrete boundaries may be distinguished, such as the tectum, and other brain areas involved in vision. At least in some cases the functional areas in turn consist of many similar subunits such as the columns of the primary visual area in the cerebral cortex. Further, there is increasing evidence for position specific properties within functional areas, both as qualitative differences and as quantitative gradients. For instance, in the chick retina there is a graded distribution of a surface component in the dorso-ventral dimension (1) and a stepwise distribution of adhesive properties in the naso-temporal dimension (2). Graded components were also found in the tectum (3). The following discussion will deal with two questions: which types of mechanisms may lead to positional specificity in the developing nervous system, and what is the role of such specificity in the generation of the neural network?

For this discussion, we will distinguish coordinates x,y for positions in the dimension tangential to the sheet of neuronal cells, and the z-coordinate describing the inside-outside dimension across the sequence of different layers. General embryological principles of spatial organization of cell sheets appear to be strikingly different for x,y-dimension on one hand and the z-dimension on the other. Let us exemplify this with Hydra as a simple model system (4). This is a primitive animal, a few millimeter long, consisting of two cell layers, ectoderm and endoderm; because of its high capacity of regeneration, it is suitable for studying mechanisms of development, in particular, the formation of spatial order within and between cell sheets. It has long been known that each piece of tissue cut from the gastric column regenerates a new animal with head and foot. First, within a few hours, there is an invisible activation of the future head area; only two days later the actual head is formed. Growth normally accompanies regeneration, but regeneration proceeds even if cell division is chemically inhibited. The primary event is the formation of a "morphogenetic field", most likely a graded distribution of a morphogenetic substance, activating the future head area within an originally near-uniform piece of tissue.

How can one explain the de-novo generation of spatial patterns on the basis of molecular interactions and movements? Turing (5) has found that spatial concentration patterns can arise spontaneously by auto- and crosscatalysis in conjunction with diffusion. We have analyzed conditions of pattern formation on this basis, showing that they are consistent with wellknown features of molecular biology and that they can account, in a straightforward manner, for the impressive "holistic" properties of developmental regulation. For instance, the size of a part - such as the regenerating head of hydra - may be regulated in proportion to the size of the regenerating tissue as a whole. Two processes are required for spatial "self-organization", short range autocatalytic activation and wider range "lateral" inhibition (6). Range is defined as mean distance between production and decay of molecules. Local activation is self-enhancing, but lateral inhibition takes care that activation somewhere occurs only at the expense of a de-activation nearby, until a stable spatial pattern is generated. The simplest form is a graded distribution specifying positional information in one dimension, but generally aperiodic and periodic patterns, in one or several dimensions, stable or pulsing in time can be produced by the same general scheme. It is not confined to multicellular tissues; similar principles may govern pattern formation within a single cell. We have shown that autocatalytic activation plus lateral inhibition or depletion are mathematically required for the simplest mechanism capable of generating concentration patterns at all, i.e. two coupled reactions with diffusion (7), and that this scheme can be generalized to a considerable extent to mechanisms involving more than two components (7,8). Only biochemistry could decide on the actual mechanism; there is no doubt, however, that primary pattern formation within the cell sheet is consistent with very conventional molecular biology. Regulatory properties such as proportion regulation, induction and polarity, can easily be accounted for on this basis.

Once a gradient has been produced within a tissue, cells may respond to local concentration values, in the simplest case, to thresholds above which a transition from one differentiated state into an other occurs. In this way boundaries are established which then can give rise to further subdivisions of the tissue. Further, cells may produce stable non-diffusable components in proportion to transient diffusable morphogenetic gradients, thus establishing stable spatially graded features across the tissue. Tissue movements may lead to new contacts between cell sheets, causing induction of new substructures. Boundaries, once established, can give rise to secondary morphogenetic gradients. The net result of the combination of such effects is a positional specification within cell sheets, expressed by a combination of qualitative and quantitative markers.

Whatever the mechanism is in detail, it is likely that spatial organization with respect to the x,y coordinates tangential to neural cell sheets follows similar rules as in non-neural tissue. As in the case of regenerating Hydra, cells within a cell sheet may communicate to acquire different, position-dependent properties without changing their neighbourhood. Positional specificity results from induction, internal "self-generation" of morphogenetic fields, cell differentiation and other cell responses (for review of primary neural induction and early neural development, see (9) and (10)).

The rules for development in the third, z dimension across cell layers of a sheet are entirely different. No gradients or other morphogenetic fields seem to be required. The spatial order may be produced and is, in any case, stabilized by contact-mediated cell interactions. Cells are often capable of moving in the z dimension until they form or find the correct layer. Let us again discuss the model case of Hydra first where we are dealing with two layers of epithelial cells, ectoderm and endoderm. We may dissociate Hydra into individual cells, form aggregates and observe that cells sort out, forming a hollow sphere of two layers: the original endoderm cells again form endoderm, the ectodermal cells ectoderm. (Only thereafter are new heads formed within the sheet - but this does not occur by sorting out of previous head cells; again, in the x,y dimension of the sheet - in contrast to the z dimension - it is cell communication across many cell diameters which leads to the activation of the subareas forming new heads).

In neural tissue, there is also evidence that the z dimension is essentially determined by contact-mediated cell-to-cell interaction. In the course of formation of multilayered sheets, differentiating cells are usually generated in one layer and migrate, subsequently, in the z dimension, to be incorporated into another layer which depends on the differentiating cell type. The spatial sequence of layers in the sheet often reflects the time sequence of production of different cell types - but there are exceptions to this rule which indicate that cell-to-cell interactions must contribute to layer specificity. The same conclusions can be drawn from experiments on sorting-out: As in Hydra, aggregates from previously isolated embryonic cells lead to histotypic configurations. Such configurations were first reported for neural tissue by Sheffield and Moscona (11). In Hydra, reconstitution of layers from clumps of cells requires the presence of cell types which are natural neighbors, ectoderm and endoderm. This suggested to us that in the retina, the natural neighbors - the pigment epithelial cells - may contribute to organizing the reformation of natural structures in aggregates. If cells obtained from embryonic retina by dissociation are allowed to aggregate among themselves, moderate degrees of order are produced by sorting. However, if pigment epithelial cells are added to the suspension of aggregating cells, they form a nucleus of aggregation; within about two weeks the full sequence of dictinct layers characterizing a natural retina of an advanced (though not fully mature) stage can be identified by a combination of staining techniques (12). The order of cell layers results from a combination of sorting, ordered proliferation and differentiation; de- and re-differentiation of pigment epithelial cells may contribute to the production of the stratified retinal tissue as well. The capacity of initially random cell aggregates for the formation of highly stratified structures supports the notion that in the z dimension spatial organization results from contact-mediated cell interaction guiding cell types to the appropriate cell layers and, possibly, the time sequence of proliferation and differentiation.

Assuming that these principles can be generalized we presume that gradients and other morphogenetic fields are involved in the x,y dimension in sheeted structures of the central nervous system and thus, for instance, in its subdivision into functionally distinct subareas, as well as in spatial specifications within subareas. On the other hand, they are not involved in the z dimension - no spatial gradients are required for programming, say, differentiation of pyramidal or stellate cells in the cortex, of ganglion or bipolar cells in the retina. As the result of brain development, cells appear to be endowed with qualitative and quantitative markers which carry information on their position in the brain.

2. CONNECTIVITY AND MAP FORMATION

So far I have discussed principles of spatial organization of neural tissue which are presumably the same as in non-neural development. The additional characteristic feature of the nervous system is the highly specific connectivity between cells which are not spatial neighbors. Part of the connections are formed within a functional area. Often, they appear to be organized in modules, as in the columnar organization of the primary visual area of the cortex. Further, there are many long distance connections. Typically, cells of one layer in one functional area extend processes which then connect with a certain layer of another functional area. In the simplest case, the spatial pattern of connections corresponds to a projection of the area of fiber origins onto the area of fiber terminals. A model system is the retino-tectal projection in birds, amphibia and fish: Cells of the ganglion cell layer of the retina extend fibers and form a nerve which crosses the brain and spreads on an area adjacent to the tectum; then, retinal axons invade the tectum and form connections there. The determination of pathways and connections is such that the ganglion cell layer of the retina eventually "maps" onto the tectum. Formally this means that if we assign coordinates u,v for the position of ganglion cells within their layer in the retina, we can transform tectal coordinates by simple squeezing, rotation and mirroring into coordinates x,y such that the target positions are x=u, y=v; in other words, all neighbourhood relations of ganglion cells in the retina are essentially conserved in the order of terminals of retinal fibers on the tectum. How can axons find the target positions corresponding to such a "map"?

Scientists in the field have proposed a set of models which presumably cover most of the spectrum of physically reasonable mechanisms. They can be ordered into a "tree of theories". The first branching is into models according to which axons carry position specific markers searching for position specific markers on the tectum (13,14,15,16), and models according to which the tectum is more or less devoid of spatial markers. In the latter class, fibers may establish order by communicating with each other directly or indirectly (17); by ordering themselves according to time of arrival at the target (18); or by establishing random positions first, to be re-arranged later by functional checks involving synaptic activity which selectively stabilize appropriate neighbourhood relations (19,20). The mechanisms are not mutually exclusive. It appears that functional checks occur, but only late in development as a corrective refining measure. Further, there is accumulating evidence (reviewed in (21)) suggesting that interaction of growing axons with position markers on the tectum does occur, particularly at early stages. Rotated parts of the tectum can lead to rotated parts of the projection (22). This indicates that spatial tectal markers matter. There are two possibilities: Either there is a mosaic of many different molecules both on the retina and the tectum, and target positions are detected by a lock-at-key mechanism in which markers of retinal axons, which are specific for their region of origin, match corresponding positional markers on the tectum; or there are only few components, but these are quantitatively graded (14,15). The mosaic theory would require extensive random search for target positions, at least for misrouted fibers, whereas graded distributions would allow fibers to assess the direction of their targets from a distance, and approach them rather directly. Experimental evidence suggests that such directional cues exist and that misrouted fibers can approach the targets across considerable distances. Often directional growth appears to be modulated by a random fluctuation of direction, leading to meandering pathways; but a meander is still essentially oriented by directional cues; self-crossing of tracks are rare, whereas they would be frequent in a stochastic, diffusion type pathway. Fujisawa (23) has found that axons may innervate newt tectum from different aspects, approaching their target positions rather directly on moderately tortuous pathways. Retinal axons are capable of distinguishing tectal cells derived from different positions of the tectum; the data suggest there is a graded distribution of guiding or affinity parameters in the tectum (3). If an obstacle is placed into the pathway of ingrowing axons on the tectum, the axons are first deflected, but thereafter they curve back towards what seems to be their target (24). Taken together the evidence on directional cues argues in favor of the involvement of graded components in fiber guidance, especially at early stages of innervation.

Gradient mechanisms for fiber guidance presume that fibers grow "up" or "down" the slope of a spatially graded distribution of some parameter p. A single monotonically graded effect in the target tissue would drive all fibers to a margin. In order to specify internal target positions two antagonistic gradients - or at least two spatially antagonistic effects arising from the same graded distribution - must be postulated for each dimension. Then their combined effect can lead to a local maximum or minimum; this would allow fibers to be guided towards internal positions. In itself, such specification of a maximum or minimum would still not suffice for generating a projection which requires that target positions depend on positions of <u>origin</u> of the fibers. The searching axons must carry markers specifying their origin; in the simplest conceivable case they would be graded with respect to position of origin of the fiber in the retina, but uniform within the axon and its growth cone. All the retinal markers have to do is to influence the relative efficiency of the antagonistic effects arising from spatial gradients in the target tissue. For instance, there might be counter-gradients in the tectum, and the retinal axon may carry receptor molecules which determine the efficiency of one of the antagonistic tectal gradients, but not of the other. Not all such interactions lead to projections, but conditions are simple and can be given in general mathematical terms (15).

A specific example is the following model based on exponential distributions. Assume that growth cones of retinal axons carry enzyme molecules of concentration $e^{-\alpha u}$, $e^{-\beta v}$, depending on the position of origin in the retina given by coordinates u, and v. Within the tectum there are graded markers distributed according to $e^{-\alpha x}$, $e^{-\beta y}$: Components of the growth cone interact with tectal components in activating and inhibitory fashion, producing products of concentration

$$q_x = e^{\alpha(x-u)};\ q_y = e^{\beta(y-v)};\ q_x{}' = e^{\alpha(u-x)};\ q_y{}' = e^{\beta(v-y)}$$

The sum $p = q_x + q_x{}' + q_y + q_y{}'$ has its minimum at $x = u$, $y = v$. Axons are assumed to grow along the slope of p until the minimal level is reached; this occurs at $x=u$, $y=v$ corresponding to a projection. The target position is approached from all directions; misrouted fibers have no difficulty in finding their respective targets.

This model is only one out of many which may easily be constructed along the same general line of thought. Another example would be a model in which production of guiding substance p is activated in a linear, and inhibited in an overlinear manner by q_x and q_y, such that

$$p = q_x/(1+q_x{}^2) + q_y/(1+q_y{}^2)$$

This leads to a maximum at $x = u$, $y = v$, again corresponding to a projection. (This scheme is the basis for the model calculation of axonal pathways shown in Fig. 2). It is not necessary that the coordinate system is Cartesian. Moreover, tectal markers need not remain at fixed levels all the time. To the contrary, if we assume that p changes slowly at a rate proportional to the density (number per unit area) and the duration of presence of the tips of invading retinal fibers, fiber terminals shift until their density has become near-uniform; in this way, regulatory effects such as extension and contraction of maps following ablations of retina, or tectum may be accounted for (16). It is likely - though not logically necessary - that synapses are disrupted and reformed during this process; regulation of development may involve "shifting connections" (25,26).

So far we have only discussed formal requirements for the formation of projections based on spatially graded components. Which mechanisms would be able to guide axons along the slope of gradients in the first place? The most common notion is guidance by differential adhesion, the idea being that axons extend microspikes which adhere more strongly in one direction than in others, thus pulling the growth cone in the direction of maximal adhesion. One may question, however, whether axons would not get stuck on the way if guidance were exclusively due to mechanical adhesion, and there are alternative possibilities to be considered which are more related to mechanisms of chemotaxis. In fact, there is evidence that both adhesion and chemotaxis may guide neural fibers - on artificial surfaces growing axons tend to avoid areas of low adhesion; but also nerve growth factor can attract certain types of axons in the direction of its source (27). One of the best studied cases of chemotaxis in non-neural biology is the guidance of cellular slime molds in the direction of a gradient of cyclic AMP (28). Some 1 % concentration difference of CAMP across the cell appears to be sufficient for activation of a part of the cell in the direction of the gradient, giving rise to pulse-like directed movement. We have suggested that the same type of mechanism proposed for intercellular pattern formation may also account for intracellular patterns giving rise to such directed growth: short range activation, in conjunction with longer range inhibition or depletion effects within the cell or on its membrane may cause a focus of activity oriented by slight directional cues (29). This principle may apply to the direction of axonal growth on the target area: A slight concentration difference of a guiding parameter p in the area of contact between growth cone and target tissue may be strongly enhanced intracellularly within the growth cone, giving rise to a focus of activation in the direction of maximal slope of p. This may lead to directed growth (Fig. 1) until the position of maximal p is reached (15). Such mechanisms do not exclude a correlation of the direction of growth and adhesivity, but the correlation is not logically necessary and, if present, it may be weak because intracellular enhancement of small differences plays a crucial role in directional guidance.

Once the area of optimal p is reached, how does the fiber sense that this is the target position? Fiber pathways documented in the literature suggest that the actual pathway is the result of superposition of directional growth and some random fluctuations of direction. In some but not in all cases there is branching in the course of fiber growth. In most cases there is a high degree of arborization near the terminal position. We may link these observations to the gradient hypothesis of directional guidance by assuming that (1) there is some random directional fluctuation superimposed on guidance by gradients, leading to a meandering pathway, and (2) whenever the proximal part of the growth cone is activated, indicating that it points into an entirely wrong direction, there is a considerable probability of branching (Fig. 2), either at the growth cone itself or at some backward position. This is consistent with some branching occurring in the course of fiber growth, but

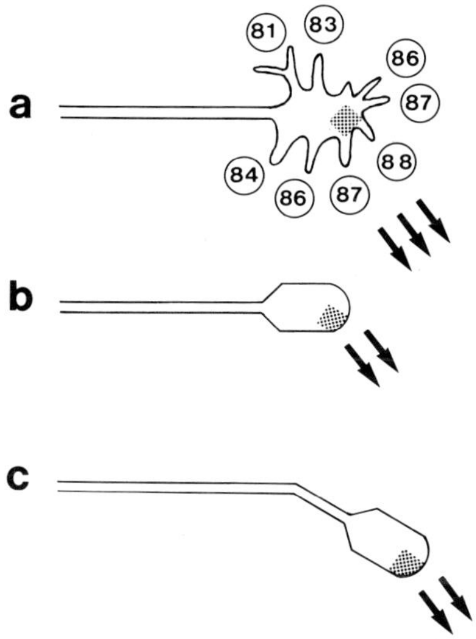

Fig. 1 Model for navigation of the axonal growth cone
(a) Guiding parameter p, generated by interaction of the growth cone with graded components of target tissue varies slightly across the area of the growth cone. The numerical values are arbitrarily chosen to illustrate the principle. Intracellular pattern formation by autocatalysis and lateral inhibition is capable of enhancing this directional cue, producing a focus of activation directed by the slope (in this version of the model towards the maximal value) of p. The arrow indicates the direction of the slope. (b,c) The focus of activation exerts a steering effect, directing fiber growth towards the target position, that is, the maximum of p.

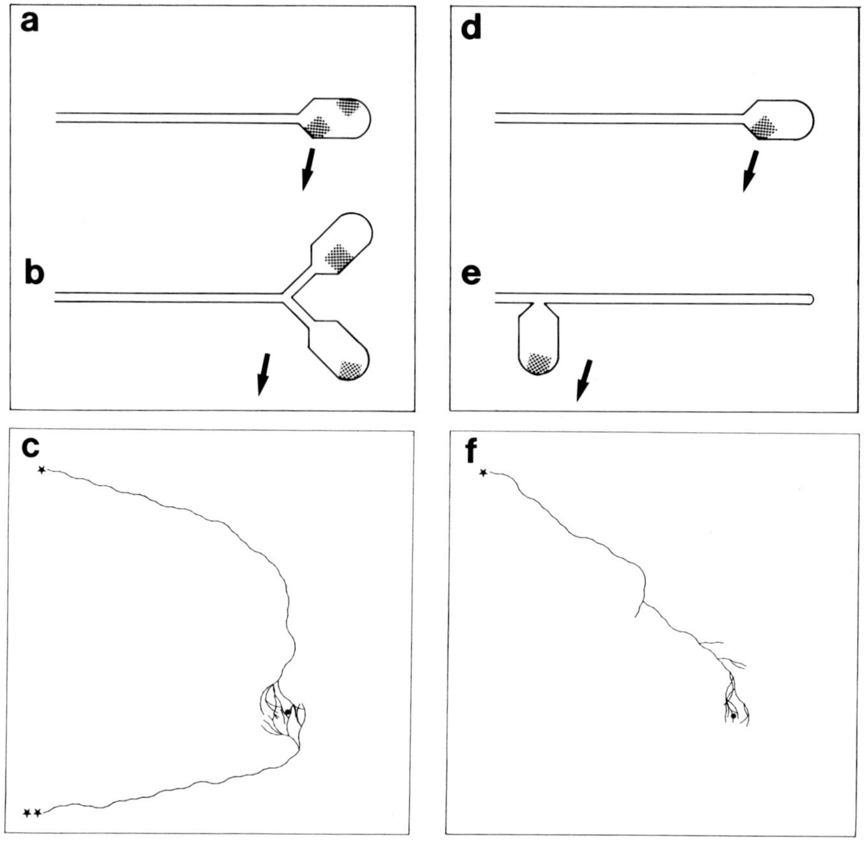

Fig. 2 Branching models:
(a-c) Forward branching. Axon navigation is assumed to be determined by the slopes of the guiding parameter p towards its maximum. A superimposed random fluctuation of direction of growth leads to meandering pathways. To account for branching, we postulate that if an axonal growth cone encounters a "downward" slope of p, and therefore the growth cone is activated in its proximal part (a), this leads, with high probability, to two focuses ▩ of activation, and this in turn causes branching (b). (Also, we introduce finite probabilities for termination of growth as well as for continuation of growth without branching in the case of proximal activation. Branching patterns strongly depend on these probability values). Such branching may but need not occur along the pathway of a meandering axon; in the area of termination where all slopes go "down", extensive terminal arborization takes place. After a certain amount of branching has occured in sequence growth is assumed to be terminated. A computer simulation of two pathways, leading from different starting positions * and ** to the same target position • is shown in Fig. c.
(d-f) Backward branching. It is assumed that proximal activation of the growth cone (d) stops further growth and causes a new branch to occur at a position not far behind the original growth cone (e). (f) Computer simulation of a pathway resulting from such backbranching.

it also explains rather directly that there is much branching once the optimal position of p is reached, because, from there, all pathways lead to "worse" positions. A certain amount of branching may then be the signal that the target area is reached with respect to the x,y dimension. (Patterns of innervation in the third, z dimension differ from one system to another - they are different in birds, amphibia and fish - and are not dealt with in this article). Details of arborization strongly depend on quantitative probability values of branching, continuation of growth, and stopping growth altogether when the growth cone senses the "wrong" direction by proximal activation.

In summary, specification of projections by gradients implies that few chemical markers can give rise to topological order of large numbers of fibers. Interactions of components which are graded with respect to positions in source and target tissue are capable of generating projections if, for each dimension of target tissue, there are two spatially graded antagonistic effects, modulated with respect to relative efficiency by axonal components which are graded with respect to position of origin. Adhesion may but need not be involved in the mechanism. The interaction of a growth cone with target tissue is assumed to produce, in their domain of contact, a slightly graded component which serves as directional cue, being strongly enhanced intracellularly within the growth cone. Curving, meandering and various types of branching are easily incorporated into the model. Let me re-emphasize that other processes may also contribute to the projections; in particular, spatially coherent impulse activity is involved in refining the map at later stages.

3. POSSIBLE RELATIONS BETWEEN PRINCIPLES OF DEVELOPMENT AND PRINCIPLES OF FUNCTION

Are there any mechanisms which are involved both in development and in function of the brain? This is conceivable, in particular, for two processes mentioned above: Shifting connections as well as the activation (or stabilization) of connections by coordinated electrical activity. Shifting connections may take part in certain processes of regeneration, such as the regeneration of fibers from a half-retina while spreading on an entire tectum; and they are also likely to occur in the formation of the primary projection onto the continuously growing tecta of amphibia and fish (25,26). In the mature brain, shifting synapses may possibly (though not necessarily) contribute to slow adaptations of functional areas within the cortex to the degree of stimulation (as described in the article by Koenderink, this volume).

Sharpening of the map in the course of development may depend on electric activity; according to current thoughts, neighbouring ganglion cells in the retina fire together, and axon terminals become stabilized if they are activated in a spatially coherent area in the tectum; this may serve as a check of correct neighbourhood relations (30,31). Solitary terminals of misrouted axons are destabilized; either they shift or they die. With respect to brain function, mechanisms of stabilizing simultaneously activated neural connections have long been invoked for processes of learning in the adult brain. A recent set of papers on such relations between development and function are collected in ref. (32). On the other hand it is to be emphasized that most important mechanisms of information processing are different from those of development.

Probably, the deepest relation between generation and function of the brain is to be found on a purely formal level. A good example is the principle of auto-catalysis in combination with lateral inhibition. It appears as a basis for the de-novo generation of spatial patterns in development, but it is also an important principle of information processing in the brain, especially in the context of visual perception (33,34). The mechanisms are different, only the second one being based on impulse activity transmitted by activatory and inhibitory synapses - but the formal similarity is close and impressive. Such relations suggest that structures which can easily be generated by biological mechanisms can often also easily be perceived. This applies for instance to textures with granularity - that is, with random positions of individual elements but systematic avoidance of small distances between them. Such textures are easily produced by lateral inhibition; they occur, in biology, for instance, as distribution of stomata in plant leaves. Textures of this type are also immediately perceived as "figure" on a ground of elements distributed entirely at random (35). We expect, more generally, that there are deep formal relationships between principles of pattern formation and pattern recognition, relationships that may also be relevant for a theoretical approach to aesthetic experience.

REFERENCES

(1) G. D. Trisler, M. D. Schneider, M. Nirenberg, Proc. Nat. Acad. Sci. USA 78 (1981) 2145-2149.

(2) W. Halfter, M. Claviez, U. Schwarz, Nature (London) 292 (1981) 67-70.

(3) F. Bonhoeffer, J. Huf, EMBO J. 1 (1982) 427-431.

(4) A. Gierer, Curr. Topics Dev. Biol. 11 (1977) 17-59.

(5) A. Turing, Trans. R. Soc. 237 (1952) 32-72.

(6) A. Gierer, H. Meinhardt, Kybernetik (continued as "Biological Cybernetics") 12 (1971) 30-39.

(7) A. Gierer, Progr. Biophys. molec. Biol. 37 (1981) 1-47.

(8) H. Meinhardt, A. Gierer, J. Theor. Biol. 85 (1980) 429-450.

(9) L. Saxen, Curr. Top. Devel. Biol. 15 (1980) 409-418.

(10) J. Cooke, Curr. Top. Devel. Biol. 15 (1980) 374-407.

(11) J. B. Sheffield, A. A. Moscona, Exp. Cell Res. 57 (1969) 462-466.
(12) G. Vollmer, P. G. Layer, A. Gierer, Neurosci. Lett. 48 (1984) 191-196.
(13) R. W. Sperry, Proc. Nat. Acad. Sci. USA 50 (1963) 703-710.
(14) S. E. Fraser, Devl. Biol. 79 (1980) 453-464.
(15) A. Gierer, Biol. Cyb. 42 (1981) 69-78.
(16) A. Gierer, Proc. R. Soc. Lond. B 218 (1983) 77-93.
(17) D. J. Willshaw, C. von der Malsburg, Phil. Trans. R. Soc. Lond. B 287 (1979) 203-243.
(18) G. Rager, Proc. R. Soc. Lond. B 192 (1976) 353-370.
(19) D.J. Willshaw, C. von der Malsburg, Proc. R. Soc. Lond. B 194 (1976) 431-445.
(20) V. A. Whitelaw, J. D. Cowan, J. Neurosci. 1 (1981) 1369-1387.
(21) F. Bonhoeffer, A. Gierer, TINS 7 (1984) 378-381.
(22) M. G. Yoon, J. Physiol., Lond. 252 (1975) 137-158.
(23) H. Fujisawa, Devl. biol. 90 (1982) 43-57.
(24) S. Thanos, F. Bonhoeffer, Neurosci. Lett. (1986), in press.
(25) R. M. Gaze, M. J. Keating, S. H. Chung, Proc. R. Soc. Lond. Biol. 185 (1974) 301-330.
(26) S. S. Easter, C. A. O. Stürmer, J. Neurosci. 4 (1984) 1052-1063.
(27) R. W.Gunderson, J. N. Barret, Science, N.Y. 206 (1979) 1079-1080.
(28) J. T. Bonner, J. Exp. Zool. 106 (1947) 1-26.
(29) H. Meinhardt, A. Gierer, J. Cell Sci. 15 (1974) 321-346.
(30) R. L. Meyer, Developm. Brain Res. 6 (1983) 293-298.
(31) J. T. Schmidt, Cell. mol. Neurobiol. 5 (1985) 65-84.
(32) Activity-dependent synaptic changes. Special issue of: Cell. Mol. Neurobiol. 5 Nos 1/2 (1985) 1-207.
(33) S. W. Kuffler, Cold Spring Harbor Symp. Quant. Biol. 17 (1952) 281-292.
(34) K. Kirschfeld, W. Reichardt, Kybernetik (continued as "Biological Cybernetics") 2 (1964) 43-61.
(35) B. Julesz, IRE Trans. Info. Theory IT. 8 (1962) 84-92.

Generation of Spatial Order in the Developing Nervous System

Discussion
edited by *Klaus-Peter Hoffmann*

Comment: The notion behind much of what you said seems to be "One should try to be economic with genetic information". The idea of the gradients practically means no genetic information.

Answer: Lower organisms such as nematodes have many more genes than nerve cells. There, in principle, it could be that most specifications have individual chemical answers. We, mice or monkeys have many more nerve cells and connections than genes and therefore the question is: How is this possible? How are far reaching connections specified? This can be divided into the problems of path finding and topological relations between order in the region of origin and the target area. The problems are solved according to simple rules requiring only few genes. Therefore, a molecular gradient scheme is attractive because very few genes would do it. On the other hand we have to incorporate many modulations into the system.

Question: What creates and maintains the gradients? Diffusion would constantly wash out the gradient.

Answer: As long as you want to maintain the gradient you have to feed in energy and products. In terms of embryology these gradients act only at relatively small time intervals and then they are canceled. Hydra seems to be an exception to this. There the gradients may be excisting for ever.

Question: Could functional mapping throughout lifetime be possible by mechanisms using chemical gradients or being formally analogue ?

Answer: It seems conceivable.

Question: Does the property of biological systems to self organise have a genetic blueprint?

Answer: It could be a set up of enzyme and receptor systems which in combination are capable of establishing patterns. Nothing in terms of preexisting spatial order in addition to this is necessary.

Question: A number of experiments (e.g. done by Michael Strycker) demonstrate the relative role of function in the organization process. Intravitreous injections of tetrodotoxin block the ganglion cell activity that in turn prevents the formation of ocular dominance stripes in visual cortex. Have such experiments been done in much younger embryos so that the entire activity of the nervous system is suppressed during the growth?

Answer: The early crude events happen without function, whereas the refinement, the sharpening is dependent on neighbourhood relation control by synaptic activity. Coherent activity in a small spot of the retina causes an entire group of activity in the tectum where the map is correct, whereas solitary activity would lead to shifts and death.

Question: Do you have a suggestion how collateral formation along the path of an axon takes place?

Answer: The navigating growth cone also can read other biochemical signals, that is signals qualitatively different from the gradient in the final target area.

Question: Do collaterals always arise from the growth cone or can they sprout later?

Answer: One could suggest two types of branching models. One, whenever the growth cone goes in the entirely wrong direction it forms two focuses and branches. Second, another model is back branching. When the growth cone comes to a wrong position it stops and gives the instruction that somewhere behind a new branch goes off in a new direction.

Comment by the speaker: Expansion of the map in the tectum of goldfish with a half retina and changes in the map of the hand representation in the monkey cortex as reported by Merzenich earlier in this meeting could be governed by similar principles. The rule of the game would be that the synaptic activity per unit area in the map is equalized. Therefore, a map expands into an empty tectum half and active skin areas occupy wider representation than non active ones. Similar algorithms could hold for expanding a map after removing half of the retina or changing the metric of the map in the tectum due to especially high activity in a particular part of the retina.

Co-Operation Between Retinal and Cortical Input in the Midbrain of Mammals

Klaus-Peter Hoffmann

Summary

A model is proposed that explains the differences in the optokinetic reflex (OKR) in various mammals by the specific interactions between retinal and cortical projections in the nucleus of the optic tract (NOT) in the pretectum. The model is based on the following assumptions.

1. A genetically prespecified retinal input reaches the NOT first during ontogeny.
2. Thereafter information flow via cortical connections is accepted only if it agrees with the complements of the retinal input.
3. After the cortico-pretectal connections have been established, the retino-pretectal connections gradually lose their influence and are replaced by cortical afferents.

This model explains why after the loss of visual cortex the optokinetic reflex is much weaker and asymmetric in cats and monkeys or absent in man and why visual deprivation or strabismus leads to a loss of binocularity in the NOT and as a consequence to an impaired and asymmetric OKR.

1 Introduction

Numerous experiments dating back to the last century have shown that visual structures in the midbrain of vertebrates play an important role in visually guided behaviour. In addition to the oculomotor nuclei themselves there are two other midbrain structures with clearly defined visuomotor functions. The superior colliculus (SC) is important for saccadic eye movements as well as head- and body- movements in the visual grasp reflex and the nucleus of the optic tract (NOT) is essential for slow eye movements in the optokinetic reflex (OKR) (1-5). Comparative neuroanatomical studies of mammals representing an ascending sample from the phylogenetic tree have shown that both the SC and the NOT in the pretectum receive a continually increasing proportion of their input from cortical visual areas as apposed to the well established direct retinal projection. In relation to the well established rule that ontogeny recapitulates in part mechanisms of this phylogeny developmental studies of these structures may offer the opportunity to examine the rules by which information from these different sources is used and integrated to create the specific properties of nerve cells in SC and NOT of adult animals. In this chapter examples will be reviewed to show that the specific filter characteristics of retinal recipient midbrain nuclei are established early in ontogeny before the arrival of cortical afferents. Thereafter information flow via cortical connections is accepted in the midbrain only if it agrees with the complements of the retinal imprint.

In adult cats and monkeys analyses of the response properties of SC and NOT cells have revealed the strong influence of the cortical input and the dependence of some specific properties on that input. For example, visual cells in the SC of the cat show a high degree of binocularity and direction specificity. These properties are lost when the cortical input is disrupted (6). Visual cells in the NOT of cats or monkeys are all direction specific over a large range of velocities and mostly (in cats) or always (in monkeys) binocular. NOT binocularity and responsivity to high stimulus velocities are lost after decortication whereas direction selectivity for lower velocities is maintained by the direct retinal input (7).

2 Properties of neurons in the nucleus of the optic tract

In the NOT of all mammals tested so far (for a review see 8) neurons with very large (up to 90° horizontal width) receptive fields and with direction specific responses to stimulus movement over a very broad velocity spectrum (0.1°/s - 100°/s) have been found. There is also an internal structural segregation according to the preferred stimulus direction. Almost all cells in the left nucleus prefer stimulus movements to the left and those in the right nucleus prefer stimulus movements to the right in the visual world (9-11). The output of these neurons goes to at least 3 sites, i.e. the dorsal cap of the inferior olive, the nucleus prepositus hypoglossi and the area of the nucleus reticularis tegmenti pontis (12-14) (Fig. 1). All these areas have been shown to contribute to the control of reflex or voluntary eye and head movements in different animals. Interestingly, recordings from NOT neurons in awake cats with implanted search coils to measure eye movements showed that the discharge rate is entirely dependent on the retinal slip of the stimulus and is not modulated during optokinetic afternystagmus or by vestibularly induced nystagmus in the dark (15). Electrical stimulation through NOT recording electrode (pulse width 1ms, frequency 60Hz, maximal amplitude 0.5mA) elicits clear nystagmus with the slow phase towards the stimulated site. Therefore, independent of whether visually or electrically driven, an increase in neuronal activity in the left NOT over that in the right NOT leads to optokinetic nystagmus (OKN) with slow phase to the left whereas higher activity in the right NOT leads to OKN with slow phase to the right. In a normal cat slow phase eye velocity during OKN was always slower than the stimulus velocity and cells in the left NOT would discharge at a rate above spontaneous activity as long as the stimulus moved leftward and vice versa (15).

3 Retinal input to the nucleus of the optic tract

The conduction time along retinal axons from the chiasma to the NOT in the cat is 3 - 7ms. By this measure, the axons are slowly conducting and must originate from W-cells in the retina (10). We sought to directly identify the type of ganglion cells in the retina projecting to NOT by recording in the retina and identifying ganglion cells which could be antidromically activated by low current stimulation in the contralateral NOT (16). From a sample of 558

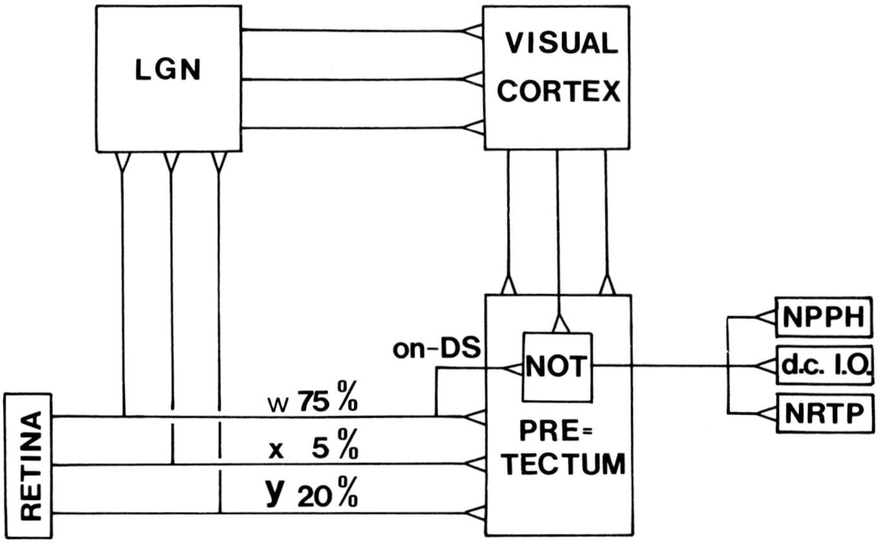

Figure 1: Visual input to the pretectum of the cat. 75% of the cells in the pretectum are activated by slowly conducting retinofugal fibers of the W-cell class, 20% by the fast conducting fibers of the Y-cells, and only 5% by fibers of the X-cells. The NOT receives a large proportion of its retinal input from W-cells with on-centre direction-selective receptive fields. The cortical input to the NOT originates from layer V pyramidal cells with "complex" receptive fields. The NOT projects to the nucleus prepositus hypoglossi (NPPH), the dorsal cap of the inferior olive (d.c.I.O.), and the nucleus reticularis tegmenti pontis (NRTP). LGN, lateral geniculate nucleus. Figure taken from (29).

retinal ganglion cells 98 were activated antidromically from the contralateral NOT including 43 Y-cells, 21 X-cells and 34 W-cells. Only some of the W-cells met both of the criteria for putative retinal afferents to the NOT: appropriate low conduction velocity and low threshold of antidromic stimulation. The rest were presumably activated by stimulation of fibers of passage in the brachium of the SC. The subgroup of W-cells includes one example or more of the following subclasses of W-cells distinguished by Stone and Fukuda (17): concentric phasic (3 cells), concentric tonic (2 cells), suppressed by contrast (1 cell), and on-center direction selective (5 cells). Of interest is the presence of these 5 on-center direction selective cells among the 11 putative NOT afferent cells because this is much higher than the proportion of these cells among W-cells generally.

These results suggest that direction selective on-center W-cells may form the major input to the NOT. The horizontal temporo-nasal preferred directions of 2 of the 5 cells correspond to the direction specificity of NOT cells. The 3 other cells, however, had oblique preferred directions (1 up and temporo-nasal, 1 up and naso-temporal, 1 down and naso-temporal). The large size of NOT receptive fields could be the result of substantial convergence from retinal ganglion cells. In decorticated cats NOT cells respond poorly to velocities higher than 20°/s but are still directionally selective for slower speeds. A similar velocity tuning was found for the direction selective on-center ganglion cells. However, NOT cells in intact cats and monkeys still respond to stimulus velocities over 100°/s. Also many NOT cells can be activated from both eyes whereas the retinal input to the NOT in the cat seems to come almost exclusively from the contralateral eye (4, 18). Thus the normal receptive field properties in the NOT of adult animals must come from other sources, most probably visual cortical areas.

4 Cortical input to the nucleus of the optic tract

Cortical cells in area 17 and 18 projecting to the NOT in the cat were identified by antidromic stimulation of the terminals of such cells in the NOT (19). In agreement with many anatomical and physiological studies these cells were shown to be layer V pyramidal cells. The response properties of antidromically activated cells were very similar to the so called cortico-tectal layer V pyramidal cells. All units had (for the visual cortex) quite large receptive fields (up to 5° in diameter) and could be equally well activated by oriented light bars as well as by large area random dot patterns moved across their receptive field. All the antidromically activated units were direction selective to some degree with a preference for horizontal movements on the average. A wide range of stimulus velocities was effective in driving these cells and most of them responded well to speeds greater than 20°/s. All cells but one were binocular and the binocular responses were always stronger than those to monocular stimulation. Electrical stimulation in area 17 and 18 in the cat or in area V1 and the middle temporal (MT) as well as middle superior temporal (MST) area in macaque monkeys activate all cells recorded in the NOT of these animals. Neurons in MT and MST are particularly sensitive to the movement of visual stimuli and are mostly direction specific. Lesion studies of Wurtz and his co-workers have recently shown that areas within the superior temporal sulcus may be involved in the control of the optokinetic reflex (20).

The comparison of retinal and cortical input to the NOT very clearly shows that the typical response profile of cells in the NOT of adult cats and monkeys reflects a very strong cortical input. As has been elaborated in a previous review (4) bilateral lesions of the visual cortex in the cat result in NOT cells that are only weakly modulated by visual input and respond only to stimuli moving at speeds less than 20°/s. Without a visual cortex all cells are exclusively driven by the contralateral eye. Clearly the functional role of the cortical visual input to the NOT added to the retinal input, is to contribute binocularity and increased responsiveness to higher stimulus velocities.

5 Ontogeny of the receptive field properties in the nucleus of the optic tract

To study the role of visual experience for the development of normal receptive field properties in the NOT we examined the effect of monocular lid suture for 6-24 months, beginning the first week after birth. In these cats binocularity in the NOT was always reduced. In most cases all NOT neurons lacked an ipsilateral input regardless of whether the deprived or the non-deprived eye is the ipsilateral one.

The neurons driven by the deprived as well as by the non-deprived eye exhibit a clear direction specificity for temporo-nasal movements of optokinetic stimuli. They differ, however, significantly in their capacity to respond to high stimulus velocities. Only cells which can be stimulated through the non-deprived eye can respond when the stimulus moves faster than 20°/s. Direction selectivity in NOT cells of the cat is thus independent of visual experience whereas responses to high velocities and binocularity do not develop without visual experience. This is consistent with the view that responses to high velocities and binocularity are mediated through visual cortex and that this structure is plastic during early life and is dependent on visual experience whereas the development and organization of the direct retinal inputs to the NOT is not.

A simple explanation can be given for the changes ipsilateral to the deprived eye (Fig. 2). The deprived eye appears to have lost its cortical connections and all neurons in the NOT are controlled by the contralateral non-deprived eye and show normal properties. The changes in the NOT contralateral to the deprived eye are quite different. Here all cells are driven by the deprived eye. The ipsilateral cortical input which should be able to mediate the influence of the non-deprived (ipsilateral) eye has not developed its normal function. The anatomical substrate for this cortico-pretectal projection is still present and can be demonstrated with electrical stimulation of the corticofugal fibers as well as with retrograde tracing methods. Electrical stimulation of the cortical areas leads to postsynaptic action potentials in NOT cells. But this cortical projection to NOT cannot change the rate of action potentials in response to visual stimulation (21). As we have to accept that cortical cells driven by the non-deprived eye send axons or axon collaterals to the NOT, the disruption may be at the cortico-pretectal synapse. We suggest that a mechanism following Hebb's paradigm (22) controls the placement of cortical terminals on NOT cells in accordance with visual experience. If, however, the retinal influence is blocked by a closed lid during the sensitive period and NOT cells discharge all the time at their spontaneous rate, the cortical axon terminals may make inappropriate or subthreshold connections in the NOT. Although the non-deprived eye remains able to drive cells in the ipsilateral cortex, the cortex is unable to exert a functional influence on the ipsilateral NOT.

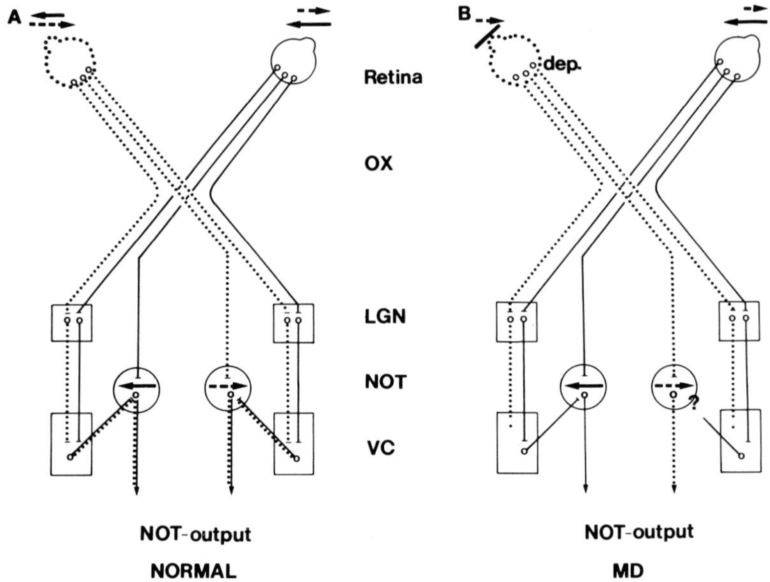

Figure 2: Schematic diagram of our model of the visual pathways mediating the horizontal optokinetic nystagmus (OKN) through the nucleus of the optic tract (NOT) for normal cats (A) and monocularly deprived (MD) cats. OX = optic chiasm, LGN = lateral geniculate nucleus, VC = visual cortex, dep = deprived eye. Projections from the left eye are represented by dotted lines, from the right eye by continuous lines. As can be seen in A, each eye in the normal cat has access to both NOTs, and the output of each NOT is thus binocularly influenced. The degree of this binocularity is a critical factor for the symmetry of OKN. Arrows in front of each eye represent the relative gain for the slow phase of OKN tested monocularly. Temporo-nasal OKN (N-OKN) has higher gain values and is generated by activation of all direction-selective cells in the contralateral NOT whereas naso-temporal (T-OKN) with the lower gain values is generated by activation of approximately half the cells in the ipsilateral NOT. In MD cats T-OKN is almost absent for the deprived eye. Accordingly the projection to the ipsilateral NOT is disrupted in visual cortex. Also N-OKN for the deprived eye is reduced due to the loss of cortical connections. The gain of N-OKN in the non-deprived eye is normal whereas the gain of T-OKN can be reduced due to non-functioning ipsilateral cortico-pretectal connections. Figure taken from (30).

6 Co-operation of retinal and cortical input during the ontogeny of the receptive field properties in the NOT

The proposed model follows closely the summary made recently by Fawcett and O'Leary (23) of the role of electrical activity in the formation of topographic maps in the nervous system (24, 25). Basically we assume that survival of retinal or cortical projections or consolidation of their synapses is dependent on the availability of a critical amount of a trophic survival factor. This survival factor is located in the target cells and the amount released is strongly dependent on the degree of depolarization of the target cells. Many inputs with the same properties releasing their transmitter at the same time will depolarize the target cell more strongly and as a consequence receive more trophic survival factor than inputs which discharge asynchronously (Fig. 3).

During early development, axons of direction selective retinal ganglion cells which respond to horizontal temporo-nasal movements grow towards and terminate within the NOT. This growth is genetically specified or due to as yet unidentified influences. The postsynaptic cell is subsequently strongly direction selective for global movement or retinal slip in temporo-nasal direction, i.e. the cell will be strongly depolarized with movements in temporo-nasal direction presented to the contralateral eye and remain uninfluenced or become hyperpolarized with movements in the naso-temporal direction. At birth, direct retinal axons from the ipsilateral eye have a much weaker influence on NOT-cells. Also, as has been shown for the rabbit (26), most retinal on-center direction selective ganglion cells prefer the opposite horizontal direction as the ipsilateral NOT (the temporo-nasal direction for the ipsilateral eye is equivalent to naso-temporal in the contralateral eye). Thus retinal terminals from the ipsilateral eye will mostly discharge when the target cell is not depolarized by its major input from the contralateral eye and the ipsilateral retinal input present at birth may even be weakened or eliminated during this shaping process. The retinal projection from the contralateral eye does not depend on experience because deprivation does not alter this connection. Rotation of an eye will also rotate the preferred direction of the NOT-cells contralateral to the rotated eye by the same amount (27). Thus a genetically

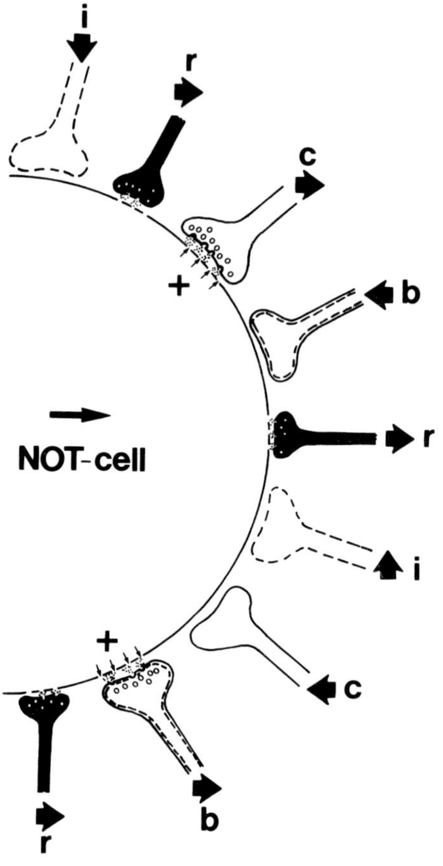

Figure 3: A possible mechanism for the formation of direction specific connections in the nucleus of the optic tract (NOT) based on the assumption that survival of retinal or cortical projections or consolidation of their synapses is dependent on the availability of a critical amount of a trophic survival factor. This factor is released from the target cell upon strong depolarization. The NOT-cell prefers movements in the direction of the arrow due to a prespecified input from direction-selective retinal ganglion cells prefering the same direction in visual space and mainly originating from the contralateral eye (solid terminals, r). Terminals from cortical axons (open terminals) which increase their discharge frequency to movements in the preferred direction of the NOT-cell and which can be activated through the eye contralateral to the NOT will receive enough survival factor. c, i, b: cortico-pretectal axon terminals which can be activated exclusively through the contralateral, ipsilateral or both eyes, respectively. Only synapses of c and b terminals which code rightward movement will be consolidated in the right NOT. They are marked by +. Figure taken from (28).

determined retino-pretectal specificity is already present before the cortico-pretectal projection matures (about 4 - 6 weeks after birth). The selection of the correct cortical axon terminals may occur according to the following rule: Cortical cells strongly activated by the contralateral eye and projecting to the ipsilateral NOT will discharge their terminals at a higher probability in synchrony with retinal axons than cells strongly activated by the ipsilateral eye. In addition only those cortical cells qualify whose preferred direction is identical to the preferred direction of the retinal axons terminating on the target cells. These cortico-pretectal terminals receive enough survival factor because this substance is released by the target cell only upon strong depolarization due to the retinal input. Some synapses will therefore be consolidated and others not. This process will select cortical axons carrying the same direction specific signal from the contralateral eye as the retinal axon. Other cells will be less successful depending on how much their preferred direction deviates from that of the retinal axons or how strongly they are influenced by the ipsilateral eye. Under the assumption that the activity of axons with information from the same retina is more correlated than the activity from different retinae, the ipsilateral retina will be connected to NOT-cells only as an accompanist of the contralateral retina, i.e. through binocular cortical cells (28).

The fate of the retinal input after its role in early development has finished is uncertain. Some retinal terminals remain throughout life. They can be demonstrated by electrical stimulation of the optic tract very well in cats and also but much less convincingly in monkeys (10, 11). Blocking the on-center direction selective cells in the retina by intravitreous injections of amino-phosphono-butyric acid (APB) had little or no effect on the optokinetic reflex (29). It is possible, therefore, that after the cortico-pretectal connections have been established, the retino-pretectal connections may gradually lose their influence and be replaced by cortical afferents. This process may occur to a small extent in cats, to a greater extent in monkeys, and completely in humans.

7 Conclusions

This model could explain why after the loss of visual cortex the optokinetic reflex is much weaker and asymmetric in cats or monkeys and absent in man. This model could explain why in the normal cat the contralateral eye dominates even the descending cortical input. Although all NOT-cells receive cortical input the contralaterally dominated cortico-pretectal axons may be selected by their higher correlation with the retinal axons driving the NOT-cells. This model could explain why in monocularly deprived cats the non-deprived eye can make no connections to the ipsilateral NOT. The NOT cells are never sufficiently depolarized by the sparse ipsilateral retinal input and the activity normally relayed by the more massive contralateral retinal input is blocked by the lid suture. This model could also explain why only contralateral input controls the NOT responses in strabismic cats. The ipsilateral eye is connected to the NOT only through binocular cortical cells and binocular cells are rare in strabismic cats. All these inferences of course could also apply to monkeys and man.

So far we have no information on the development of this pathway in primates. The similarity in the ontogeny of the optokinetic reflex in cats and monkeys suggests, however, that similar developmental mechanisms exist in all higher mammals, including man.

8 References

(1) K. Akert, Helv. Physiol. Acta, 7, (1949), 112-134.

(2) R. H. Wurtz, J. E. Albano, Ann. Rev. Neurosci., 3, (1980), 189-226.

(3) W. Precht, Functional basis of ocular motility disorders, Pergamon, (1982), pp. 291-302.

(4) K.-P. Hoffmann, Spatially oriented behavior, Springer, (1983), pp. 135-153.

(5) H. Vanegas, Comparative neurology of the optic tectum, Plenum Press, (1984).

(6) B. G. Wickelgren, P. Sterling, J. Neurophysiol., 32, (1969), 16-23.

(7) K.-P. Hoffmann, Progress in oculomotor research, Elsevier, (1981), pp. 443-454.

(8) J. I. Simpson, Ann. Rev. Neurosci., 7, (1984), 13-41.

(9) H. Collewijn, Studies of brain function, Springer, (1981), Vol. 5, pp. 75-106.

(10) K.-P. Hoffmann, A. Schoppmann, Exp. Brain Res., 42, (1981), 146-157.

(11) K.-P. Hoffmann, C. Distler, Adaptive processes in visual and oculomotor systems, Pergamon, (1986), pp. 261-266.

(12) W. Precht, P. G. Montarolo, P. Strata, Neurosci. Lett., 17, (1980), 39-42.

(13) M. Magnin, J. H. Courjon, J. M. Flandrin, Exp. Brain Res., 51, (1983), 298-303.

(14) J. Lannou, L. Cazin, W. Precht, Brain Res., 301, (1984), 39-45.

(15) K.-P. Hoffmann, H. P. Huber, Soc. Neurosci. Abstr., 9, (1983), 1048.

(16) K.-P. Hoffmann, J. Stone, Exp. Brain Res., 59, (1985), 395-403.

(17) J. Stone, Y. Fukuda, J. Neurophysiol., 37, (1974), 722-748.

(18) I. Ballas, K.-P. Hoffmann, H. J. Wagner, Neurosci. Lett., 26, (1981), 197-202.

(19) A. Schoppmann, Brain Res., 223, (1981), 1-17.

(20) R. H. Wurtz, Soc. Neurosci. Abstr., 12, (1986), 361.

(21) K.-P. Hoffmann, Exp Brain Res., 51, (1983), 236-246.

(22) D. O. Hebb, Organization of behaviour, Wiley, (1949).

(23) J. W. Fawcett, D. D. M. O'Leary, Trends Neurosci., 8, (1985), 201-206.

(24) J.-P. Changeux, A. Danchin, Nature, 264, (1976), 705-711.

(25) W. M. Cowan, D. D. M. O'Leary, Medicine, science and society, Wiley, (1984), pp. 643-668.

(26) C. W. Oyster, H. B. Barlow, Science, 155, (1967), 841-842.

(27) K.-P. Hoffmann, M. Cynader, unpublished

(28) K.-P. Hoffmann, Imprinting and cortical plasticity, Wiley, (1987).

(29) K.-P. Hoffmann, Progress in brain research, Vol. 64, (1986), Elsevier, pp. 75-84.

(30) C. Markner, K.-P. Hoffmann, Exp. Brain Res., 61, (1985), 117-127.

Co-Operation Between Retinal and Cortical Input in the Midbrain of Mammals

Discussion
edited by *George L. Gerstein*

The Midbrain System for Control of Eye Position

The midbrain (pretectal) neurons under discussion have strong output to the muscles of the eye that cause horizontal movement. Input to these neurons differs in "high" and "low" mammals. In "low" mammals there is mainly a strong projection directly from direction and movement sensitive retinal ganglion cells in the retina. In "high" mammals there is a much stronger additional projection coming from cortex. This explains why the system is little perturbed in "high" mammals when APB, a known blocker of on-pathways is applied intra-ocularly. In contrast, the same manipulation on a "low" mammal abolishes directional selectivity in the pretectal nuclei, and is accompanied by spontaneous nystagmus.

Most of the cortical projection to the pretectal neurons comes from higher visual areas: MT (medial temporal) and MST (medial superior temporal). Those neurons have broad receptive fields that are highly selective for stimulus movement direction, but little selective for details of spatial structure, size, orientation etc. of the stimulus. The projection is extremely strong when tested by electrical stimulation of MT. The population of axons reaching the pretectal neurons from MT represents information from both eyes, and all directional preferences.

The pretectal neurons themselves respond only to horizontal components of the stimulus motion, and only through the contralateral eye. There is no response to the ipsilateral eye. There may be two classes of pretectal neurons: one class with sensitivity to a very broad range of (horizontal) movement velocities, and the other class with similar sensitivity at relatively low (horizontal) velocities, but inhibition by saccadic velocities in any direction.

Ontogeny

In higher mammals, the pretectal system develops through the critical period. Initially the direct retinal input is important, but by the time the animal is adult, there is very little direct retinal input, while the cortical input has become dominant. Therefore, lesion experiments on the cortical sources have different effects on adults as compared to young animals.

Interfering with the system

Lesions of MT in the adult produce a defect in tracking horizontal movements towards the site of the lesion (experiments of Wurtz). However, there is some controversy over whether the defect spontaneously disappears in the course of a few weeks.

Interference with the system is possible during development: deprivation or artificially induced strabismus can both produce ocular instability. Albinism is associated with misdirected fibers and wrong connections: such humans often show strong spontaneous nystagmus.

Electrical stimulation of the pretectal area produces strong nystagmus, so that its output goes to the eye muscle motor region. With "incorrect" sensory input to the pretectum, it is therefore relatively easy to get into positive feedback, as expressed by nystagmus.

Possible Mechanisms

This topic aroused considerable questions and suggestions. The basic property to be explained is how the pretectal neurons select only a portion of the information coming to them from the cortical projection (i.e. the selection is only for horizontal movement and from the contralateral eye), and how this selection develops from the initial dual input (retinal and cortical).

There seems to be no evidence for some kind of feedback mechanism from the motor system for horizontal eye movement.

There is no evidence for a local inhibitory process in the pretectal area which would involve GABA as a transmitter: there is no change in pretectal neuron properties when a GABA antagonist is used. The local pretectal process could be excitatory, but there is no direct evidence.

The direct retinal input to the pretectal neurons is completely contralateral. Could there be some requirement for correlation in the internal structure of the spike trains that helps form the selectivity of the pretectal neurons? This seems unlikely in any detailed sense, because the direct retinal input originates with W ganglion cells, while the cortical input comes via Y ganglion cells. Coarse correlation from the stimulus is possible. Experiments addresssing this question have not been done.

There may be a number of different selection mechanisms at work. Removal of cortical input during the critical period (while the direct retinal input is still strong) does not prevent the pretectal cells from forming horizontal movement preference.

Sources

This summary was made from auditory tape recordings of the actual discussion made by Ulrich Leinhos. In this condensation process, the identity of the actual discussants has deliberately been eradicated, although I have attempted to keep the spirit and essence of the interchanges.

Comments:

Concepts for Changing Neural Network Topology

by *Rolf Eckmiller*

The preceding paper by Hoffmann gives a very interesting example for the adjustment of the neural network topology in the vertebrate visual system by means of a presumably cooperative and irreversible process during early stages of ontogeny. Such a cooperative adjustment may involve pre- and post-synaptic changes, as well as the formation of new synapses and the degeneration of existing ones.

The tacit assumption behind such concepts is, that they all involve the specification of a given neural network topology during a 'teach in' phase early in ontogeny and that henceforth the 'hardware' is permanently fixed (disregarding mechanisms of neural plasticity) to match a specific task of transforming input signals into output signals.

I would like to draw attention to additional mechanisms that may exist for dynamical changes of the neural network topology, namely by means of a specific class of command neurons specifically assigned to this task of topology control. For example area V1 of the primate visual cortex has at least nine cortical layers (1) with input neurons, intracortical interneurons, and output neurons. The horizontal spread (within the layer) of intracortical neurons for example in layer 4B extends over more than 5 mm corresponding to several degrees of visual angle (2). On the other hand area V1 is obviously capable of distinguishing much smaller visual patterns, in the range of minutes or even seconds of arc (3). The <u>concept of command neurons</u> implies that some cortical input signals do not carry visual (or other sensory) information, but rather specify and continuously modify the cortical network topology, by means of influencing membrane thresholds and synaptic weighting factors of selected cortical neurons depending on the desired cortical function.

Schlag and Schlag-Rey (4,5) studied the neurophysiology of the internal medullary lamina (IML) in the central thalamus of alert macaque monkeys. They suggested that the IML with its massive projection to the visual cortex serves as a central controller of the network topology in the visual cortex, rather than as a data processor. In fact, central thalamic lesions were found to cause a neglect syndrome, as would be expected from an impairment of a putative central controller represented by command neurons (6,7). It is theoretically conceivable that the functional network topology is transiently altered by means of such a central controller in order to optimize various visual and oculomotor functions.

Another brain region where a role for command neurons has been suggested, is the posterior parietal cortex in primates as initially proposed by Mountcastle and co-workers (8). Their work provoked considerable discussions and led to important new studies (9,10).

The term 'command neuron' is used slightly differently in studies on various rhythmic motor program generators (e.g. for respiration or locomotion in vertebrates (11) and invertebrates (12), where the command neuron provides the essential input (command) signal for motor program generation, rather than modifying the data processing functions of a neural network by changing its topology.

Presently available 'massively parallel' computers (e.g. Connection Machine, Transputer, BBN Butterfly, Ncube) have a fixed topology, represented by the hardwired connections among its node processors. Considerable efforts are currently being made to develop computers with self-organizing network topology in order to overcome the need of programming the system by means of a sequentially operating host computer. It may be useful to examine possible applications of the command neuron concept, to optimize the programming procedure of novel parallel computers with flexible neural network topology.

REFERENCES

(1) Eckmiller, R.: The neural control of pursuit eye movements. Physiol. Rev. 67, (1987), 797-857.

(2) Gilbert, Ch.: Microcircuitry of the visual cortex. Ann. Rev. Neurosci. 6, (1983), 217-247.

(3) Westheimer, G.: Panum's phenomenon and the confluence of signals from the two eyes in stereoscopy. Proc. R. Soc. B 228, (1986), 289-305.

(4) Schlag, J., Schlag-Rey, M.: Visuomotor functions of central thalamus in monkey. II. Unit activity related to visual events, targeting, and fixation. J.Neurophysiol. 51, (1984), 1175-1195.

(5) Schlag, J., Schlag-Rey, M.: Role of the central thalamus in gaze control. In: Oculomotor and Skeletalmotor System (Prog. Brain Res. vol. 64), H.-J. Freund, U. Büttner, B. Cohen, J. Noth (eds.). Amsterdam: Elsevier, (1986), 191-202.

(6) Orem, J., Schlag-Rey, M., Schlag, J.: Unilateral visual neglect and thalamic intralaminar lesions in the cat. Exp. Neurol. 40, (1973), 784-797.

(7) Watson, R. T., Heiman, K. M.: Thalamic neglect. Neurol. 29, (1974), 690-694.

(8) Mountcastle, V. B., Lynch, J. C., Georgopoulos, A., Sakata, H., Acuna, C.: Posterior parietal association cortex of the monkey: command functions for operations within extrapersonal space. J. Neurophysiol. 38, (1975), 871-908.

(9) Hyvärinen, J.: The parietal cortex of monkey and man. Berlin: Springer, (1982).

(10) Kupfermann, I., Weiss, K. R.: The command neuron concept. Behav. Brain Sci. 1. (1978), 3-39.

(11) Grillner, S., Wallén, P.: Central pattern generators for locomotion with special reference to vertebrates.
Ann. Rev. Neurosci. 8, (1985), 233-261.

(12) Selverston, A. I. (ed.): Model neural networks and behavior.
New York: Plenum Press, (1985).

Review:

The Visual System

by *Martin I. Sereno*

SUMMARY

I discuss selected aspects of the anatomy and physiology of the visual system that may be of interest to theorists attempting to model real neural circuitry. Though a great deal is already known about the visual system, our knowledge is still quite incomplete, particularly with regard to the principles governing interactions among different types of neurons. I therefore consider some strategies for building stronger links between models and the kinds of data we can now obtain.

1 INTRODUCTION

Neural modelers often construct simple abstract networks out of "units", and then analyze the dynamical behavior of these networks in detail, considering only afterwards their possible relevance to real neural circuits. A similar strategy has at times proved very successful in the physical sciences. A turning point in the development of mechanics at the time of Galileo and Newton, for instance, came with the idea of constructing and studying the physics of a thoroughly abstract and simple "absolute space" lacking friction, gravity, magnetism, and so on. Only after this could the physics of more complex and cluttered real worlds be articulated. Similarly, a neural modeler tries to find the basic mathematical forms (presumably, rather more complex ones) that get fleshed out but also made less obvious in the particular networks that happen to have been patched together in the evolution of real brains. I am in sympathy with this approach and only want to make sure that the basic abstract forms are near enough in complexity to real networks that it becomes possible to begin relating the two. It is often suggested that neural-like "units" need not always be construed as neurons, but instead sometimes as modules or even groups of modules, especially when higher level functions are considered. The use of similar models across several levels of organization, however, would seem to imply the existence of strong between-level isomorphisms unlike anything seen in other biological domains. It is difficult to think, for instance, of any isomorphisms across protein structure, cytoplasmic organization, and multi-cellular assemblies. To avoid glossing over significant between-level differences, I think it is essential to be as explicit as possible in describing how a model relates to biological neurons and neural assemblies. To do this, it may be necessary to start out with more realistic neuron-like elements and connection schemes than is customary in network modeling. There are several longer reviews covering some of the topics discussed here (1, 2, 3, 4, 5).

2 VISUAL SYSTEM ANATOMY

The visual system, broadly defined, is spread across 40 or more nuclei and laminated structures in the pons, midbrain, thalamus, basal forebrain and cortex. The first four brain regions are often artificially set to the side or completely omitted in discussions of the cortex. This is unfortunate since every visual cortical area has direct access to the motor system via these structures. It seems likely that the nature of the activity in cortical neural networks has a lot to do with this thoroughly distributed sensorimotor linkage. Having said this, I shall nevertheless have to focus on primate cortex for the sake of brevity. My anatomical notes begin with areal features and then consider successively smaller subdivisions.

2.1 Topography of Visual Areas

Primate visual cortex consists of at least 20 distinct, adjoining regions or "areas" that occupy as a group about half of the total area of the neocortex (2, 3, 4, 5, 6, 7, 8, 9). A working list of visual areas in the macaque monkey followed by their probable and possible counterparts in the owl monkey is as follows: V1 (=V-I), V2 (=V-II), V3 (=DM?), V4/VA (=V4?/VA), V4t/VOT (=DL proper), MT (=MT), VP (=VP), PO (=M), V3A (=DI?), MST (=STv), FST (=?), LIP (=FS?), VIP (=PP?), 7a (=TP?), DPL (=?), PITd (=ITcd?), PITv (=ITcv?), CIT (=?), AIT (=ITr?), TF (=ITml?), TH (=ITmm?), TG (=ITp?). This list is incomplete; but the total number of cortical visual areas is certainly no more than 30. The remaining cortex consists mostly of somatosensory, auditory, and motor areas. Cortical areas have extensive direct interconnections within, but not between, modalities. Between-modality connections in monkeys are made in large part via nuclei in the amygdala and areas in frontal cortex (10, 11).

Visual stimuli are tranduced by photoreceptors, processed by 40 to 60 different cell types in each local patch of the retina (12), and passed along about one million retinal ganglion cell axons into the CNS where precise topological maps are established in the superior colliculus, the dorsal lateral geniculate nucleus (dLGN), and in several other smaller brainstem structures. The projection of the dLGN to striate cortex (V1) is as orderly as the retinogeniculate pathway, and V1 has the largest, finest-grained retinotopic map of any visual area; it occupies about 8% of the total neocortex in macaque monkeys (about 17% in owl monkeys, and 1-2% in humans). In primates, the other, so-called "extrastriate" visual areas receive much of their visual input from V1, either directly, or via intermediate cortical areas and also via cortico$_1$-thalamo-cortical$_2$ links involving six or more topographic visual nuclei in the pulvinar (13, 14). (In some mammals [e.g., cats] every extrastriate area receives a direct input from the dLGN). A second source of input is the superior colliculus, which projects to some of the extrastriate-projecting nuclei in the pulvinar and is capable of visually driving extrastriate cortex in the absence of V1 (15). Cortico-cortical connections and lesion studies have defined somewhat separate parietal and inferotemporal processing streams (see next section).

As one moves by stages into the system, retinal receptive fields grow larger and the retinotopy becomes less clear. However, the absolute topological fidelity of different interareal projections does not vary greatly, and much of the increase in receptive field size can be attributed to the smaller size of most extrastriate areas (from 30% to less than 5% as big as V1) and to the effect of successive slightly divergent remappings. Non-topological projections between adjacent stages in the cortical areas hierarchy appear to be uncommon but have been clearly demonstrated in several feedback projections from areas (CITd, CITv, TF, TH) in inferotemporal and parahippocampal cortex (9). Notably, connections between the thalamic and cortical areas most remote from V1 are topographic (16).

Laminar connection patterns of corticocortical projections can be used to group the 20 or more visual areas into five or six rough hierarchical levels by counting the number of "feedforward" interareal projections one must pass through to get to an area. It is possible to subdivide these levels further. For example, two areas B and C that receive feedforward input from area A might be placed on separate sublevels if B has a feedforward projection to C but not vice versa. In general, there are a number of different feedforward routes to a given cortical area, and these routes often differ in length (i.e., in the number of cortico-cortical projection stages). There are many different routes to higher areas. "Feedforward" projectionssynapse mainly on layer 4 of the target area (the main input layer of cortex--see below), while "feedback" connections avoid layer 4, synapsing above and below it. Most interareal connections are reciprocal. These hundred or so cortico-cortical bundles, each containing a large number of axons, interpenetrate in the white matter below visual cortex, which also contains a hundred or so input and output bundles involving subcortical centers. Limited studies of the latency of response to a flashed stimulus confirm the anatomical hierarchy. Striate cortex is activated with a sharp rise time about 30 milliseconds after the onset of a flashed bar (most of the initial delay occurs in the retina) and each successive interareal feedforward link consumes approximately an additional 10 milliseconds (17, 18) (most of the interareal delay reflects within-area processing). Thus, over 100 milliseconds typically passes before the areas at the top hierarchical levels become active (19).

There are numerous outputs from the visual cortex. The higher areas project into the limbic system. Areas in the top three hierarchical levels project to several nuclei in the amygdala. The areas at perhaps the highest level (TF, TH) project to the entorhinal cortex, which projects into the first of a sequence of areas in the hippocampus (in order: dentate gyrus, CA3, CA1, subiculum). Some of these last projections (e.g., dentate to entorhinal, and CA3 to CA1) appear to be topographic (20). In contrast with the upper level projections into the limbic system, outputs to the motor system emerge from all levels in the hierarchy. Each visual area projects to several of the following: frontal eye fields, caudate nucleus and putamen, intralaminar nuclei (which project to motor cortex), superior colliculus, pontine nuclei. In many cases (e.g., cortical areas to the caudate nucleus and superior colliculus) these projections are topographic.

Some areas contain an almost complete, topologically continuous map of the contralateral visual hemifield (e.g., V1, MT). Other areas, however, introduce a discontinuity into the visual field representation at the horizontal meridian (e.g., the upper and lower field V2 representations are separated), lack a representation of part of the visual field (VP lacks a complete lower field map), distort or stretch their representation (e.g., V2, V3, VP, and VA and others are elongated in the central to peripheral direction), emphasize central vision or peripheral vision (inferotemporal and parietal areas, respectively), or have callosally supported representations of part of the ipsilateral visual field (present only in highest areas). The majority of areas are topographic, and when viewed from the surface, can be divided into areas that have a mirror image transformation of the visual field (e.g., V1, V3, VP, DL) and areas that have an untransformed representation (e.g., V2, V4, MT, M).

2.2 Subareal Features

There are a number of macroscopic (or nearly so) area-specific features of the cortex defined by vertical and tangential groupings of cell types and their dendritic and axonal processes. These were first studied in detail by Brodmann and other architectonicists at the turn of the century; this is how a number of cortical areas were initially recognized. The vertical or laminar organization of all of the cortex follows a general plan, and is discussed first. A newly discovered largely tangential sub-parcellation of V1 and V2 is then summarized.

The cortex is divided into six main layers with layer 1 being the most superficial and layer 6, the deepest. The general input/output wiring scheme is the same in every cortical area (see 21, 22, 23, 24, 25, 26 and the next section for visual cortex). Layer 4, the middle layer in most cortical areas, is the "feedforward input" layer, receiving input from thalamic nuclei, the claustrum (a part of the basal ganglia), and other cortical areas. Cortico-cortical feedback inputs avoid this layer. Layer 4 neurons project mainly to neurons above and below them in the same area. Layer 4 is particularly well developed in striate cortex where four subdivisions are recognized (from the top: 4A, 4B, 4C-alpha, 4C-beta). The output from a cortical area comes mainly from layers 2, 3, 5, and 6. Layers 2 and 3, the "feedforward output" layers, project primarily to layer 4 of other cortical areas. Layer 6, the "feedback" layer, projects back to thalamic nuclei, the claustrum, and cortical areas that each provide a layer 4 input. Layer 5 is the main "motor output" layer, projecting to the spinal cord, red nucleus, and reticular formation (from sensorimotor cortex), intralaminar thalamic nuclei, the pontine nuclei, the superior colliculus and pretectum, and the caudate and putamen. Layer 1 contains many long tangential axons from pyramidal neurons and a sparse population of cells with horizontal axons; both contact the apical dendritic tufts of cells from all layers.

Since the early studies of Hubel and Wiesel in cats and primates, it has been known that area V1 contained tangentially arrayed gradients of orientation tuning angle intercalated into its retinotopic map. More recent work suggests that different sorts of visual information in V1 and V2, including a non-orientation-tuned component, are partially segregated into three parallel "streams" for each region in the visual field (27, 28, 29, 30, 31). This work recapitulates and extends work in the cat visual system that grew out of the X, Y, and W retinal ganglion cell classification (32, 33). In monkeys, the three streams have been labeled the parvocellular, magnocellular, and interlaminar pathways after their presumed relay layers in the dLGN. In the cortex, these pathways are segregated vertically in different cortical layers, and tangentially in regions--the blobs and interblobs in V1, and the thick stripes, thin stripes and interstripes in V2--that can be visualized with a stain for the metabolic enzyme, cytochrome oxidase. The V2 stripes extend from its posterior border (the vertical meridian) to its anterior border (the horizontal meridian) thus forming 90 degree arcs when back-projected into the visual field.

The magnocellular pathway begins with Y-like retinal ganglion cells, distinguished by their larger receptive fields, more transient response, sensitivity to low contrasts, lack of color selectivity, and sensitivity to higher velocities. It passes through the two magnocellular layers of the dLGN on the way to V1 where the first stage is layer 4C-alpha and the second is layer 4B, which projects directly to the V2 thick stripes, V3, and MT. The V2 thick stripes project to MT as well. The subsequent projections in this pathway are primarily to parietal visual areas. The parvocellular pathway begins with X-like retinal ganglion cells, distinguished by their smaller receptive fields, more sustained response, lack of response to contrasts below 10%, color selectivity, and lower velocity cutoff. It passes through the four parvocellular layers of the dLGN, and then to V1, where the first stage is layer 4C-beta, the second, layer 4A, and the third, the interblob regions of layers 2 and 3, which project to the V2 interstripes. The interstripes in turn project to V4. Subsequent projections involve mainly inferotemporal areas. The interlaminar pathway begins with W-like ganglion cells, presently a heterogeneous class known to contain a number of distinct types. Some of these project to the six interlaminar layers of the dLGN, which project directly to the blobs in layer 3 of V1. Layers 3 and 2 then project to the thin stripes in V2, which also project to V4. Again, subsequent projections go to inferotemporal cortex.

2.3 Cell Types

There are a large number of different cell types in the cortex--probably at least 50 (based on present anatomy), but probably less that 500 (based on estimates of dendritic "tiling"). I will simply mention some of the the better-known types found in primate V1. Primate V1 is one of the most derived cortical areas in an evolutionary sense, and probably not the best area to use for demonstrating the general plan of cortical organization. For example, almost all cortical areas in mammals (including the extrastriate visual areas) have about the same number of cells per unit surface area--80,000 per square millimeter (41); primate V1, by contrast, has about two and a half times that number. It is, nevertheless, one of the most extensively studied areas.

Cortical cell types are divisible into excitatory types, "standard" inhibitory types, and peptide-containing types. The major distinctions noted are local (i.e., intracortical) versus white matter (corticocortical) axons, and small-field (less than 100-200 microns diameter) versus large-field (1,000 microns or more) axon arbors. Known or presumed excitatory types have spiny dendrites and are further divided into stellate and pyramidal types (21, 22, 25, 26, 34, 35, 36). Spiny stellate types include: layer 4C-beta cells with small-field local axons to layers 4A, 3, and 5; layer 4C-alpha cells with small-field local axons to layers 4B, 3, and 5; layer 4B cells with wide-field local axons in layer 4B plus a white matter axon; layer 4A horizontal dendrite cells with local axons to layers 3 and 5. There are many pyramidal cell types: layer 4A cells with local axon to layers 3 and 5; layer 4B cells with local axon to layer 3 plus white matter axon; small layer 5 cells with local axon to layers 4A and 3; small layer 2 and 3 cells with local axon to layers 2, 3, 5; larger layer 2 and 3 cells with local axon to layer 5 plus white matter axon (these are typical feedforward cells); large layer 5 cells with local axon to layers 2 and 3 plus white matter axon to tectum, and/or basal ganglia and pons; layer 6 cells with local axon to 4C-alpha or 4C-beta and white matter axon to magnocellular or parvocellular dLGN, respectively; layer 6 cells with local axon to layer 4B and magnocellular dLGN; layer 6 cells with white matter axon to claustrum; and finally, layer 1 Cajal-Retzius neurons with widely branching tangential axons contacting pyramidal cell apical dendrites in layer 1. The local axon arbors of large pyramidal cell types are considerably larger than their dendritic fields; some smaller pyramids and stellate cells have more nearly comparable axonal and dendritic arbors.

There are many "standard" inhibitory types known or presumed to use GABA as a neurotransmitter that usually have smooth dendrites without spines and lack interareal axon branches entering the white matter (25, 26, 34, 35, 37, 38, 39). Some of these are: medium-sized layer 4C-alpha cells with 4C-alpha arbor and branch to layers 4A, 5A and 6; layer 6 cells back to layer 4C-alpha; layer 4C-beta cells to layers 4A and 3B; layer 4A cells back to layer 4C-beta; layer 4C-alpha cells with local axon spreading into layer 4B; layer 5A cells to layer 4C-alpha, 4A, 5A and 6; layer 5A cells with axon to only 5A and 3; basket cells in layers 3, 4, and 5 with beaded dendrites and wide-field local axons; medium-sized layer 4C-alpha cells with small field basket-like axons in layers 4C-alpha, 3B, 5A and 6; chandelier cells in layers 2, 3, 4, and 5 with small-field local axons terminating mostly on the spike-initiating part of pyramidal cell axons; double bouquet cells in layers 3 and 4 with extremely narrow vertically elongated local axon arbors (sometimes less than 20 microns wide) spreading across all layers; very small "clewed" cells with both dendrites and primary axon arbors restricted to layers 4C-alpha or 4C-beta; small cells with dendrites restricted to layer 4C-alpha or 4C-beta, but with axons spreading across both these layers; medium layer 4C-beta cells with dendrites spreading across layers 4C-beta and 4C-alpha but with axons restricted to 4C-beta. The axon arbors of inhibitory cell types may be similar size or much larger than their dendritic fields and in a number of cases, appear to provide points of contact between the cortical streams discussed above.

Finally, there are peptide-containing cells (40), presumed to have slow modulatory actions on post-synaptic cells. They have no clear axon and their vesicle-filled dendritic varicosities do not make conventional synapses. Some peptide-containing cells also appear to release GABA, a "standard" fast neurotransmitter. There are many neuropeptides (e.g., VIP, somatostatin, peptide Y, CCK, dynorphin, substance P) but it seems that there may be only two types of peptide cells--those containing VIP and those containing (simultaneously) several of the other peptides plus GABA. Most GABA cells do not contain peptides.

Most of the cell types described are present in other visual areas (42) and in somatosensory (34) and auditory areas. The main difference is probably that non-primary areas do not have as many cell types in layer 4. However, other areas probably have more cells types in other layers. Also, other areas may have cell types that have restricted tangential as opposed to laminar distributions (e.g., stripe-specific cells in V2).

2.4 Synapses

Our knowledge about the interconnections among different neuron types in the visual system at the level of the synapse is quite incomplete. Here I shall only make some general points about synaptic efficacy and connectional patterns based partly on more complete information available in other systems.

An excitatory or inhibitory cortical synapse normally involves a presynaptic "bouton" and a postsynaptic site on a cell body, on the flat surface of a dendrite, on a dendritic spine, or on the initial segment of an axon. A single bouton often makes one synapse, but large boutons can make several. The axon of a single cell typically branches extensively to give rise to at least hundreds and perhaps as many as ten thousand boutons, and a target cell typically receives synapses from thousands of boutons. Recent studies in the spinal cord and brain stem suggest that both excitatory and inhibitory boutons behave in a probabilistic, all-or-none fashion, emitting either a single "quantum" of neurotransmitter or nothing at all when invaded by a spike (43, 44, 45, 46, 47, 48). (This contrasts with the better understood neuromuscular synapse, which typically releases several hundred quanta at once; however, the endplate is much larger than a bouton, and it drives a much larger target.) A single quantum produces a depolarizing (excitatory) or hyperpolarizing (inhibitory) postsynaptic potential of 100-200 microvolts in the target cell, measured at its cell body. Thus, a minimum of 50-100 of the excitatory boutons contacting a cell would have to be simultaneouly active (within about 5 milliseconds) to overcome the typical 10 millivolt threshold for generation of a spike. Models of the passive electrical properties of dendrites have suggested that the cell body potentials produced by distal dendritic synapses are attenuated and temporally broadened in comparision with potentials produced by otherwise equivalent proximal synapses. Consequently, distal synapses have traditionally been thought to be less efficacious in bringing a cell to threshold. However, the studies cited above suggest that distal synapses produce potentials at the cell body that are similar in size to those of proximal synapses, implying that distal synapses produce a larger local potential deflection. Since the distally-generated potentials have a longer time course at the cell body, they may in fact be quite efficacious.

These studies are significant because they allow us for the first time to attach rough numbers to anatomical pictures. Boutons are easily distinguished, for example, in HRP-filled axon arbors, even at the light microscopic level. A bouton distribution is difficult to assess, however, without knowledge of its postsynaptic targets. Fortunately, it has been possible in a few instances to inject both pre- and post-synaptic partners with HRP so that all contacts between the two can be visualized (43, 44, 45, 48, 49). The number of contacts between a pair of cells varies but has often turned out to be quite small. Connections that have been studied in this way include: Ia afferent contacts on motoneurons (one afferent makes 1-10 contacts per motoneuron--median 4); inhibitory cell synapses on Mauthner neurons (3-60 contacts--median 9); tectal output axons onto cells in the midbrain reticular formation (1-15 contacts--median 3); hair follicle afferents to spinocervical tract cells (1-60 contacts--a few for afferents supporting receptive field edges but many for afferents supporting center). Probably the most patent connection between cells is found in the cerebellum. Inferior olive axons branch to form several "climbing fibers" that each make hundreds of excitatory synapses onto single Purkinje cell dendrites. So far, a similar morphology has not been seen in the cerebral cortex.

Information about connections between neurons at the single cell level is scarce in the visual thalamus and cortex, and existing wiring diagrams have been constructed using mostly indirect data. Probably the best known connection involves the highly specialized chandelier cells or "axo-axonic" cells, which make 5-10 synapses with the axon initial segment of each of several hundred nearby pyramidal neurons (39). Up to 5 chandelier cells synapse on some pyramidal neurons. Thus, although a single chandelier cell might not be able to "veto" the output of a well-driven pyramidal cell (cf. ref. 1), a group of them probably could. Recently, the dLGN to V1 connection was examined at the single axon/single dendrite level, with some unexpected results (50). Single geniculate afferents provide only a few synapses to any given cortical neuron cell body or dendrite (estimated to amount to 1-8 contacts total per cell), but apparently contact very many of them (perhaps 400-800) in layer 4 of V1 in the cat. (Single monkey dLGN afferents in V1 have an order of magnitude fewer boutons than cat dLGN terminals; the number of contacts per cell they make is unknown).

Previously, it was thought that V1 cortical receptive fields were constructed from a much smaller number of dLGN cell inputs. It is perhaps not unexpected that a single axon in one of the highly non-topographic pathways from the tectum to the reticular formation (48) should be unable to single-handedly drive postsynaptic cells. The tectoreticular pathways mediate an activated-locus-to-firing-rate sensorimotor transformation that requires divergent/convergent connectivity in many parts of the reticular formation, and it may be impractical for one cell to make enough synapses. Surprisingly, however, low cell-to-cell connection strengths characterize even highly topographic projections like the one from dLGN to V1. In this case, single input axons make all their synapses in a small volume, but the synapses are not confined to few enough cells to single-handedly drive any of them.

In general, an axon in the cortex distributes boutons to a number of different cell types. Perhaps because of this, a cell of a given type does not usually make many connections with others like itself. Boutons from dLGN axons in layer 4 synapse on both excitatory and inhibitory cell types. They contact stellate and other non-pyramidal cells in layer 4 and pyramidal cells from other layers on spines and shafts and on basal and apical dendrites. Often the contact consists of a single bouton. Most inhibitory cells contacted by dLGN axons are in layer 4 because these cells tend to have smaller dendritic fields. The within-area targets of a few cortical cell types have been studied at the single-cell level, but the total number of synapses made by an axon on any given target cell is generally not known. Excitatory and inhibitory cortical axons, like dLGN axons, contact both excitatory and inhibitory cells, but the proportions and types of contacts vary. For example, the local collaterals of layer 6 pyramidal neurons contact the shafts of significant numbers of putative inhibitory cells in layer 4 (51), while the patchy wide-field local collaterals of large layer 3 pyramids make 90% of their contacts on layer 3 and 5 pyramidal neuron spines but only 5% of their contacts on inhibitory GABA cells (52). Inhibitory types also show quantitative differences. Chandelier cells only contact pyramidal types (39) (they may be unique in this), basket cells contact mostly excitatory types (pyramidal cell bodies, shafts and spines) but also some inhibitory types (26), and double bouquet cells contact mostly inhibitory types and only a few spines (38).

Since excitatory postsynaptic potentials and dendritic time constants are rather short, the typical firing rate of cortical input axons (usually under 100 Hz) is probably not fast enough to generate within-fiber temporal summation (50). Such a phenomenon may occur with inhibitory synapses since their potentials have longer time courses than those of excitatory synapses (20-40 milliseconds versus about 10 milliseconds). This asymmetry of time course may partly explain why there are fewer inhibitory cells and synapses than excitatory ones in the cortex. Since cells often receive 100 times the number of excitatory synapses needed to fire them, a given cell can probably be fired by many different input coalitions. Factors affecting the temporal coincidence of different inputs may thus powerfully alter the effective connectivity of a local cortical network.

3 VISUAL SYSTEM PHYSIOLOGY

3.1 Excitatory Receptive Field Properties

The classically-defined excitatory receptive field of a cell is the region of the visual field from which a simple visual stimulus will elicit an excitatory response. The smallest receptive fields are found in V1 in the representation of the center of gaze and span less than a sixth of a degree of visual angle in macaque monkeys. Receptive field size increases with eccentricity in most areas. Receptive fields in secondary areas span several degrees, in tertiary areas, roughly five to ten degrees, in quaternary areas, over ten degress, and in upper level areas, even more. Even at the highest levels, however, few cells have excitatory receptive fields covering the entire visual hemifield. Receptive fields usually have sharply defined edges, even when they are quite large; almost invariably their borders are convex.

The original Hubel and Wiesel classification of orientation-tuned striate cortex cell types recognized three putatively serial stages (simple, complex, and hypercomplex cells), and has been reexamined in many studies from many different angles. Simple cells have spatially segregated subregions that respond to either the onset or offset of a small spot. Recent studies suggest that there are X-like (parvocellular) simple cells and Y-like (magnocellular) simple cells (53, 54). Some simple cell receptive fields have many "on" and "off" regions arranged as thin parallel strips (up to 8 total) so that they respond best when several cycles of a luminance grating are presented (54). Complex cell receptive fields do not have subregions when plotted with a spot; however, as with simple cells, some prefer more than one cycle of a grating (55). There may also be X-like and Y-like complex cells (53). It is known, now, that some complex cells are driven monosynaptically from the dLGN (56). The hypercomplex property (endstopping) is a graded property and may appear on the ends of both simple and complex cells. This property quite unexpectedly turned out to depend on the cortico-claustral loop (57). So far, few correlations between physiology and anatomy have emerged in cat V1. For example, there is no clear relation between receptive field size and dendritic field size and no obvious anatomical correlates of simple and complex cells. In fact, the one inhibitory cell type that has been characterized physiologically (the basket cell) can be either simple or complex, and be mono- or polysynaptically activated by either X- or Y-like inputs; physiologically, then, these cells are grossly indistinguishable from the various pyramidal cell types (26).

The responses of cells in a number of extrastriate visual areas have been tested with a variety of basic stimulus parameters (e.g., contrast, bar orientation, spatial frequency, direction of motion, speed of motion, wavelength, size, interocular disparity). Except for some simple cells in V2 (55), the receptive fields of cells in most extrastriate areas tend to be "complex" in the Hubel and Wiesel sense with respect to basic stimulus parameters, and most cells are binocular. In general, some cells in each area respond to changes in every parameter, although the proportion of cells responding to a particular parameter changes from area to area. In many cases, the sharpness of tuning for a parameter resembles that observed for this parameter in V1. Contrast affects the response of cells in all areas. All areas also contain cells sensitive to bar or grating orientation, including the highest non-topographic areas in inferotemporal cortex (58). Most areas, again including inferotemporal cortex, contain cells tuned to spatial frequency (IT neurons often prefer many cycles within their receptive fields). The thick V2 cytochrome stripes, MT, V3 (DM), and their targets seem to be especially concerned with analyzing direction of motion (3, 4, 29, 30, 31). Cells in all areas have a broad tuning for stimulus speed (59). Isoluminant edges are specifically <u>not</u> detected by cells in MT and associated areas. Area DL shows many more cells tuned to the size of a stimulus than areas MT, DM, and M (the best stimulus size was often much smaller than the excitatory receptive field) (60). Cells in V2, V3, VP, and MT (at least) respond to horizontal (and in MT, also vertical) disparities (59, 61).

The sensitivity to several more complex stimulus parameters, and the effect of behavioral variables in awake animals has been examined, but usually only in one or two areas at a time. Some cells in V2 (but none in V1) respond to illusory contours generated by real contours just outside the excitatory receptive field (62). Some cells in MT are tuned to the overall direction of motion of a checkerboard pattern while other MT cells and all V1 cells are tuned only to the direction of movement of its component gratings (63). Some cells in area MST (but none in MT) respond to one or the other direction of rotation of a dot field, anywhere within their large receptive fields (64). Cells in MST and parietal cortex (but none in MT) respond to either contracting or dilating patterns (64, 65). Cells in MST respond during smooth pursuit eye movements, even when the target is temporarily extinguished (MT cells are silent without the target) (66). Attention to a simple spot stimulus, whether or not it is used as a target for a saccade, enhances the response to stimuli in much of the hemifield in parietal areas but not in inferotemporal areas (66, 67). Attention to one of two stimuli in a V4 or inferotemporal (but not a V1) receptive field suppresses the response to the other stimulus (68) (it is not clear whether this depends on attention to spatial location or to stimulus properties). Eye position signals can be used to sensitize different parts of the large receptive fields in area 7a (69) (it is not known if spatial attention can have such a localized effect in parietal areas). Finally, in a visual-visual or somatosensory-visual orientation match task, many cells in V4 increase their background firing and respond better to visual target stimuli after presentation of a cue to look for a particular orientation (different cells have different "cue preferences"), whether the initial oriented cue is presented visually or by touch (70). These results suggest that internally-generated activity in visual cortical areas (roughly, visual imagery) may appear quite early in the cortical areas hierarchy.

3.2 Receptive Field Surrounds

Silent inhibitory surrounds are well known at many levels in the visual system; they are "silent" since they become apparent only when the excitatory receptive field center is simultaneously stimulated. Typically, surrounds have been described as having an area comparable to that of the the excitatory center (see e.g., the hypercomplex end-zone). Recently, however, it has become clear that cortical neurons in all visual areas have much more extensive <u>stimulus-specific</u> inhibitory surrounds (71, 72, 73, 74, 75). In MT, random dots moving though the surround in the same direction as an optimally directed center stimulus can completely suppress the center response, while antagonistic surround movements enhance the response. The surround is typically 50-100 times the area of the excitatory center, and exerts its effect with about a 40 millisecond delay (71; cf. 53). Similar interactions have been demonstrated with respect to speed, color, spatial frequency, and orientation of center/surround pairs. This suggests that although most cortical neurons have local receptive fields, they do not signal purely local events except when given extremely sparse displays (like those in neurophysiology experiments).

3.3 Problems of Multiple Selectivities and Controls

The firing rate of a single neuron typically reflects selectivity to several different receptive field parmeters and to the nature of stimuli in the surround. It is intriguing to consider how subsequent processing stages might use such radically ambiguous information. A very large number of non-optimal combinations of center and surround parameters will give similar outputs. For instance, an MT neuron may significantly increase its firing rate if the direction or speed or contrast of either the center or the surround stimulus is changed. Furthermore, these parameters may interact nonlinearly in the neuron's response. From this perspective, it would be nice to have complete firing rate maps and also to know which parameters the neuron is actually _in_sensitive to. Unfortunately, single neurons can rarely be held for the long periods (days--e.g., 5 factors by 5 levels each with 5 repetitions at 5 sec/repetition = 22 hours) such a map requires, and so such information is usually not available.

A related problem arises when more complex stimuli are employed, typically in higher areas. A reasonable strategy is to try to determine the 'simplest' explanation for why a cell is responding (e.g., orientation or spatial frequency content of a stimulus; simple spatial attention) and then invoke sensitivity to more complex stimulus configurations and behavioral states when simple explanations fail. As stimuli become more complex, the number of possible simple-explanation controls multiplies rapidly, and in most cases, we must be satisfied with a few spot checks.

4 STRATEGIES AND SUGGESTIONS FOR CONNECTING MODELS AND DATA

4.1 A Comparative Approach to the Data

In examining real nervous systems, it is important to realize that few structures are likely to exhibit their function transparently. Biological networks reflect the operation of many developmental and phylogenetic as well as purely functional constraints. Developmental programs are not infinitely plastic, but instead can be viewed as defining groups of possible network morphologies. Evolution does not produce the "best of all possible networks" for a particular task, but rather, ones that are just good enough, and ones that work within the developmental means of the brain region at hand. A five centimeter thick, 150 layer cortex may be ideal for certain visual tasks, but may not be developmentally practical. A corollary of this notion is that some aspects of the structure of a network may be non-functional--developmental excresences unavoidably generated in the process of making the functional parts, and persisting simply because they do little enough harm (cf. parasitic transistors in integrated circuit design). The notion that every structure or activity in an organism must have a function has been criticized before, though mostly with respect to non-neural morphology, as the "adaptationist fallacy" (76). None of this is to deny that some aspects of a neural circuitry represent remarkable functional adaptations.

To tease apart some of these influences, it sometimes helps to adopt an explicitly comparative perspective on function, comparing brain areas both within and between species. For example, the amygdaloid nuclei are not laminated like the cortex, and they have no vertically aligned elements like pyramidal cells. Nevertheless, the amygdala contains chandelier cells (77), which in the cortex, specifically inhibit the initial segments of pyramidal cells. Part of the explanation for why chandelier cells also appear in the amygdala (or conversely, in the cortex) is that the amygdala and the neocortex both develop from the pallium, the dorsal subdivision of the forebrain, and these two regions probably rely on similar mechanisms for determining cell types, and so on. The presence of these cells may not indicate that the "need" for the "chandelier" function in the amygdala (or the cortex) resulted in chandelier cells evolving there--another developmentally unrelated brain region might do it another way--but simply that the chandelier-generating machinery was handy. We thus might not want to provide a strictly functional explanation for their presence.

On the positive side, we can use parallel morphologies in unrelated brains or brain regions to attempt to argue that a particular function does in fact require a particular structure. One example concerns the tendency for evolutionarily derived ('advanced') brain regions to develop many layers and many cell types. The midbrain auditory nucleus torus semicircularis (=inferior colliculus) of a turtle, an animal with servicable though not acute hearing, is a small, unassuming structure with few layers and few cell types. By contrast, the same structure in gymnotid electric fish, which have a well-developed electric sense, has 12 layers and at least 50 cell types, and complex interlaminar connections (78) (the auditory and electrosensory sytems are both based on hair cell receptors and have closely related central pathways). Similarly, the cortex of 'primitive' mammals like insectivores and bats, but also dolphins (probably the first placental mammalian group to reenter the water) is thin and poorly laminated and appears (on quick inspection) to contain fewer and less differentiated cell types (79). The cortex of advanced mammals like seals, carnivores, and primates, by contrast, is much more clearly laminated and has many distinct cell types.

Other comparisons provide an interesting perspective on the function of the cortical streams discussed above. It has been suggested on the basis of finding double-opponent color-selective cells in the blobs of V1 in the macaque monkey that the blob system is specialized to process color. However, both owl monkeys and galagos have well-differentiated V1 blobs, in spite of their nocturnal habits and very poor color vision. Furthermore, the early primate ancestors of these groups were also probably nocturnal. Thus, it appears that color processing was added onto a previously existing system that did and may still mainly do something else (e.g., analysis of coarse shading) (80).

Clear contrasts, on the other hand, argue that a particular function identified in area A might not exist in area B. For example, it has been suggested that the cortex might perform computations based on fine-grained temporal interactions. When brain regions known to distinguish small time differences (less than 100 microseconds) are examined anatomically, however, they often show extremely specialized forms, including axons that synapse entirely on cell bodies (calyces of Held), turnip-like cells with extremely short dendrites, and electrical synapses, which all presumably function to preserve temporal information. The lack of such structures in the cortex suggests that operations with a similar temporal "grain" are not performed there.

4.2 Some Specific Suggestions

Subject to the cautions above, I list some general principles of organization in cortical visual systems. At the level of a single cortical area, there are many anatomical cell types (at least 50-100) with a great variety of local axonal arbor sizes and laminar specificities. Local collaterals of a cell may spread across several percent of an area. Within each area, there are many layers of cells and axon terminals (10 or more). In general, dendritic fields are an order of magnitude less widespread than local axon fields. Excitatory cell dendrites tend to cross more layers than inhibitory cell types. Most connections are between, not within cell classes. Cell to cell connections are weak (amounting to 1-5% of the firing threshold). Cells typically receive inputs from thousands of other cells. Excitatory and inhibitory types get both excitatory and inhibitory input (reflected in similar receptive field properties).

At the next level or organization, there is an interconnected hierarchy of distinct areas (20 or more) each characterized by a particular (probably quantitative) variation on the general pattern of within-area connectivity. Interareal connections are generally "acyclic" in the sense that there are (feedforward) paths to a given area of different lengths (i.e., that pass through different numbers of areas). Each area shows retinotopically-indexed excitation of increasing coarseness as one ascends, and wider field stimulus-specific inhibition. Between-area connections are strictly excitatory, usually topological (even at the highest levels), and they generate convex excitatory receptive fields at all levels. Each area contains cells with several different receptive field properties, but different areas have different mixes of cell properties. In general, lower areas in the hierarchy appear to have little "memory" (at least after development), reflected in their prompt, stable, non-habituating responses. Cells with very complex properties (e.g., "place cells" in the rat hippocampus that fire when the animal senses it is in a particular place in a familiar environment) may be only a few stages away from the input area (with place cells--V1, 18b, Ent, DG, CA3).

There are already instances where data begins to contact neural models more specifically. First, one class of abstract network models uses symmetric connectivity matricies (81), and the ubiquity of cortico-cortical feedback might at first seem to be an example of this. However, a round trip between areas requires considerable time (as much as 25 milliseconds if feedback influences have the same minimum latency as feedforward influences), and the return pathway is probably multisynaptic (some aspects of <u>within</u>-area connectivity may be more nearly symmetric). Such a feedback delay may in fact lead to interesting temporal pattern sequences (82) with the persistence time of each pattern equal to the feedback delay. So far, such transition and hold patterns have not been obvious in recordings from early, stimulus-bound levels in the areas hierarchy, although no one has explicitly looked for them. Given typical cortical neuron firing rates, each pattern could be represented by only a few spikes in a single neuron. Second, a number of models result in neural-like units being driven to saturation (or silence) as the network approaches a stable state. This is not usually seen in the cortex (cells tend to show tuning curves), but as before, it may occur in less well studied higher areas, where the cells sometimes tend to burst and habituate. Third, there are a variety of models of how motion information is used to infer structure and other things. The experiments with translation, dilation, and rotation of dot fields in parietal cortex suggest that this is done in at least two stages, with the first stage (MT) sensitive only to translation and the next stage (MST, 7a, and others) sensitive to all three. This data argues against one stage models. Finally, there has not yet appeared evidence of explicit z-axis information at the single cell level in any cortical area, in great contrast to the many explicit representations of variously blurred x- and y-axis information. This argues that models with explicit z-axis coordinates are too abstract to make contact with cortical mechanisms.

These notes are very brief, but they may suggest that a good plan might be to use known neuron properties as starting point. A particular cortical area doesn't "know" where, say, the direction selectivity or end-stopping in its input came from, and is thus a bit like a neural modeler. By using what is already known at one level in the cortical hierarchy, it may be possible to suggest plausible response properties at the next level that neurophysiologists could search for in the brain. It seems unlikely that a blind empirical approach to higher areas is going to produce major insights by itself.

4.3 Other Analogies

A class of network models has been shown to be formally similar to physical systems known as spin glasses, of which a cool magnet is an example. Analogies with other physical systems may be possible. For instance, there is an interesting code-based physical system in living cells that controls a great variety of chemical reactions and distributions in a thoroughly 'parallel' fashion (cell metabolism) through the use of high-specificity reaction-controlling devices (enzymes and other proteins). Although spin glasses may have a "rich statistical physics" (83), there are no living magnets. Living physical systems are predicated on making and breaking strong covalent bonds in addition to exploiting weak cooperative interactions like those seen in spin glasses. This suggests that we might do well to look for analogues to strong (covalent) bonding and bond-making and breaking processes between stable 'molecular' firing patterns (perhaps embodied in strongly interacting cell subassemblies) in neural networks. The metabolic network in code-based living chemical systems, of course, depends heavily on objects whose detailed three-dimensional configurations are determined by cooperative relaxation (protein folding). But protein self-folding occurs only after the arena of possible weak interactions has been defined by the serial synthesis of a 'one-dimensional' chain of peptides using strong covalent bonds.

It may be possible to build up an analogous 'metabolism' of patterns in a network controlled by other serially assembled and then 'folded' patterns. In any case, it will be necessary to go beyond the simple stimulus-response framework that characterizes most present day network models (and neurophysiology experiments). Eventually, we need to approach questions of how, for instance, one integrates successive glances in scene perception or successive words in language comprehension, and how the integrated scene or discourse representations interact appropriately with their previously constructed fellows in the brain.

5 CONCLUSION

A great deal of data had to be very briefly summarized here. Even so, it is clear that our knowledge of many basic parameters in the visual system is rudimentary. And the suggestions given on what to do with some of this information are, to be polite, preliminary. But it is emphatically necessary to start somewhere. Neural modelers are fond of pointing to the futility of trying to determine how a large computer functions by taking 'single-unit' recordings from it (81). I think such pronouncements underestimate human experimental and theoretical ingenuity (and luck). We will get better theories and data by trying to connect the two. Some of the intermediate stages in this development are bound to turn into curiosities known only to future historians of science. We shall not, however, be able do without intermediate stages.

6 ACKNOWLEDGEMENTS

I thank T. Sejnowski for helpful discussions and C. Hochenedel for preparing the manuscript. Supported by NIH grant F32 EY05887.

7 REFERENCES

(1) F. Crick, C. Asanuma, In Parallel Distributed Processing, Volume 2: Psychological and Biological Models, MIT Press, Cambridge 1986, pp. 333-371.

(2) G.A. Orban, Neuronal Operations in the Visual Cortex, Springer-Verlag, Berlin 1984.

(3) D.C. Van Essen, In Cerebral Cortex, Volume 3: Visual Cortex, Plemun Press, New York 1985, pp. 259-329.

(4) J.H.R. Maunsell, W.T. Newsome, Ann. Rev. Neurosci. 10 (1987) 363-402.

(5) J.H. Kaas, Ann. Rev. Psych. 38 (1987).

(6) J.F. Baker, S.E. Petersen, S.E. Newsome, J.M. Allman, J. Neurophysiol. 45 (1981) 397-416.

(7) R.E. Weller, J.H. Kaas, J. Comp. Neurol., 256 (1987) 137-172. See references within.

(8) M.I. Sereno, C.T. McDonald, J.M. Allman, Neurosci. Abstr. 12 (1986) 1181.

(9) D. Felleman, J.J. Knierim, D.C. Van Essen, Neurosci. Abstr. 12 (1986) 1182.

(10) M.M. Merzenich, J.H. Kaas, Prog. Psychobiol. Physiol. Psych. 9 (1980) 1-42.

(11) D.N. Pandya, E.H. Yeterian, In Cerebral Cortex, Volume 4: Association and Auditory Corticies, Plenum Press, New York 1985, pp. 3-61.

(12) R.W. Rodieck, In Comparative Primate Biology, Volume III, Neurosciences, Alan R. Liss 1987 (in press).

(13) C.-S. Lin, J.H. Kaas, J. Comp. Neurol. 187 (1979) 655-678.

(14) J. Graham, C.-S. Lin, J.H. Kaas, J. Comp. Neurol. 187 (1979) 557-580.

(15) H.R. Rodman et al., Neurosci. Abstr. 11 (1985) 1246.

(16) C. Asanuma, R.A. Andersen, W.M. Cowan, J. Comp. Neurol. 241 (1985) 357-381.

(17) J.H.R. Maunsell, In Matters of Intelligence, Reidel Press, Dordrecht 1987 (in press).

(18) S.E. Petersen, F. Miezin, J. Allman (1987, in press).

(19) B.J. Richmond, R.H. Wurtz, T. Sato, J. Neurophysiol. 50 (1983) 1415-1432.

(20) D.G. Amaral, In Handbook of Physiology, Williams and Wilkins 1987 (in press).

(21) J.S. Lund, J. Comp. Neurol. 147 (1973) 455-496.

(22) J.S. Lund, R.G. Boothe, J. Comp. Neurol. 159 (1975).

(23) D. Fitzpatrick, J.S. Lund, G.S. Blasdel, J. Neurosci. 5 (1985) 3329-3349.

(24) G. Blasdel, J.S. Lund, D. Fitzpatrick, J. Neurosci. 5 (1985) 3350-3369.

(25) F. Valverde, In Cerebral Cortex, Volume 3: Visual Cortex, Plenum Press, New York 1985, pp. 207-257.

(26) K.A.C. Martin, In Cerebral Cortex, Volume 2: Functional Properties of Cortical Cells, Plenum Press, New York 1984, pp. 241-284.

(27) R.B.H. Tootell, M.S. Silverman, R.L. DeValois & G.H. Jacobs, Science 220 (1983) 737-739.

(28) M.S. Livingstone, D.H. Hubel, J. Neurosci. 4 (1984) 309-356.

(29) S. Shipp, S. Zeki, Nature 315 (1985) 322-325.

(30) D.H. Hubel, M.S. Livingstone, Nature, 315 325-327.

(31) E.A. DeYoe, D.C. Van Essen, Nature 317 58-61.

(32) C. Enroth-Cugell, J.G. Robson, J. Physiol. 187 (1966) 517-552.

(33) Y. Fukada, J. Stone, J. Neurophysiol. 37 (1974) 749-772.

(34) E.G. Jones, J. Comp. Neurol., 160 (1975) 205-268.

(35) T.L. Davis, P. Sterling, In Models of Visual Cortex, John Wiley, New York 1985, pp. 524-532.

(36) C.D. Gilbert, T.N. Wiesel, J. Neurosci. 3 (1983) 1116-1133.

(37) J.S. Lund, J. Comp. Neur. 257 (1987) 60-92.

(38) P. Somogyi, A. Cowey, In Cerebral Cortex, Volume 1: Cellular Components of the Cerebral Cortex, Plenum Press, New York 1984, pp. 337-360.

(39) A. Peters, In Cerebral Cortex, Volume 1: Cellular Components of the Cerebral Cortex, Plenum Press, New York 1984, pp. 361-380.

(40) E.G. Jones, S.H.C. Hendry, Trends Neurosci. 9 (1986) 71-76.

(41) A.J. Rockel, R.W. Hiorns, T.P.S. Powell, Brain 103 (1980) 221-244.

(42) J.S. Lund, A.E. Hendrickson, M.P. Ogren, E.A. Tobin, J. Comp. Neurol. 202 (1981) 19-46.

(43) J.J. Jack, S.J. Redman, K. Wong, J. Physiol. 321 (1981) 65-96.

(44) S. Redman, B. Walmsley, J. Physiol. 343 (1983) 117-133.

(45) S.Redman, B. Walmsley, J. Physiol. 343 (1983) 135-145.

(46) E.A. Neale, P.G. Nelson, R.L. Macdonald, C.N. Christian, L.M. Bowers, J. Neurophysiol. 49 (1983) 1459-1468.

(47) H. Korn, A. Mallet, A. Triller, D.S. Faber, J. Neurophysiol. 48 (1982) 679-707.

(48) M.I. Sereno, J. Comp. Neurol. 233 (1985) 48-90.

(49) A.G. Brown, Organization in the Spinal Cord, Plenum Press, New York 1981.

(50) T.F. Freund, K.A.C. Martin, P. Somogyi, D. Whitteridge, J. Comp. Neurol. 242 (1985) 275-291.

(51) B.A. McGuire, J.-P. Hornung, C.D. Gilbert, T.N. Wiesel, J. Neurosci. 4 (1984) 3021-3033.

(52) Z.F. Kisvarday, K.A.C. Martin, T.F. Freund, Z. Magloczky, D. Whitteridge, P. Somogyi, Exp. Brain Res. 64 (1986) 541-552.

(53) J. Bullier, M.J. Mustari, G.H. Henry, J. Neurophysiol. 47 (1982) 417-438.

(54) L.A. Palmer, J.P. Jones, W.H. Mullikin, In Models of the Visual Cortex, Wiley, New York 1985, pp. 273-280.

(55) D.A. Pollen, K.H. Foster, J.P. Gaska, In Models of the Visual Cortex, Wiley, New York 1985, pp. 281-291.

(56) G.H. Henry, In Cerebral Cortex, Volume 3: Visual Cortex, Plenum Press, New York 1985, pp. 119-155.

(57) H. Sherk, S. LeVay, J. Neurosci. 3 (1983) 2121-2127.

(58) D.A. Pollen, M. Nagler, J. Daugman, R. Fronauer, P. Cavanagh, Vision Res. 24 (1984) 233-241.

(59) A. Burkhalter, D.C. Van Essen, J. Neurosci. 6 (1986) 2327-2351.

(60) S.E. Petersen, J. Baker, J.M. Allman, Brain Res. 197 (1980) 507-511.

(61) J.H.R. Maunsell, D.E. Van Essen, J. Neurophysiol. 49 (1983) 1148-1167.

(62) R. von der Heydt, E. Peterhans, G. Baumgartner, Science 224 (1984) 1260-1262.

(63) J.A. Movshon, E.H. Adelson, M.S. Gizzi, W.T. Newsome, Exp. Brain Res. (1987, in press).

(64) H. Saito, M. Yukie, K. Tanaka, K. Hikosaka, Y. Fukada, E. Iwai, J. Neurosci. 6 (1986) 145-157.

(65) M.A. Steinmetz, B.C. Motter, C.J. Duffy, J. Neurosci. 7 (1987) 177-191.

(66) R.H. Wurtz, B.J. Richmond, W.T. Newsome, In Dynamic Aspects of Neocortical Function, Wiley, New York 1984, pp. 195-217.

(67) S.E. Petersen, D.L. Robinson, W. Keys, J. Neurophysiol. 54 (1985) 867-886.

(68) J. Moran, R. Desimone, Science 229 (1985) 782-784.

(69) R. Andersen, G. Essick, R. Siegel, Science 230 (1985) 456-458.

(70) P.E. Haenny, J.H.R. Maunsell, P.H. Schiller (1987, in press).

(71) J. Allman, F. Miezin, E. McGuinness, Perception 14 (1985) 105-126.

(72) J. Allman, F. Miezin, E. McGuinness, Ann. Rev. Neurosci. 8 (1985) 407-430.

(73) J.I. Nelson, B. Frost, Brain Res. 139 (1978) 359-365.

(74) S. Zeki, Neurosci. 9 (1983) 767-781.

(75) J. Moran, R. Desimone, S.J. Schein, M. Mishkin, Neurosci. Abstr. 9 (1983) 957.

(76) S.J. Gould, R.C. Lewontin, Proc. R. Soc. Lond. B 205 (1979) 581-598.

(77) A.J. McDonald, J. Comp. Neurol. 212 (1982) 293-312.

(78) C.E. Carr, L. Maler, J. Comp. Neurol. 235 (1985) 207-240.

(79) P.J. Morgane, M.S. Jacobs, A. Galaburda, In Dolphin Cognition and Behavior, Lawrence Erlbaum, Hillsdale, NJ 1986, pp. 5-29.

(80) J. Allman, M. Sereno, Dahlem conference (in press).

(81) J.J. Hopfield, D.W. Tank, Science 233 (1986) 625-633.

(82) H. Sompolinsky, I. Kanter, Phys. Rev. Lett. 57 (1986) 2861-2864.

(83) J.J. Hopfield, D.W. Tank, Science 235 (1987) 1228-1229.

Design for a Sensorium

Jan J. Koenderink

SUMMARY

Two problems in brain theory as related to the structure of sensoria are discussed: how to construct a topology of the visual field and how to embed differential geometry in the representation in order to enable the construction of shape measures. Possible embodiments in terms of the known physiological structure are investigated.

A topology can be based on the correlation structure of neural activity. Under certain constraints this can be shown to lead to a neat triangulation, i.e. the dimension and the connectivity of the space are obtained from scratch.

Sets of receptive fields (RF's) can act as local "n-jet's": truncated Taylor series that locally approximate the retinal illuminance distribution. The RF's can be considered "fuzzy partial derivatives". In such a view many apparently multilocal properties (such as boundary curvature) can be considered simple point properties, thus diminishing the level of geometrical expertise required from sensorimotor routines addressing the sensorium.

1 AIMS AND SCOPE

The brain is the most complex system that science has to come to terms with. Experiments yield a higher datastream than our concept formation can keep up with. Modern technology merely increases the data overload, one can expect no understanding from just that. On the other hand brain theory is bedevilled by the mind-brain problem which induces a veritable babylonean chaos. What is needed are fresh concepts that enable one to order and structure the facts and understand them simultaneously at multiple logical levels. New concepts are forthcoming, be it less readily than data. In order to be useful, concepts have to be abstract and general. They are not merely meant to model arbitrary sets of data but to help understand how to describe what goes on and how to formulate new questions. The present theory is about a tiny facet of perception although it may be interpreted in a much wider context. It may be used to put some order on existing data in a general sort of way. Whether it is worth anything will depend on the uses to which it will be put.

The structure of this paper is as follows: First a rather simplified view of how to understand the function and structure of sensoria is presented. Then two examples taken from the field of vision are (rather summarily) presented. These are used to adstruct certain general principles which will be (over-) stated at the conclusion.

2 A LOGICAL STRATIFICATION OF BRAIN SCIENCE

A sharp distinction of different logical strata involved in our scientific description is no philosophic pedantry but of crucial importance in brain science. This is not particular to brain theory but applies to many fields of endeavour. In order to make points clear I will often draw examples from electronics and computer science. The reader may be assured that this does not mean that I imply that the brain is at all like our electronic computers. Far from it.

At the most basic level of scientific description we are in the domain of physics (and chemistry). We observe processes that may be termed energy conversions. For example a photon is absorbed by a rhodopsin molecule in a visual receptor and we observe an electrochemical disturbance. This is the pre-syntactic domain, i.e. the notion of "signal" has no place here.

The next stratum is the syntactic domain of signals and signal processors. It is important to see that this is a disparate stratum that is coupled to the physical domain by essentially arbitrary conventions. Signals are equivalence classes of physical states and signal processing maps signals on other signals. For example if you buy a chip with logic AND gates, you may look in the datasheets for the definition of what the in- and output signals are. E.g. voltages should be within certain limits, output impedance of the source should be at least lower than something, etc. Only if you use this arbitrary convention does the manufacturer guarantee that the device "is a logic AND gate". Other arbitrary definitions are possible, e.g. you may use much higher voltages and currents and take the temperature of the device as the output, etc. In fact logic INVERTERS are often used as amplifiers for the "wrong" signals. I call the interface between the physical and the syntactic domain a "physical format". We can only guess at the physical format for neural "signals" since there are no datasheets. Any convention is possible, though some will no doubt be more convenient than others (e.g. action potentials) (note 1).

If we describe neural processes as "signal processing" we have (often implicitely) adopted such an arbitrary format.

The signals have no meaning. An example from computer science may help to see this: the same memory address may be taken to contain an integer, a character, a set of booleans,... This depends on the calling program and is effectively defined through a "logical format" ("declaration", "format statement"). Thus semantic content is independent of the syntactic structure and depends again on an arbitrary convention. The semantic domain is a disparate stratum and we can only link it to the syntactic domain through an act of choice (note 2).

Again, in brain/mind theory the choice is ours (the scientist). In this paper I will denote semantic processors as "sensorimotor routines". The sensorium is understood as a part of the body that is treated syntactically and assumes states that depend on the environment through physical interactions at the body surface. The semantic processors regard the sensorium as a medium that can be "read" (note 3).

This means the following: the processor issues a read command that consists of two distinct parts, namely _where_ to read (an address) and _what_ to read (the logical format that defines the meaning). The semantic content is the syntactic structure interpreted in terms of the format.

3 THE SENSORIMOTOR LOOP

We conceptually divide the object of study into an "environment" and an "organism". This may be interpreted in terms of world and animal, but we may also think of brain "modules" interfaced with the rest of the brain (note 4) (1).

On the "surface" of the organism we distinguish sensor and motor regions as interfaces or physical formats. (Figure 1). Via the _motorium_ the organism may act on the environment (I will refrain from analysis of the logical levels involved here). Via the _sensorium_ the environment acts on the organism, a dynamic syntactic structure is maintained in the sensorium that may be regarded as essentially a corner of the environment in machine readable form, i.e. written in a uniform format. This means that the sensorium does not "process data" and that its structures are pre-categorical, non-iconic and strictly meaningless. Because of that the sensorium is objective and can never be wrong. (The latter condition is a logical impossibility just like the world can't be wrong even if our perceptions often are.)

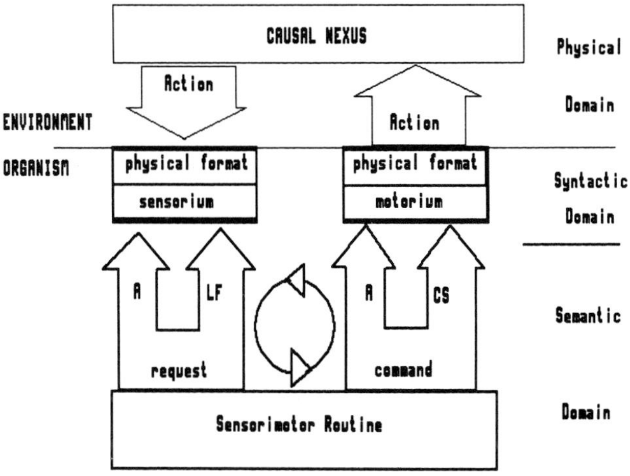

1. Generalized scheme of sensorimotor loops. The box diagram is to be understood as a logical rather than a hardware decomposition. Note how the stream of information is a closed loop that encompasses both the environment and the organism.

A : Address

LF : Logical format

CS : Command signal.

Sensorimotor routines (hierarchies/heterarchies of them) read from the sensorium and extract their own meaning from maybe the same structure. They issue commands to the motorium. Through the causal nexus of physical laws these motor commands eventually effect the sensorium, thus the sensorimotor process is an informationally closed loop. Through this loop meaning is generated and engrained in the structure of the sensorimotor routines. The organism can tune itself (i.e. synchronize its behaviour) with the structure of the environment. I have no occasion to amplify on this in the context of this paper, however.

4 DESIGN PRINCIPLES FOR SENSORIA

I have defined a sensorium as a syntactic structure that represents (in a very concrete way) the physical structure of the environment in machine readable form. Basically physical structure is "sensed" (the physical format), formed (signal processing) and distributed, i.e. placed into bins where it can be found by interested sensorimotor routines. The problems to be dealt with are those of selection, representation and distribution. The criteria to be met are generality (many routines must be able to put the same structure to different use), speed, i.e. explicit representation (check of single "flag" to elicit a flight reaction would be good) and selectiveness (what is not needed should not be represented at all).

Some of the criteria are clearly contradictory. A hybrid solution seems optimal: explicit, stereotyped representations for dumb "automatic" routines and an abstract, general representation for unforeseeable problems.

The latter system is especially interesting here: suppose you don't know all the problems the system will ever have to solve. (Maybe the long term goals will change.) Then the representation had better be universal such that if it is needed any data can be extracted, albeit slow.

Universal representation means necessarily non-"concrete" (i.e. of the "go-no go"type) but highly abstract representation. In the case of the visual system we think of the basis levels of geometrical structure on which structures of arbitrary complexity can be grafted such as topological relations and basic differential geometry. I have picked these two possibilities as examples. Topology is really the skeleton for all geometry. Differential geometry sews the flesh on the bones, it basically sustains all notions of shape.

5 TOPOLOGICAL STRUCTURES IN SENSORIA

Point set topology is basically the theory of neighbourhood relations, connectivity relations, dimension, etc. For instance, you might well ask the question of how the brain finds out that the visual field is two-dimensional on the basis of a few million signals that arrive simultaneously over the optic nerve fibers (2). Some people will answer that the famous somatotopic maps solve the problem. I don't think so because these maps exist only for the external observer (electrophysiologist). We need a topology without homunculus as was already clearly indicated by Lotze in his theory of "local sign" (3).

The following reasoning leads to a possible solution (4). Consider a set of units [a,b,c,...]. (You may think of receptive fields, of local neural assemblies,...) Let there be a symmetric relation between them, i.e. a set of unordered pairs of units. If (a,b) is in this set then "a is related to b". (You may think of correlation, either for spatial or temporal patterns, or both.) This structure may be interpreted in pseudogeometrical language, e.g. (a,b) means that a and b overlap in some sensory space. From the notion of overlap we may construct many other notions by combinatorial processes. For instance, we say that "a is included in b" iff for any c such that c is related to a it is also true that c is related to b (Figure 2).

From this notion of inclusion we may go to higher constructs such as the "hull" of two units. (Figure 3). I call P the hull of u and v if it contains any unit x

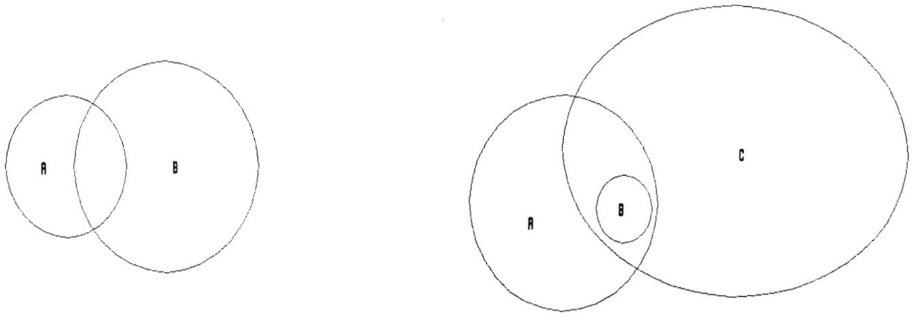

2. The upper picture illustrates the symmetrical relation "A is related to B" in terms of geometrical overlap.
The lower picture illustrates the definition of "included in". The unit C stands for <u>any</u> unit that is related to B: if it follows that it is also related to A it is said that "B is included in A". In terms of the geometrical picture this is obvious.

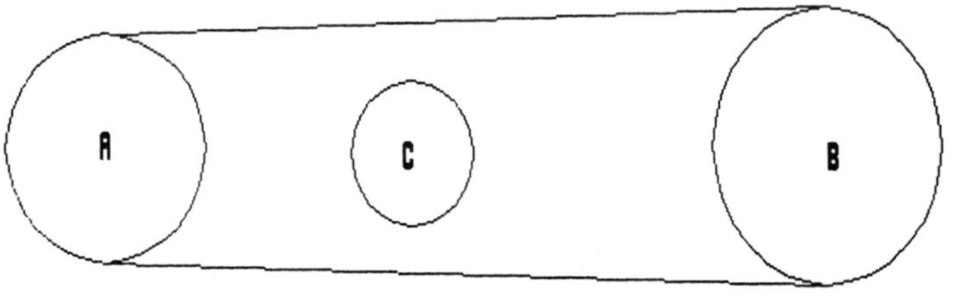

3. C is included in the convex hull of A and B if the set of units consists of ovoid shapes of arbitrary size.

that is contained in all units that contain both u and v. The reader may easily verify that if we use a mathematical model in which the units are circular discs in the plane the hull of two points (arbitrary small circles) is a line segment. Similarly the hull of three units is an abstract triangle. Here is where we can actually compute the dimension of the space: if we are able to find a triangle that contains a unit that is not contained in one of its sides we may conclude that the space is at least two-dimensional, etc. (Figure 4).

It is not at all guaranteed that this process of construction of geometrical objects on the basis of a symmetrical relation on pairs always leads to neat geometrical constructs. One easily devises models for which this is not the case (e.g. one-dimensional spaces that are not topological lines or circles but are unordered, contain branches, etc.). Examples with nice behaviour are well known, however: The set of all circles in the plane has half a century ago been shown by Huntington to be isomorphic to Euclidean geometry. One general requirement is "functional convexity": the abstract overlap relations should be isomorphic to the geometrical overlap relations for some set of convex regions in the plane. If we think of a representation in terms of receptive fields then it is sufficient (but not necessary!) that their shapes be convex for the external observer. Moreover, there should be sufficiently many of them. If the necessary requirements are met we can actually produce a triangulation of the sensory space (Figure 5) (4). No need to say that this is an abstract space (a semantic construct) and not e.g. retinal space. For instance, the brain may not be able to decide between spatial overlap in the retina and correlation between fibers from left and right eye: the latter two are in the same sensory space. There are many indications from neurology that this kind of semantic structure is actually there (note 5).

6 A RECEPTIVE FIELD DIFFERENTIAL CALCULUS

If you consider the structure of well known receptive fields it is very striking indeed that they are like approximate partial derivatives or combinations of those.

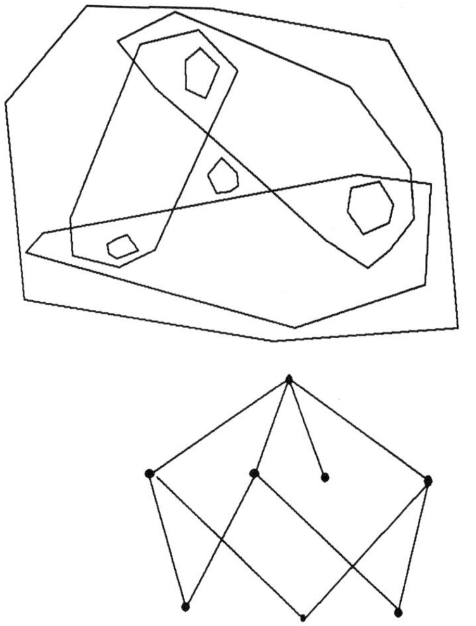

4. A "triangle". Upper picture the geometrical layout of units, lower picture the corresponding Hasse diagram.

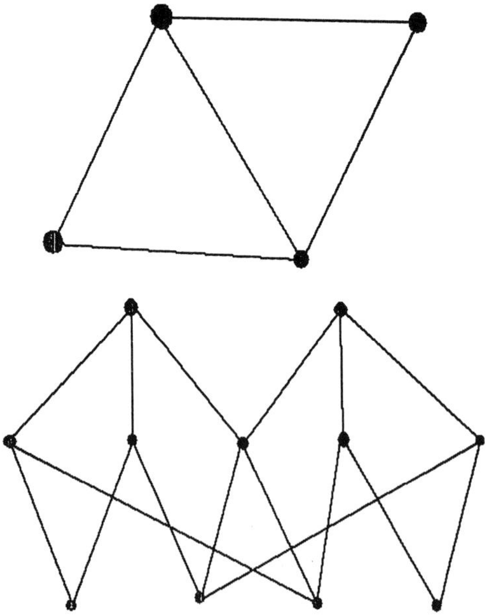

5. An example of a small "triangulation". In the upper picture a schematized geometrical depiction is given; the lower figure specifies the Hasse diagram.

(Figure 6). For example the Mexican hat profile is like a Laplacean operator, a "line detector" is like a second derivative in one direction, an "end stopped cell" is like a mixed derivative, twice in one direction and once in an orthogonal one. This suggests strongly that one should think of given assortments of RF's as a set of partial derivatives. This would mean that a single datum is not a RF signal, but a record (set with labelled members) of such signals that locally determine e.g. the truncated Taylor series of retinal illuminance in the neighbourhood of a point. If one could make such a notion precise one would essentially establish a "RF-differential calculus" (5).

One fact that has to be taken into account is that RF diameters can vary widely, even for a single type. This is in fact a hint towards a solution: we could conceive of RF's as exact partial derivatives of blurred versions of the retinal illuminance pattern. This makes good sense because the visual system has to entertain a representation on multiple levels of resolution anyway in order to solve the "forest and the trees"-problem. Recently the problem of how to describe such representations mathematically has been solved. We know that there exists only one rational way to blur (only the Gaussian kernel yields a causal structure), moreover, we know how to build multiresolution space into a linked datastructure, i.e. how to relate representations at different resolutions (6). The partial derivatives idea fits beautifully into this structure. We can easily design RF profiles for arbitrary mixed partial derivatives of the representation at any level of resolution. Considered as linear systems they yield exact results. (Figure 7).

These RF profiles are extremely convenient for the theorist because they have very simple concatenation properties (putting them in series produces more of the same with altered resolution and order) and because all rules of calculus concerning mixed partial derivatives now can be applied to the RF combinations. This is a most elegant and powerful tool. The empirically minded will hardly be impressed by the elegance although they may be irritated by the introduction of yet another theoretical RF profile. They can put their mind at rest by noting that the RF

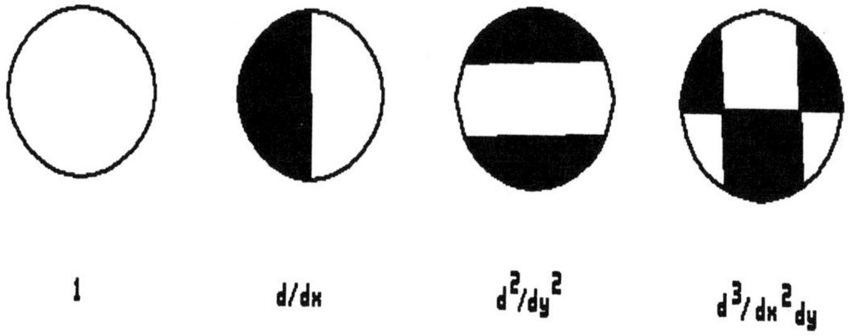

6. Neighbourhood operations that approximate partial derivatives. The second partial derivative would be classified as the "line detector", the third mixed derivative as an "end stopped cell" by the electrophysiologist. The first derivative is at least implicated by psychophysics; the constant (zeroth derivative) does not occur.

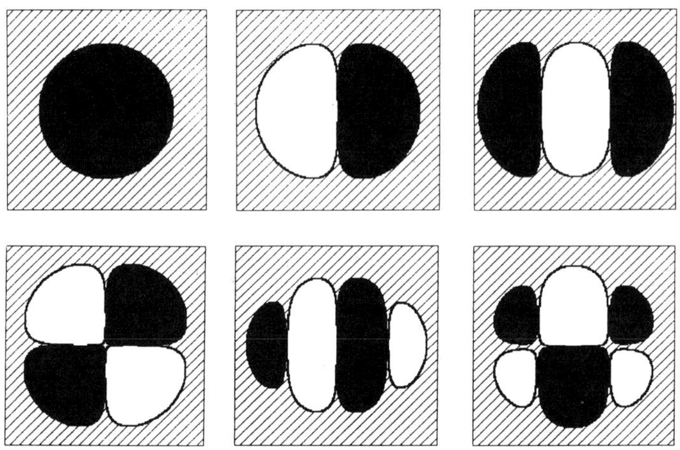

7. Some low order mixed partial derivatives. The areas where the absolute value of the weighting function exceeds a certain level has been given a black tone for positive, a white tone for negative levels.
From left to right, top to bottom you have:
1 (zeroth derivative), first derivative in horizontal direction, minus the second derivative in the horizontal direction, the mixed derivative with respect to both directions, the third derivative in the horizontal direction and a mixed third derivative (two times horizontally, once vertically).

profiles in question are very much like some of the commonly accepted types (DOG, Gabor functions, you name it). (Figure 8). In empirical studies one needn't care about the differences, but conceptually the differences are crucial.

Regarded as spatial sampling windows these RF's are like partial derivatives and inherit all their theoretical properties. Regarded as "spatial frequency filters" they can be regarded as local gratings. (E.g. the partial n-th order in x and m-th order in y is like a grating with 2n cycles in the x and 2m cycles in the y-direction.) This latter view is indeed sometimes profitable because it suggests that we group derivatives of even and odd order as sine-like and cosine-like pairs. Than we can easily define an envelope (amplitude) which defines the spatial window through which the RF's look at the retinal illuminance and a phase which is in fact a useful linear position signal.

7 THE REPRESENTATION OF THE RETINAL ILLUMINANCE

The mathematical construction that models the RF calculus globally is the "jet-extension" (7). A "jet" is an equivalence class of functions that agree in all their partial derivatives at the point of application up to the n-th order. The truncated Taylor series up to the n-th order may be taken as a representative. Concretely, we just have a labelled set of numbers, namely the coefficients of the series or the partial derivatives. This record is to be regarded as a single datum. If one replaces the value of the illuminance at every point by the n-jet one gets the "n-jet extension". In practice one won't take "every point" but a number of samples. (The sampling density is a technical matter (note 6).)

Each sample effectively approximates the pattern in the neighbourhood of its point of application. The structure is very similar to that of the cortical columnar structure.

The representation of a jet can vary greatly. Instead of taking the mixed partial

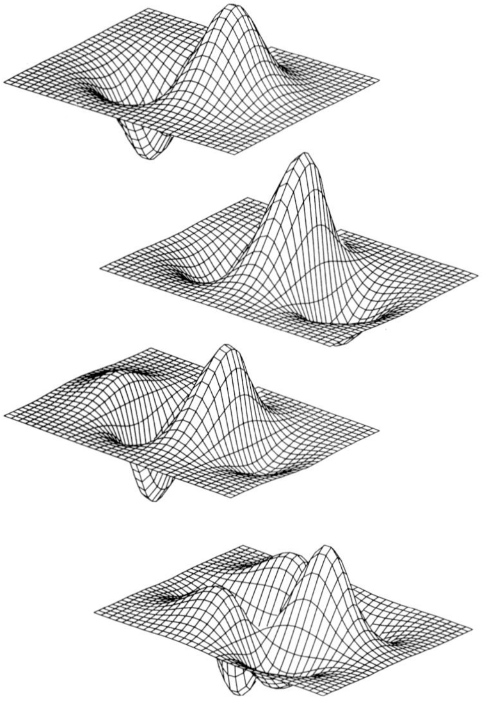

8. First, second and third derivatives with respect to one direction and (lowest) a mixed third derivative.

derivatives with respect to two Cartesian axes one might consider ordered sets of congruent RF's in different orientations. For example the three RF's for the Cartesian second order terms can be replaced with a series of "line detectors" of various orientations. Thus a complete orientation cycle of "line detectors" effectively represents the second order terms of the jet. Similar considerations apply to the other orders. If we do this we indeed obtain structures that have a remarkable likeness to the cortical representation as described in neurophysiology.

As said a jet is to be considered a single datum. This means that it has no geometrical structure ("a point is that which has no extension"). A sensorimotor routine addressing the jet need have no geometrical expertise if it can compute its output on the basis of a single jet. Yet the outcome of such a calculation might best be described in geometrical terms, e.g. "boundary curvature". I call such routines "point processors", the properties they compute "point properties". It is important to note that e.g. boundary curvature is a point property of the 2-nd order jet but a multilocal property of the 1-st order jet. A computation of a multilocal property usually does imply a geometrical expertise (e.g. topological such as knowing where the neighbours are or even a "connection", i.e. the ability to compare orientations at different places) on the side of the processor.

In order to establish the "glue" for multilocal processes it would make sense to consider the jets as the "units" in the local sign computation. The "relation" could then be a correlation of local pattern via the jet representation. Moreover, once neighbourhood has been established a comparison of mutual phase can be used to mensurate the positions of the jets, i.e. to put a metric on the topological triangulation.

8 THE "MEANING" OF A RF SIGNAL

If only the jet as a whole is a single datum, the RF signal as such can have no fixed semantic content. The following example illustrates this vividly. The example is about "curvature". We consider the curvature of two different geometrical objects: the "boundary curvature" in an illuminance gradient (basically the curvature of an equi-illuminance curve) and the "line curvature" of an elongated "ridge" of locally higher (or lower) luminance. (Of course this distinction applies only to a certain level of resolution. E.g. at a higher level the "line" will show up as a "patch" and has edges to which the boundary concept applies. There is no space to consider such interesting effects here.) It is rather easy to compute these curvatures from elementary differential geometry. Transposed to the neural domain the curvatures are the ratios of RF outputs. What happens is this:

- for the boundary curvature one first detects the direction of the gradient with the "gradient detectors" (first order directional derivative). Then one looks at the output of a "line detector" (second order directional derivative) whose orientation is found via the gradient detector and looks at its output. It is proportional to boundary curvature. (Figure 9).

- for the line curvature one first detects the orientation of the line with the "line detectors". Then one looks at the output of an "end stopped cell" (third order mixed partial derivative) whose orientation is found via the line detectors and looks at its output. It is proportional to line curvature. (Figure 10).

The lesson to be drawn from this is the following: in both cases we use "line-detectors" but they appear in different ways. Once as "line detector" and once as "curvature indicator". The line detector is no "feature detector" (i.e. "line detector") with a fixed meaning at all. Instead, the calling routine (boundary curvature detector, line curvature detector) defines the meaning via a logical format.

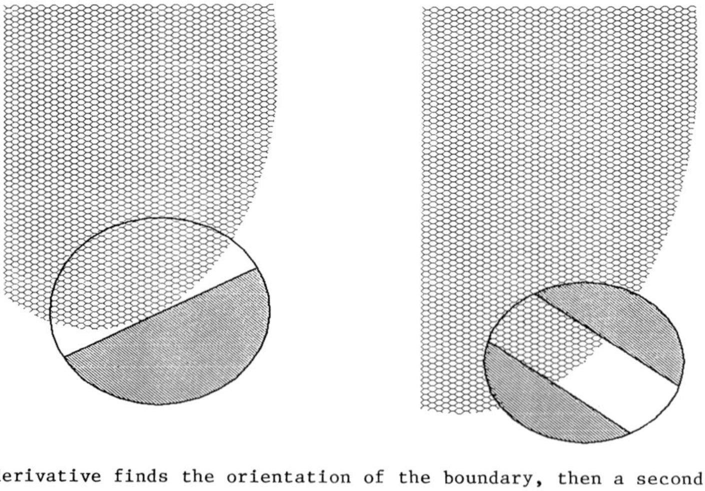

9. A first derivative finds the orientation of the boundary, then a second derivative in the perpendicular direction yields a signal that is proportional to boundary curvature.

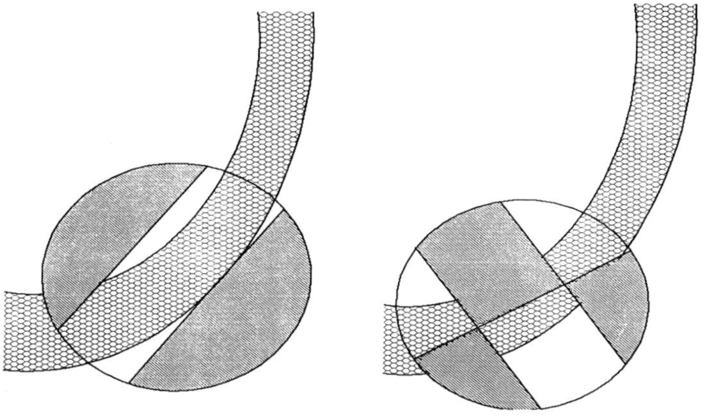

10. A second order derivative ("line detection") finds the orientation of the line, then a suitably positioned mixed third derivative yields a signal that is proportional to line curvature.

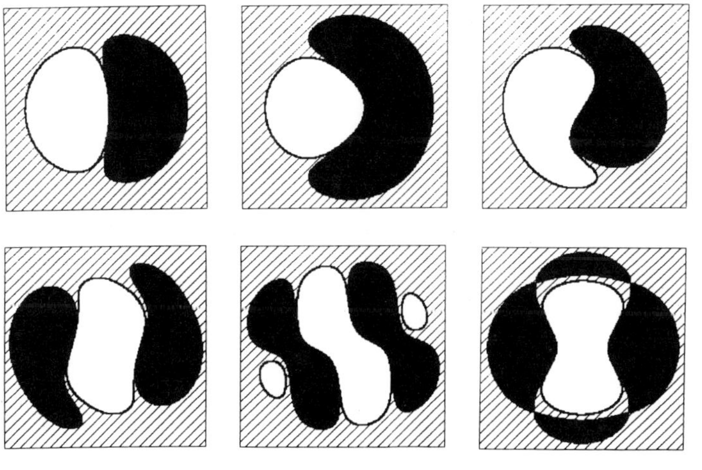

11. Some simple linear combinations. Upper row: boundary curvature slight and strongly; a boundary inflexion. Lower row: a line inflexion, a Vernier offset and a "neck detector".

9 DIFFERENTIAL GEOMETRY IN THE NEURAL DOMAIN

The present interpretation allows us to transpose the whole apparatus of differential geometry to the neural domain. The power of this move can hardly be overestimated: it means that it becomes at once trivial to find out the abtract structure of the visual routines concerned with problems of shape. One just solves the corresponding problem in differential geometry (which is easy as this field has been highly developed) and transposes the result to the neural domain. The results are partly amenable to empirical verification. For instance, we now predict which types of RF's the routine has to access, whether it will be a point or a multilocal processor, and we know the abstract structure of the processor itself. Moreover, we can predict relations between different routines in so far as complicated shape properties can usually be broken down into modules concerned with simpler properties. This is merely a matter of algebraic factoring in the mathematical picture. For multilocal processes we can indicate the degree of mathematical expertise required, which again predicts that certain structures should be present.

One may now use differential geometry to devise classification schemes for RF's (e.g. linear combinations of partial derivatives up to a certain order, etc.). Not only does this produce a useful labelling, more interestingly still these may be used to "predict" certain types (e.g. "missing" terms of a n-th order jet, invariants, etc.).

10 PROGRAMMABLE RECEPTIVE FIELDS

Because local shape properties can always be written as algebraic combinations of mixed partial derivatives, we can actually design programmable RF's. We represent the partial derivatives with RF's and build a network that combines a set of them in linear or nonlinear ways via synaptic connections whose weights can be set, either slowly, or fast via switching. Then we obtain RF's that can either be switched to different modes or adapt and "learn" to assume certain forms, e.g. to detect commonly presented patterns. Some important patterns could be hardwired in and might e.g. form the substrate for the "textons" of preattentive vision (note 7).

Even simple linear combinations of RF's can be used to build interesting complex RF's. (From a n-th order jet you can effectively construct arbitrary local patterns of as many degrees of freedom ("pixels") as there are terms in the truncated Taylor series. If you take the 3-rd order jet for instance you can construct RF's for boundary curvature, edginess, endedness, wedginess, etc.; there is information to specify a local 3 x 3 pixel "local image".) (Figure 11).

11 CONCLUDING REMARKS

In this paper I have essentially illustrated a few principles. The mathematical background has to be found elsewhere. I nevertheless hope that the exposition has served to make the following remarks at least worth pondering. They are intentionally overstated in order to evoke interest, if stated more cautiously they might be dismissed as the trivial truisms which they really are.

- sensoria do not compute perceptions, they are corners of the environment that fit a certain physical format and can be addressed by sensorimotor routines;

- somatotopy has no functional meaning unless it is between the receptors and effectors. Semantically it only exists for the external observer. In the syntactic domain it serves wiring economy, etc.;

- the semantic domain only applies to the complete sensorimotor loop. The sensorium does not represent meaning, only structure;

- "meaning" is a useful scientific concept in so far that it is defined relative to some essentially arbitrary convention (e.g. hypothetical "goals");

- the same part of the physical world can (even simultaneously) be described as a physical system, a signal processing device, an "agent" (implying semantic structure). Many alternative descriptions are always possible. They can be equally valid because they may belong to incommensurate logical domains. Only economy principles and arbitrary decisions can decide which one suits us best.

- the concept of "feature detector" only makes sense in terms of a limited paradigm. It is counterproductive if we use it to understand what goes on in sensoria because it negates the possibility of fractional/multiple meaning;

- the definition of "neural signal" is a real problem and is essentially arbitrary; i.e. cannot really be "decided" but merely agreed upon on the basis of economy arguments;

- we really don't know which neural activity patterns are most fruitfully thought of in terms of single data items on the semantic level;

- it is very well possible to claim whole fields of mathematics for neurphysiology (e.g. topology via partially ordered sets and lattices, differential calculus and differential geometry). This lets us look at the data from novel perspectives.

12 ACKNOWLEDGEMENTS

This research has been supported by the Dutch Organization for the Advancement of Pure Research (ZWO) and a NATO grant.

13 NOTES:

1. Strong nonlinearities often induce "natural" choices. For instance the action potential is a natural choice because "the same" action potential can stand for an enormous universe of histories of graded potentials, etc. However, this is merely a matter of degree. When we take a better look at action potentials we will find that no two of them are identical. Although a natural choice it is still arbitrary.

2. Again, some choices appear more "natural" than others. For instance, it is natural to relate the "meaning" of a sensorimotor action to the chances of survival of the organism. However, this remains an effectively arbitrary choice.

3. It makes no sense to have a semantic processor "write" to the sensorium. This would only be "tampering with the evidence". It may be useful to send signals to the sensorium to set global amplification factors, etc. Such commands are in principle not different from e.g. adjusting the optical strength of the eye lense. (The processor need not even know where the commands go to! It only commands and the meaning of a command is the expected result which will show up via changes in the sensorium.) We consider them basically "motor commands" even if no mechanical transduction is involved.

4. In this picture "brain modules" are effectively independent agents and their relation to the rest of the brain is not different from the relation between animal and world. "Independent" doesn't mean "decoupled", just as the animal is a mirror of the environment in many ways, so the modules contain representations of each other. The notion has a certain affinity to Leibnitz's system of "monads".

5. A famous example is due to an observation by Helmholtz (8): Patients are often unable to point out whether a bad tooth is in the upper or the lower jaw. The teeth are perceptually in the same place! Helmholtz explains this in terms of the neural correlation structure.

6. The total amount of data to be represented need not be much larger than the amount needed to represent the illuminance pattern itself as estimated from Nyquist's sampling theorem.

7. A natural inference is that textons are point properties which can be taken in at one gulp, then the possible "textons" are limited by the order of the jet extension. It is possible to characterize them for any given order. This might be a method to estimate the order psychophysically.

14 REFERENCES:

1. J.J. Koenderink, A.J. van Doorn
Logical stratification of organic intelligence.
In: Cybernetics and Systems 1986, ed. R. Trappl. D. Reidel Publishing Company 1986, pp. 871-878.

2. J.J. Koenderink
Simultaneous order in nervous nets from a functional standpoint.
Biol. Cybern. 50, 35-41 (1984).

3. J.J. Koenderink
The concept of local sign.
In: Limits in perception, eds. A.J. van Doorn, W.A. van de Grind, J.J. Koenderink. V.N.U. Science Press, Utrecht, 1984, pp. 495-547.

4. J.J. Koenderink
Geometrical structures determined by the functional order in nervous nets.
Biol. Cybern. 50, 43-50 (1984).

5. J.J. Koenderink, A.J. van Doorn

Representation of local geometry in the visual system.

Biol. Cybern., in press (1986).

6. J.J. Koenderink

The structure of images.

Biol. Cybern. 50, 363-370 (1984).

7. J.J. Koenderink

Image structure.

Nato Conference: Mathematics and Computer Science in Medical Imaging. September 21-October 4, 1986. Il Ciocco, Italy. Eds. M. Viergever, A.Todd-Pokropek.

8. Helmholtz, H. von

Handbuch der physiologischen Optik.

Voss, Leipzig, 1896.

Design for a Sensorium

Discussion
edited by *Günther Palm*

Comment: Your discussion of local operators assumes an undistorted mapping between the visual field outside and the cortex inside. In fact the magnification factor in this mapping changes with excentricity. This distortion may produce different operators.

Answer: I know that.

Question: In the introductory part of your talk you stressed the unity composed of sensory and motor systems. In the second half of the talk you split this unity, you only discussed the internal representation of visual information. How does this fit together? Can visual representation be discussed at all independently from the motor tasks it is need for? How can one decide, where sensory representation ends and where motor processing starts?

Answer: This split is in our description of the object, not in the object itself.

Comment: Your description of local differential operators amounts to a kind of pattern transformation or feature extraction, which reminds me of the Hubel-Wiesel-type detectors or even of the grandmother cell.

Question: Why should there be a topological representation?

Answer: You need it to be able to detect spatial structure.

Comment: In olfaction it seems to be different from vision: the "shape" of an odour has no meaning.

Comment: You could probably improve your system by introducing feedback and by working from low resolution operators down to high resolution.

Answer: The second idea is exactly what we are implementing now.

Comment: The jet-representation you propose is extremely redundant.

Answer: That is true but I believe that our visual system cannot significantly reduce the redundancy.

Neural-Like Graph-Matching Techniques for Image-Processing

Elie Bienenstock

SUMMARY Some problems in pattern recognition may be phrased as graph-matching tasks. Indeed, graph-theoretic, i.e., relational descriptions provide a powerful theoretical tool for solving recognition problems which require invariances with respect to various transformations of the image. Relational approaches may also be used for image segmentation and figure/ground separation. From the computational point of view, conventional notions of graph-matching, which are generally related to graph-isomorphism, are in most cases impracticable. An alternative notion of graph-matching is proposed, based on the definition of a cost-function, the minima of which provide the solutions to the matching problem. This alternative notion of a match is less restrictive than graph-isomorphism, and more appropriate for problems of pattern recognition. It is also more convenient computationally. A similar formulation is used for the problem of figure/ground separation or segmentation. This uniform treatment makes it possible to interface in a natural way high-level and low-level processing, which may then be carried out simultaneously and interactively. Thus, a 3-layer model for segmentation-and-recognition is sketched; the model is appropriate for processing images which are too degraded to allow data-driven segmentation. These approaches to graph-matching and to segmentation suggest a ***relational paradigm***, which may be relevant in both artificial and natural vision. The implementation of this relational approach in the brain is briefly discussed.

1 INTRODUCTION

This paper is concerned with a family of problems which arise both in digital image processing, i.e., computer vision, and in natural vision. In its simplest form, the problem is to recognize a pattern or shape belonging to a family of known patterns. In computer vision, this often includes a classification problem (1). For instance, optical character recognition is a 26-category classification problem. In general though, classification is only one aspect of the task. One may for instance also have to decide how many characters there are and where they are in a given image. This task of figure/ground separation is often performed before, and to a large extent independently of the classification itself.

On the other hand, the competences of animal and human visual systems are much broader than those of current artificial vision systems. Indeed, our everyday sensory-motor behavior involves the analysis and interpretation of natural scenes which include varied and unforeseeable elements, and are in general extremely complex and cluttered, when compared to the images that man-made vision systems --even the most sophisticated of those-- are able to process. Natural perception is thus a truly intelligent behavior. It involves several high-level cognitive functions, which cannot be easily dissociated from each other. Under these circumstances, it is hardly possible to reduce object recognition to a classification problem: the very attempt to circumscribe and classify in a useful way the objects that we are able to recognize, manipulate and use, seems extraordinarily difficult. The main issue seems to be that of the format of knowledge representation and knowledge organization in our brains. This issue is, at the present day, largely unsolved.

The approach adopted here consists in isolating problems in pattern recognition which capture the essence of some of the tasks solved with great efficiency by our visual system. These problems are cast in a framework of artificial vision, yet we shall look for ***neural-like*** solutions.

We shall focus the discussion on two image-processing problems: invariant pattern recognition and figure/ground separation or segmentation. Invariant pattern recognition is phrased as a task of graph-matching (Section 2), and we propose a mathematical formulation based on the notion of a ***cost-function*** to be minimized (Sections 3 and 4). One advantage of this formulation is that it makes it easy to construct a multi-layer model in which the various levels of processing interact with each other in a natural way: figure/ground separation, which is a low-level task, is carried out simultaneously with the actual recognition process (Section 5). The various tasks at the different levels are indeed all incorporated within a single cost-function. Section 6 outlines a possible neurobiological substrate for this computational theory of graph-matching. The main idea is that fast-varying synaptic weights may embody the dynamical links required for the implementation of the theory.

This paper describes work in progress, done in collaboration with Christoph von der Malsburg, who elaborated the theory of brain function outlined in Section 6, and with Stuart Geman, who suggested much of the three-layered image-processing model described in Section 5. If anything should appear inadequate in this presentation, the present author would claim the entire responsibility for it.

2 PATTERN RECOGNITION AS A PROBLEM OF GRAPH-MATCHING

One of the most remarkable features of our visual system is its ability to cope with a great many transformations of the sensory input. In particular, we have an amazing capacity to recognize and identify an object in spite of the virtually infinite variety of retinal images generated by the object at different positions, orientations, distances etc. The ability to "invert these transformations" is often referred to as a phenomenon of ***perceptual invariance.*** To a certain extent, this ability is also part of our faculty to generalize from examples: an object seen once at a given distance, orientation etc, will be recognized without any effort at other orientations and distances.

Our solution to the invariance problem relies on the notion of an invariant representation. A representation of an object is said to be invariant with respect to a given family of transformations if it remains unchanged whenever the object undergoes one or several of these transformations. For instance, a position-invariant representation of a 2-D shape is a description which does not depend on the absolute position within the plane: this description is based on the positions of parts of the shape relative to each other, that is, on ***relational information***. One may also be interested in size changes and in more complex transformations, such as non-isotropic distortions. These can be handled by extracting appropriate topological invariants.

Invariant representations are thus essentially relational: they describe how various elements of a shape or object relate to each other. We shall build a framework for pattern-recognition based on relational descriptions.

The simplest abstract relational structure is a graph. A graph G is a set of ***vertices*** or ***nodes***, along with a set of ***edges*** or ***links*** between the nodes. Yet graphs are too abstract to be used as such for representing objects. More structure needs to be incorporated in the graph-representation. This may be done by labelling the nodes and/or the links of the graphs with various (local) features, or labels. Labels may be of relatively low-level and universal applicability, or on the contrary they may be high-level and domain-specific.

For instance, the labels affected to the nodes of the graph could simply be the actual grey-level values of individual pixels in the original unprocessed intensity image. A more sophisticated label-set may consist of local visual features extracted from the image, e.g. local intensity gradients as are commonly used in image processing. Contrasting with these relatively low-level and almost universal label-sets, one may use descriptors which are specific to a well-defined class of objects. For instance, when the application is character recognition, an adequate set of labels may consist of various types of straight or curved strokes, crossings, Y-shaped bifurcations etc. One may also use descriptors related to geometrical or topological properties of the image, e.g. connected components or convex regions, possibly with specifications of size and shape of these regions (see 2 for a review of relational descriptions in computer vision).

The common feature in all these representations is the crucial role played by the edges or links of the graph. These links embody the relational aspects of the description. Again, the meaning of the links may vary depending on the application. Simplest of all is a link which indicates a neighborhood relationship between two elements (nodes) in the image. Next, the link between nodes i and j may carry a label specifying the actual distance between i and j, and/or the geometrical relationship between them, e.g. node i is above node j. Other more specific link-labels may refer to the specific labels carried by the nodes. For instance, a link may indicate the ratio between the sizes of the two image regions (e.g. connected components) corresponding to the two nodes.

To summarize, labelled graphs offer a broad range of possibilities for constructing relational descriptions of objects and scenes. Clearly, the nature and the degree of the invariances afforded by these relational descriptions critically depend on the nature of the labels used both for the nodes and for the links in the graph. For instance, if the presence of a link between nodes i and j simply indicates that i and j are neighbors in the image and if the node-features are rotation-invariant, the resulting description will be fully invariant with respect to all rigid transformations, i.e., shifts and rotations. If on the other hand the links also specify geometrical relationships beween the nodes, the description will be invariant with respect to shifts, but not rotations.

How are these relational descriptions actually used? Typically, one will be faced with the problem of comparing labelled graphs with each other: a labelled graph extracted from the observed image will be compared to several ***prototype-graphs*** which constitute our knowledge --or are part of our knowledge-- about a given world of images. One seeks the best possible match for this observed graph: this match identifies the object or shape under consideration.

3 VARIOUS NOTIONS OF GRAPH-MATCHING

Given two graphs G and G', matching these two graphs means constructing a map from the nodes of G into the nodes of G' in such a way that any two nodes i and j in G which are linked to each other by an edge of G, map into two nodes k and l in G' which are linked to each other by an edge of G', and vice versa. This means simply that the map preserves the structure of the graph.

Thus, the strongest notion of graph-matching is graph-isomorphism. A graph-isomorphism is a bijective, i.e., a one-to-one map which preserves accurately the existence or non-existence of edges between pairs of nodes. This is indeed the strongest possible notion of match: isomorphic graphs --like any other isomorphic structures-- are really identical up to a change of notation.

This notion of match is not appropriate for pattern-recognition. First, it is too restrictive. We would like to accommodate various distortions and degradations of the image. If we require that the observed pattern be strictly isomorphic to one of our "memory-graphs", we would probably never recognize even a single real pattern. In addition, graph-isomorphism is computationally very inconvenient. Mathematicians have indeed investigated the computational complexity of the problem of finding a graph-isomorphism or deciding whether two graphs are isomorphic to each other. These problems are essentially ***intractable*** (3), which means that the time required for finding an exact solution by any known algorithm implementable on a conventional sequential machine, grows prohibitively fast --faster than any power function-- as the size of the graphs increases. This is due to the unfortunate fact that in order to find a strict isomorphism, one has to essentially scan through the space of all possible solutions in a sequential manner. The ***search-space*** in this case contains all the one-to-one maps from G to G'.

We therefore need a ***weaker notion of graph-matching,*** which would accommodate various distortions and degradations that affect real images, and which would also lend itself to much faster computations. The graph-matching notion that we shall use may actually be regarded as a generalization, or approximation of graph-isomorphism.

Notice that a map from a graph G to a graph G' is an isomorphism if it satisfies many individual, one may say local requirements. These individual requirements are of two types. First, the map should be one-to-one, which means that each node in G should correspond to exactly one node in G' and vice-versa. This constitutes the first family of local requirements, or ***constraints.*** Second, the map should preserve the existence of links. For each link in G or in G', this constitutes an additional constraint.

The basic idea is to relax somewhat these strict requirements. This means that we shall use the same constraints, but in a weaker form: the constraints can now be violated, but each violation incurs a small penalty. We add all the penalties together, and term the sum a ***cost-function***. The lower the cost, the better the match. For instance, if a node i in G maps into 2 nodes in G', this violation of bijectivity contributes a small penalty, say α, to the global cost of the solution. If i is connected to 3 nodes, the penalty incurred will be somewhat larger than α. The parameter α is the same for all nodes. A map with minimum penalty, i.e., with zero global cost, is a perfect one-to-one map.

Similarly, the requirement of preserving the links may also be viewed as the "sum" of many individual constraints. Notice that a configuration of two connected nodes i,j in G which map into two connected nodes k,l in G' is a ***4-cycle***, i.e., a closed circuit of length 4. This is illustrated in Fig. 1. Thus, the more 4-cycles of this sort we can find in a map from G to G', the better the match. We may now incorporate these individual requirements in the global ***cost-function***. Yet rather than using penalties for unsatisfied local constraints, we shall use ***rewards*** for satisfied constraints: we reward the 4-cycles, which means that we add a small negative term, say $-\beta$, for each individual 4-cycle in the map.

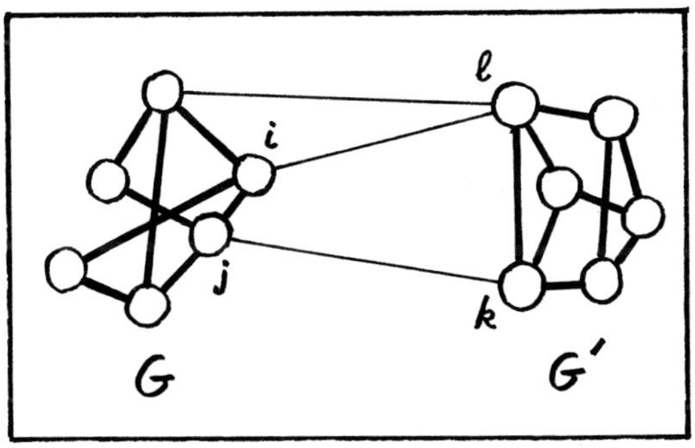

Figure 1. A map between graph G and graph G', including a ***4-cycle*** (i,j,k,l). The more 4-cycles the map includes, the better the match between the two graphs. Notice that *if* an isomorphism between G and G' exists, the isomorphism is precisely the bijective map which includes the maximum number of 4-cycles of type (i,j,k,l).

The global cost-function thus obtained is a sum of many individual, or "local" contributions, of two different types: rewards for local matches, penalties for violations of the bijectivity constraint. The quality of the match is determined by the cost: the lower the cost, the better the match. Finding good matches, i.e., maps of lowest cost, is an optimization problem. Notice that the cost-function that we have built embodies two antagonistic "forces". The first is associated with the penalties for violating the bijectivity requirement. This force tends to keep the number of links from G to G' small, as the best configuration actually includes exactly one such link per node. The second trend is akin to a "growth process": it tends to favor many links, as it favors the occurrence of 4-cycles. Thus, the nature of the best match according to this definition will depend on the relative weight of the two terms, i.e., on the ratio between the parameters α and β. Depending on this ratio, the best solutions may be sparse maps, or on the contrary many-to-many maps.

To summarize, we have defined a notion of graph-matching based on a "cost-function" which is a measure of the quality of the match. This notion is, in a sense, a generalization of the usual definition of graph-isomorphism, yet it differs from it in several important respects.

First, it tolerates errors, and the nature of these errors depends on the adjustment of numerical parameters in the cost-function. Thus, we really have not one definition but a family of definitions of graph-matching, and we may pick the most appropriate one for each particular application.

Second, since we now have a quantitative notion of the quality of the match, we may content ourselves with near-optimal solutions, i.e., maps which are not quite the lowest-cost states, yet are sufficiently close to provide the right answer to many pattern-recognition problems. A near-optimal solution is in many cases much easier to find than a strict graph-isomorphism, if the isomorphism exists at all.

Finally, as we shall see below, this optimization problem may be solved by using parallel-realizable algorithms which have the neural-like flavor mentioned earlier.

4 MATHEMATICAL FORMULATION

Let w_{ij} denote the value of the link beween nodes i and j. The variable w_{ij} is binary-valued: $w_{ij}=1$ means that the link is ON, that is, present, whereas $w_{ij}=0$ means the link is OFF. Specifying a graph G on a given set of nodes is thus equivalent to specifying a set of values w_{ij} for all pairs of nodes (i,j). We now define:

$$H(w) = -\beta \Sigma_{i,j\in G\ k,l\in G'}\ w_{ij}w_{jk}w_{kl}w_{li} + \alpha\ \Sigma_{j\in G}\ (\Sigma_{k\in G'}\ w_{jk} - 1)^2.$$

H(w) is the cost function with its two terms discussed above. The first term is the number of 4-cycles including one link w_{ij} in G and one link w_{kl} in G', times a constant β. The second term is a quadratic term which vanishes if and only if each node of G is linked to precisely one node in G', and is positive in all other cases.

The links of G and the links of G' are of course kept fixed. Only the "inter-graph" links are allowed to switch on and off. These links define a ***connectivity state*** $(w_{jk})_{j\in G,k\in G'}$ and the problem is the following: given two graphs G and G', defined by a set of fixed w_{ij}'s for $i,j\in G$ and w_{kl}'s for $k,l\in G'$, minimize H as a function of the inter-graph connectivity state $(w_{jk})_{j\in G,k\in G'}$.

The function H we have just defined is the simplest of a large family of cost-functions for graph-matching. Other functions can be obtained by various modifications of one or both of the two terms, and by including additional terms. In particular, when dealing with labelled graphs, a term may be included to penalize links between nodes carrying different labels. Again, this is a weak form of a strict constraint. Also, if the links are labelled, one may restrict the summation in the first term of H to those 4-cycles in which the G-edge w_{ij} and the G'-edge w_{kl} carry the same label.

Notice that a 4-cycle (i,j,k,l) with $i,j\in G$ $k,l\in G'$ is actually a graph-isomorphism between (i,j), an elementary subgraph of G, and (k,l), an elementary subgraph of G'. Thus, our definition of graph-matching is equivalent to the "sum" of many graph-isomorphisms between elementary subgraphs of G and elementary subgraphs of G'. This readily suggests a further modification of H, which consists in using small subgraphs including more than just two nodes and one link.

Cost-functions are sometimes termed "energy-functions", by an analogy with statistical mechanics. The individual w_{ij}'s are then regarded as "particles" which interact with each other. The system of particles subject to interactions is a fictitious physical system, and we may pretend that it is subject to the usual laws of thermodyamics. Examples of physical systems composed of many individual particles interacting with each other include gases and magnets. For instance, in the well-known Ising model of magnetism, the particles are elementary magnets called spins.

In statistical physics, the states of lowest possible energy are of particular interest, as are the lowest-cost states in our optimization problems. Indeed, at low temperatures, physical systems assume low-energy values. This is more accurately stated in a proper thermodynamical framework. Temperature introduces thermal noise, which results in a random motion of the state. This means that the state is distributed over a large portion of the state space. The equilibrium behavior of the system at a given temperature is thus characterized by a probability distribution function, the well-known Gibbs-Boltzmann distribution. At low temperatures, this distribution is concentrated on low-energy states. One may say that the system spends almost all its time in the low-energy regions of the state space. On the other hand, at high temperatures, thermal noise is so important that the distribution is virtually flat: the system spends on the average an equal amount of time in all the regions of the state space.

The interactions included in the energy function determine the aspect of the "energy landcape" of the problem, that is, the shape of the energy as a function of the configuration of the system. Since we deal with very large systems, i.e., systems including huge numbers of particles, the state spaces are always of very high dimensionality. The shape of the energy function, the energy landscape, is therefore difficult to visualize. One may nevertheless derive some important aspects of it. For instance, some functions are smooth, other are chaotic. The behavior of the system, particularly at low temperatures, depends critically on the smoothness or roughness of the energy landscape.

Clearly, working with a chaotic energy landscape is more difficult than with a smooth one. This is true in particular when the goal is to find the low-energy states, or "valleys". The simplest technique for finding the low-energy states is to follow the lines of steepest descent in the landscape. If there are many communications between the valleys, one will eventually find a "good solution", i.e., a low-energy state. However, it may happen that the landscape contains many local energy minima. The gradient descent technique will perform rather poorly in this case: the state will soon get trapped in a local minimum, which may stand quite high on the energy scale. Alternative techniques should then be used.

For instance, instead of rolling down the slopes in a "deterministic" fashion, one may introduce an element of randomness, which can be viewed as resulting from thermal noise in the statistical physics analogy. This is equivalent to going uphill from time to time instead of only downhill. Of course the amount of noise, i.e., the "temperature" of the system, should be carefully adjusted. Typically, if the temperature is low, the probability of escaping from a local energy minimum is very small; the time required for finding a good solution may then be prohibitively long. Put in technical terms: the convergence to the equilibrium distribution is very slow. On the other hand, if the temperature is high, then, as said already, the distribution becomes flat: the state goes uphill almost as often as downhill, and never stays in low-energy regions.

Some more sophisticated techniques have been devised, and these are particularly useful when almost nothing is known *a priori* of the energy landscape. For instance, one may start at high temperature and progressively and very slowly lower down this temperature: as time goes by, the probability of going uphill instead of downhill becomes smaller and smaller. This process achieves a good compromise between the fast convergence at high temperature and the good quality of low-temperature solutions. This algorithm is called simulated annealing (4) because it is analogous to the process of annealing used in metallurgy. Simulated annealing has been applied successfully to various optimization problems and to image-processing (5).

In our problem of graph-matching and its applications to pattern recognition, the shape of the "energy-landscape" defined by the cost-function will of course depend on many details, e.g. on the parameters α and β, and on other parameters that may occur in additional terms of the function. The main point is that this family of cost-functions provides a definition of graph-matching which is both computationally tractable and appropriate to the problem of pattern recognition. Indeed, when the parameters are correctly adjusted, the energy-landscape is sufficiently well-behaved, and a satisfactory solution can be found in reasonable time. The weak constraints used in our cost-functions also accommodate various distortions and degradations of interest in pattern recognition (6,7).

5 APPLICATION TO DIGITAL IMAGE-PROCESSING

In this section, we outline a computer vision algorithm partly inspired from the graph-matching technique presented in Sections 3 and 4. Consider again the problem of character recognition, where the task is the localization and identification of characters or numerals within an image. As was already mentioned in the introduction, this is not merely a classification problem: one also needs to segment the image, i.e., separate the figure(s) from its (their) background.

In many cases, segmentation may be performed as a first pre-processing step, ***before and independently*** of the computations involved in figure-identification proper. Figure/ground separation is then a strictly low-level task, which is carried out in a purely data-driven fashion, and does not require any higher-level knowledge. For instance, if the figure is always much brighter than the background, then using an appropriate grey-level threshold to binarize the image will unambiguously separate the figure from the background.

On the other hand some images cannot be as easily pre-processed. In many cases, pure grey-level information is not sufficient to separate figure from ground. This may be due to an intrinsic variability in the generation of the figures, or to various sorts of noise which may affect the image, and render the process of segmentation based on low-level information much too hazardous to be used in isolation from high-level information. Degradations of the image caused by reflections, shading, dust, scratches etc., will in general result in highly correlated noise. Unlike uncorrelated "white noise", correlated noise has a structure of its own. It creates "spurious shapes" which are hard to distinguish from the real signal, in this case the characters or the numerals. One therefore needs higher-level information, i.e., knowledge about the shapes of the characters, to disambiguate such a noisy image. Higher-level knowledge may help separate the signal from the noise. Clearly this calls for a partially "top-down" approach, where high-level knowledge interacts with low-level processing.

We shall outline an approach to this class of image-processing problems, partly inspired from the graph-matching paradigm described above. The other source of inspiration is the Bayesian approach to image-processing introduced in 5. We shall formulate the problem in such a way that the segmentation and the recognition are carried out simultaneously: instead of first segmenting the image and then only matching it to the prototypes, one directly looks for the best possible "segmentation-and-match". The process of finding the match, that is, identifying the character in the image, will then guide the process of segmentation.

The model has three layers. The bottom layer, called L_0, is the observed grey-level image. The intermediate layer, L_1, contains the tentative segmentation of the grey-level image: L_1 is an array of same size as L_0, yet it is binary-valued: each pixel in it is either -1 or +1. Finally, the upper layer, L_2, contains a labelled graph which is the prototype to be compared to the segmented image.

The cost-function in this model is the sum of two terms. The first is the graph-matching cost-function studied in the previous sections, with L_1 and L_2 playing the roles of the two graphs G and G'. Thus, the L_1-L_2 interaction is governed by a cost-function which embodies the search for the best possible match between the prototype and the segmented image. Depending on the type of labelled graph used in the model, one may obtain various types of invariances in the L_1-L_2 correspondence. The second part of the cost-function concerns the interaction between the observed image L_0 and the segmented image L_1.

For the sake of illustration (Fig. 2), we use here a very simple type of labelled graph in L_2. This labelled graph is generated from a perfectly undegraded {-1,+1}-valued shape (character). The nodes correspond to individual pixels, some of them belonging to the character itself (in which case the label is +1), others to the background (the label is -1). The edges indicate the geometrical relationships between these pixels. Thus, node-labels are binary-valued, while edge-labels carry relative positional information. In general, one does not need to include all the pixels and all the edges in the graph. Various heuristic methods can be used to pick an appropriate set of pixels and edges. Clearly, this representation is only shift-invariant. Invariances with respect to other transformations may be obtained from other labelled-graph representations.

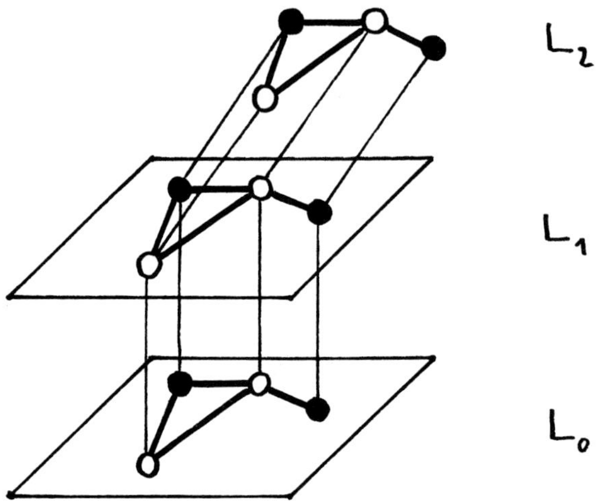

Figure 2. A 3-layer relational model for image-segmentation and pattern-recognition. See text for details.

There are two types of variables in this model: the binary-valued pixel-variables in L_1, defining the segmented image, and the L_1-L_2 connectivity variables, defining a correspondence between the segmented image and the prototype-graph. The segmented image in L_1 is viewed as a labelled graph, and the aim is to find a map between L_1 and L_2 which is the best possible match between the two graphs in these two layers, in the sense of the previous section. Simultaneously, the segmentation, i.e., the L_1-state by itself, is allowed to vary, so that one really looks for the best possible segmentation-and-match. The algorithm is partly sequential: given an observed image, i.e., an L_0-state, one first attempts to interpret it under the hypothesis that it contains the character "A", that is, one looks for the best segmentation-and-match with prototype "A" in L_2. One then iterates with the other prototypes in L_2, namely "B", "C", etc. The final decision is given by the best result reached.

We shall not give here the explicit form of the graph-matching cost-function. We just recall that this function embodies the various constraints mentioned in the previous sections: (i) sparseness of the map, here the L_1-L_2 correspondence; (ii) consistency of the map with the graph structure, expressed in terms of the number of 4-cycles; (iii) consistency of the map with the node-labels; (iv) consistency of the map with the edge-labels. The meaning of the last requirement in the present case is that in order for a 4-cycle (i,j,k,l) to be allowed to contribute to the energy, the geometrical relationship between the pixels i and j in L_1 has to be the same as the relationship between pixels k and l in L_2.

The second part of the cost-function in this model contains the interactions between layer L_0, the raw image, and layer L_1, the segmented image. The two arrays L_0 and L_1 have same size, and the correspondence is strictly one-to-one and "retinotopic": a pixel in L_1 carries the binarized value of the grey-level of the corresponding pixel in L_0. The part of the cost-function corresponding to the segmentation task is again made of many individual weak constraints. These can be chosen in several different ways.

The simplest, already mentioned segmentation scheme consists in picking an appropriate grey-level threshold μ, and affecting value +1 (in L_1) to pixels with grey-level (in L_0) above μ, and value 0 to pixels with grey-level below μ. Notice that in such a scheme, the constraints embody ***pixel-to-pixel interactions:*** each pixel in L_1 "sees" only the corresponding pixel in L_0. One may also base the segmentation on averaged grey-level values in the image, or on other local informations such as descriptors of texture, instead of using only single-pixel information. These schemes are essentially local: each pixel in L_1 is connected in a "retinotopic" fashion to a small region in L_0.

We shall use a different segmentation scheme, in which the L_0-L_1 interaction is ***link-to-link*** rather than pixel-to-pixel. This improves considerably the behavior of the model, in particular its robustness to various degradations. The cost-function now includes ***one contribution for each pair of neighboring pixels*** in the image, with the following effect. If the difference of grey-levels in any two neighboring pixels in L_0 is large, the two corresponding pixels in L_1 tend to have different labels, while if the difference is small, the two corresponding pixels in L_1 tend to have identical labels.

This type of interaction may be easily implemented in the form of a quadratic term. It is interesting to note that the resulting cost-function is then identical to the interaction energy in the already mentioned Ising model of magnetism. Indeed, the energy in the Ising model is:

$$H(\boldsymbol{s}) = - \Sigma_{ij} T_{ij}s_i s_j .$$

The binary-valued spins s_i correspond to the L_1-variables (the discretized image), and the interaction constants T_{ij} are determined by the raw-image (L_0) instead of being true constants as in the Ising model: T_{ij} is large if the grey-levels of pixels i and j in L_0 are alike, and small if these grey-levels are very different. This results in an attractive (ferromagnetic) force in case the grey-levels are close to each other, and in a repulsive (antiferromagnetic) force if the grey-levels are far apart. This type of interaction achieves a correct segmentation.

This new segmentation scheme is quite different from a simple thresholding algorithm. The interaction between the two layers L_0 and L_1 concerns now only relationships between pixels. The advantage of this ***relational segmentation scheme*** is, evidently, a new type of invariance. For instance, the model is now insensitive to the sign of the figure/ground contrast. In other words, it is not necessary, when using this cost-function, to know in advance whether the figure is brighter than the background or vice-versa. More importantly, the model performs equally well when the figure is for instance brighter than the background in the top half of the image, and darker in the bottom half. In such an image, a discontinuity exists between the top and the bottom halves, but this discontinuity is "of measure 0": it is a single line, hence the number of pairs of neighboring pixels across the discontinuity is negligible when compared to the pairs of neighboring pixels in the whole image. As a result of this, the lowest-cost state for the global cost-function including both the L_1-L_2 interactions and the L_0-L_1 terms, is indeed achieved by the correct segmentation: the figure will be uniformly labelled +1 and the background -1 (or vice-versa), since the resulting violations on the discontinuity line incur only a small penalty when compared to the global cost and reward contributed by the rest of the image.

So far, in our relational approach to the segmentation problem, we have included interactions between nearest-neighbors only. Yet it is equally easy to build a cost-function based on long-range interactions instead of, or in addition to nearest-neighbors interactions. We then obtain, as before, an Ising-like energy for the segmentation problem, with the interaction constants T_{ij} derived as before from the grey-level difference between pixels i and j. The only difference is in the interaction graph: instead of using only nearest-neighbor pairs (i,j) as in the conventional Ising model, one includes now pairs of pixels (i,j) arbitrarily distant from each other.

From the point of view of computational complexity, the best strategy for choosing pairs of interacting pixels (i,j) for the purpose of segmentation, consists very often in picking them at random. (A random choice generates mostly distant interactions, if the array is large.) Indeed, due to theoretical properties of random graphs that we shall not dwell upon, the L_0-L_1 energy landscape based on a random interaction graph has virtually no local minima. This means that the best segmentation can be found extremely rapidly and reliably, by using essentially a deterministic gradient-descent method. This is in sharp constrast with the usual nearest-neighbor scheme, which results in a much more chaotic energy landscape: as in the conventional Ising model, the nearest-neighbor interaction energy has many local minima, which make it much difficult to find the lowest-energy state, that is, the best segmentation.

We conclude this section with a remark concerning the relationship between the present segmentation-and-recognition model and the Bayesian approach to image analysis (5). Notice that the full interaction energy H, which includes the L_0-L_1 terms as well as the L_1-L_2 terms, may be used to generate, through the usual Gibbs-Boltzmann formula, a probability distribution function on the joint L_0-L_1-L_2-Γ_{12} state, where Γ_{12} denotes the L_1-L_2 connectivity state. (The prototype-graphs in L_2

are given equal probability.) In a Bayesian perspective, this distribution is an ***a priori probability,*** or "prior", embodying our knowledge about a given world of images. Now for a given ***observed image,*** i.e., a fixed L_0-state, the ***posterior probability*** is obtained by conditioning this prior on the observation. Thus, the minimum of H given the observed image, which is the solution to our segmentation-and-recognition problem, is actually the usual MAP (Maximum A Posteriori) estimate, i.e., the mode of the posterior distribution. In short, the present multi-layer relational model may be regarded as an extension of the Bayesian approach to image processing (5), incorporating higher-level knowledge in the prior distribution.

6 IMPLEMENTATION IN THE BRAIN

Let us summarize the preceding sections.

(i) We have outlined an approach to invariance based on the notion that the relevant information in an image is essentially of relational nature.

(ii) Thus, we have proposed an approach to invariant pattern recognition based on an appropriate notion of graph-matching.

(iii) Similarly, we have outlined a scheme for figure/ground separation characterized by the fact that it exploits only relational information in the image.

(iv) These approaches to pattern-identification and to figure/ground separation are both implemented in the form of a ***cost-function*** whose minimum realizes the solution of the problem under consideration.

(v) Finding the minima of the cost-function can be done by a variety of algorithms, some of which are inspired from statistical physics.

(vi) We have sketched a 3-layer architecture in which the figure/ground separation task and the figure-identification task are carried out simultaneously and interactively. This is made possible by the uniform treatment using energy functions. This model is appropriate for the processing of images which are too degraded to be segmented in a data-driven fashion.

(vii) Regarding the extraction of relational information from an image, we mentioned the fact that it is in many cases advantageous to use long-range, in particular randomly chosen links, rather than short-range ones. This may considerably simplify the ***energy-landscape*** of the problem under consideration.

These propositions may be regarded as an outline of a ***relational paradigm*** for vision. In this section, we shall briefly discuss a possible implementation of this paradigm in the nervous system.

Three issues at least should be discussed. First, the format of representation of relational information in the nervous system should be specified. The question is: How are the link-variables in our graph-theoretic approach physically coded in the brain?

The second issue concerns the dynamics of the system. In the mathematical formulation we have used, solving a figure/ground separation and/or pattern recognition problem amounts to minimizing an appropriately defined cost-function. The question is then: What are the mechanisms underlying this optimization process in the neural-network implementation?

The third and last issue is that of the relationship between the fast processes of optimization used in problem-solving, and the slower memory-like processes required to store knowledge in the network.

We first address the coding issue. The simplest scheme for coding information in the brain is known as ***rate-coding***. In this scheme, the variables which carry the information are activity rates of individual neurons or assemblies of neurons. The presence of a given object in our visual field is signalled in the brain by a configuration of activity, specific to this particular object. Similarly, attributes of the object and its relationships to other objects present in the scene are coded in the form of specific patterns of activity. This view about brain function is a time-honored one (8,9), and it is probably fair to say that it is still largely prevailing in the literature (10,11).

One should also mention that activity rates are generally understood as ***time-averaged rates.*** Indeed, individual neural activity has a conspicuous stochastic character, which is viewed by many authors as a nuisance to brain function. A simple way to overcome this nuisance is to average the activity over time, a typical figure for the averaging period being a few hundred msec.

On the other hand, several authors have expressed the idea that information could be encoded in the brain in more sophisticated ways. A variety of such suggestions have been made, including several forms of ***statistical coding*** (12,13). In this alternative, or complementary scheme for knowledge representation, the accurate time-sequences of action potentials play a determining role. The relevant time-scale is in the range of the milliseconds, which is consistent with the properties of spike propagation, synaptic transmission, and spatio-temporal integration on dendritic membranes. Information is carried essentially by the statistical relationships between action potentials in different fibres. Such information is of course hidden from the neurophysiologist equipped with a single recording microelectrode.

Another possibility for encoding relational information in the brain relies on the notion of ***fast synaptic plasticity*** (14). In such a scheme, the synapses themselves are the locus of fast and reversible processes, which affect their transmission efficacy. These processes can be as fast as the time required for the propagation of nervous activity from one neuron to another, i.e., about 1 msec. This scheme clearly provides the most straightforward implementation of a graph in the nervous system: a dynamical link, with its two possible states $w_{ij}=0$ and $w_{ij}=1$, is simply implemented by a fast-modifying synapse, which can switch instantaneously from the ON-state to the OFF-state and vice-versa. The dynamical state of the network is thus a ***connectivity state*** rather than, or in addition to an activity state.

The relational paradigm outlined in the present paper clearly militates in favor of statistical coding and fast synaptic plasticity. Indeed, it seems hard to object to the argument that the most natural way to represent external relationships, i.e., relationships between objects and sub-objects in the world, consists in using internal relationships, i.e., relationships between neurons in the brain, thereby preserving the very structure of the information itself. This argument has been developed in detail elsewhere (15,16,17).

Of particular interest is the conjunction of the last two schemes, namely statistical coding and fast synaptic plasticity. In this mixed scheme, relationships are encoded in the form of two types of complementary variables: temporal relationships between activity in neuron i and activity in neuron j, in particular co-activity; strength of the synapses between neuron i and neuron j.

The advantage in using both ***co-activity and connectivity*** becomes clear as soon as one raises the second issue mentioned earlier, namely the dynamics issue. As has been proposed elsewhere (16), the dynamics of the joint co-activity and connectivity state may be characterized by positive feedback: co-occurrence of activity in the pre- and post-synaptic neurons would enhance the efficacy of the synapse (a very fast "hebbian" synaptic modification rule), while it is of course always true that an increased excitatory synaptic weight enhances the probability of co-activity in pre- and post-synaptic neurons.

Positive feedback provides stability and diversity: a dynamical system endowed with positive feedback will generally have several stable states. In the present case, stability and diversity are needed to explain how different temporal patterns of co-activity and connectivity may come into existence in a neural network depending on the task currently performed by the system, and how these patterns may be temporarily stabilized. In particular, positive feedback between co-activity and connectivity provides a simple implementation of the optimization process proposed in connection with graph-matching. Remember that the cost-function in the graph-matching problem had two main terms, which resulted in two different trends in the minimization process. The first was a reward term or growth term which had the effect of favoring the creation of 4-cycles. The second was a penalty term which prohibited the growth of too many links at any node.

With the links of the graph implemented in the form of excitatory synaptic weights, the rule of positive feedback just described is clearly a growth mechanism. This mechanism favors and stabilizes specific configurations, closely related to the

4-cycles on which the graph-matching paradigm is based. The argument is briefly summarized here.

In order to trigger an action potential, or a burst of action potentials in a given neuron, the activity in the afferent fibres to this neuron has to be synchronized on a relatively short time-scale: membrane time-constants are typically of the order of 5 msec. Neurons are therefore sometimes termed "coincidence-detectors" (13). Now a closed cycle (i,j,k,l) may be viewed as containing two alternative pathways from neuron i to neuron k, namely the pathway i → j → k, and the pathway i → l → k. More generally, configurations containing many short closed cycles also contain many short alternative pathways between neurons. Clearly, such configurations create synchrony in the network. Hence, they will be favored in the positive feedback dynamics, as required.

We next briefly discuss the implementation of the connectivity-penalty term in the cost function. The mean level of activity in cortex is strictly regulated, which implies in particular that co-activity and connectivity should never be allowed to exceed certain critical values. This constraint, which can be implemented by an appropriately wired inhibitory system, is virtually identical to the connectivity-penalty used in the graph-matching cost-function.

We finally address the third issue, concerning the relationship between long-lasting memory and fast working memory in this "graph-coding" approach. There is little debating that the physical locus of memory in the brain is essentially the configuration of synaptic connections, with their respective strengths. Thus, the present model should include at least two different processes of synaptic plasticity: (i) a fast reversible process, which is part of the actual computations performed at any instant; (ii) a long-lasting, essentially irreversible process, leaving a permanent memory trace.

In a computational perspective, the memory is roughly the equivalent of a computer program, including various stored data such as prototype patterns for pattern-recognition. In our approach, using the very same format, namely labelled graphs, for the permanent memory and the working memory, makes it extremely easy to incorporate new data in the program whenever required: the fast co-activity and connectivity pattern is simply "printed" in the network of permanent connections. Physically, this amounts to a purely local transcription from a reversible trace to a permanent one. The advantage of this scheme when compared to the learning algorithms in many connectionist models is quite appreciable (see 17 for a more detailed discussion).

7 CONCLUSION

We have outlined a computational approach to pattern-recognition and image-segmentation based on a ***relational paradigm:*** these problems are formulated as a task of minimization of a cost- or "energy-" function, which embodies interactions between relationships in one image and relationships in another image. In other words, the interactions between the two images, e.g. an observed grey-level image and the corresponding binarized image, or the binarized image and a prototype image, is a ***link-to-link interaction*** rather than a ***pixel-to-pixel interaction.*** Formally, the objects we manipulate are labelled graphs.

Applications of this approach to several image-processing problems are currently investigated. The range of applications is of course limited by the fact that conventional machines use sequential algorithms, which are bound to ***simulate,*** in a sometimes rather inefficient way, distributed processes. It is however possible to outline two main features that would be required from a parallel machine, for a better implementation of the distributed optimization processes underlying our relational approach. (a) High connectivity. This should include at least two patterns of connectivity: local, short-range connections in a low-dimensional space, e.g. on a regular 2-D lattice; and long-range, essentially random connections. Random connectivity is in a sense equivalent to very high dimensionality. (b) Fast, on-line modification of the connectivity.

These "anatomical" and "physiological" features may actually account for much of the superiority of brains over computers: they could be the substrate of a format of knowledge representation and organization largely unknown to artificial computing systems.

8 REFERENCES

(1) Duda, R.O., and Hart, P.E. (1973) *Pattern Classification and Scene Analysis.* John Wiley and Sons, New-York.

(2) Ballard, D.H., and Brown, C.M. (1982) *Computer Vision.* Prentice-Hall, Englewood Cliffs, New-Jersey.

(3) Garey, M.R., and Johnson, D.S. (1979) *Computers and Intractability.* W.H. Freeman and Co., New-York.

(4) Kirkpatrick, S., Gelatt, Jr., C.D, and Vecchi, M.P. (1983) Optimization by Simulated Annealing. Science, **220**, 671-680.

(5) Geman, S., and Geman, D. (1984) Stochastic Relaxation, Gibbs Distributions, and the Bayesian Restoration of Images. IEEE Trans. Pattern Anal. Machine Intell., vol. PAMI-6, 721-741.

(6) Bienenstock, E., and von der Malsburg, C. (1987) A Neural Network for Invariant Pattern Recognition. Europhys. Lett., in press.

(7) von der Malsburg, C., and Bienenstock, E. (1987) A Neural Network for the Retrieval of Superimposed Connection Patterns. Europhys. Lett., in press.

(8) Hebb, D.O. (1949) *The Organization of Behavior.* Wiley, New-York.

(9) Barlow, H.B. (1972) Single Units and Sensation: A Neuron Doctrine for Perceptual Psychology? Perception, vol. 1, 371-394.

(10) Hopfield, J.J. (1982) Neural Networks and Physical Systems with Emergent Collective Computational Abilities. Proc. Natl. Acad. Sci. USA, **79**, 2554-2558.

(11) Ackley, D.H., Hinton, G.E., and Sejnowski, T.J. (1985) A Learning Algorithm for Boltzmann Machines. Cognitive Science, **9**, 147-169.

(12) Sejnowski, T.J. (1981) Skeleton Filters in the Brain. In: *Parallel Models of Associative Memory.* G.E. Hinton and J.A. Anderson, eds. Erlbaum Associates, Hillsdale, N.J.

(13) Abeles, M. (1982) *Local Cortical Circuits: An Electrophysiological Study.* V. Braitenberg, ed., Springer-Verlag, Berlin.

(14) von der Malsburg, C. (1981) *The Correlation Theory of Brain Function.* Internal Report 81-2. Max-Planck Institute for Biophysical Chemistry, Department of Neurobiology, Göttingen, West-Germany.

(15) von der Malsburg, C. (1987) Synaptic Plasticity as a Basis of Brain Organization. In: *The Neural and Molecular Bases of Learning.* Dahlem Konferenzen. J.-P. Changeux and M. Konishi, eds. John Wiley and Sons, Chichester.

(16) von der Malsburg, C., and Bienenstock, E. (1986) Statistical Coding and Short-Term Synaptic Plasticity: A Scheme for Knowledge Representation in the Brain. In: *Disordered Systems and Biological Organization.* E. Bienenstock, F. Fogelman-Soulié, and G. Weisbuch, eds. Springer-Verlag, Berlin.

(17) Bienenstock, E. (1987) Connectionist Approaches to Vision. In: *Models of Visual Perception: from Natural to Artificial.* M. Imbert, ed., Oxford University Press, in press.

Neural-Like Graph-Matching Techniques for Image-Processing

Discussion
edited by *Gert Hauske*

First question was whether the system has to go through all possible mappings to reach its final state or whether there would be a perhaps more intelligent way e.g. to use a priori information. As already mentioned in the lecture one way of using a priori information is to confine the mapping to small neighborhood ranges which makes a solution much more tractable. But the essential point is that the system properly operates also under the worst case of no a priori information. In this case it makes a global decision starting at high "temperatures" which breaks the symmetry leading to a rough global match. When the "temperature" is decreased the details of the mapping emerge and you come closer to a perfect isomorphism between image and interpretation.

A critical remark referred to the concept of transformation energy in connection with pattern recognition systems. Experiences with a practical system showed that for larger distortions images were often transformed into false categories even with low transformation energies. Such a behaviour is not unexpected in systems of that kind but the use of transformation energies seems to be a little questionable.

The question of how many possibilities of connections are actually needed in the model system did not lead to a precise quantitative estimate. It was explained, however, that a minimum of connections is necessary to have significant effects on self organization in terms of synchrony relations between neurones.

Whether an analytical solution is conceivable for simple cases with definite distortions was not completely rejected, e.g. if computer vision problems are considered. On the other hand the class of real world transformations is not very well defined so that the direct approach using a rubber sheet transformation seems to be the best so far.

The problem of the dimensionality of a-priori-information under genetic aspect led to a more general consideration of how much is learned and how much is hard-wired. Although a definite ratio cannot be given the important role of both aspects is evident.

In another question the relation was considered between the biologically given effect of synchronization within a system of plastic synapses on one hand and the concept of cost functions and annealing processes on the other. The response was that different mathematical formulations can be given of the fundamental role played by statistical relationships embodied in a interconnected system and the above mentioned concept represents only one of them.

The final question was whether the system has to know exactly what it is looking for. The answer was no; the only thing the system needs is a good prior distribution before it starts interpreting.

In a concluding remark the necessity was explained to differentiate between structure and the more functional aspect of connectivity which may exist in different kinds of processing.

The Temporal Integration of 3-D Structure From Motion: A Computational and Psychophysical Study

by *Norberto M. Grzywacz, Ellen C. Hildreth, Victor K. Inada* and *Edward H. Adelson*

SUMMARY

This paper approaches the problem of how humans utilize motion cues in the process of recovering the three-dimensional structure of objects, by a combination of computational and psychophysical methods. At the computational level, we explore a theory recently proposed by Ullman (18), the incremental rigidity scheme. This theory postulates that the recovery of structure from motion may be accomplished by the visual system through the maintenance of an internal model of the structure of the object, which is transformed from one instant to the next in such a way that its shape remains as rigid as possible (but not strictly rigid), while still being consistent with the image data. This scheme was shown to be able to recover the structure of objects in both rigid and nonrigid motions. Here, we report our studies on two properties of Ullman's scheme. The first property is that the rate of convergence of the internal model and quality of the recovered structure decrease as the scheme compares image frames that are temporally closer. This result is surprising, because more views of the motion are inspected by the visual system when it is sampled at higher frequencies. We found, however, that as the time distance between frames decreases the transformations performed by the algorithm become ill-conditioned, i.e. less robust against noise, analogous to what happens in numerical differentiation. In fact, one can show that a continuous formulation of the algorithm, which may be devised in the limit of infinitesimally small time intervals between inspected image frames (i.e. when velocity instead of positional information is used), is unstable. This result raises the hypothesis that the human visual system needs to process simultaneously long periods of image data in order to obtain a reliable recovery of structure from motion. The first psychophysical experiment reported in this paper measured the integration time for this recovery and found it to be on the order of 1 second. This is a remarkably long time if one considers that the integration time for velocity estimation in the human visual system is at most about 100 *msec* for the range of velocities used in this paper (44). The second property of Ullman's scheme that we report here is that the asymptotic behavior of the algorithm depends on the initial internal model used. In particular, we found that if the initial internal model is stretched compared to the real structure of the object it will not find the correct solution, but rather remain relatively stretched. On the other hand, if the initial model is flat, a correct convergence is guaranteed under certain conditions. We found psychophysically, that humans too have difficulty initially in recovering the correct structure of a moving object if they are forced to have an initial percept that is stretched compared to the real object, but quickly perceive the correct structure if the initial percept is flat. We argue that this coincidence between the algorithm and human performance strongly constrains theories for the structure from motion mechanism used by the visual system. We suggest that the human visual system prefers a long temporal integration of the images using all possible *a priori* information about the object's structure, even if this sometimes leads to the incorrect structure, to a "sure" temporally local method, which nevertheless is not robust against noise.

1 INTRODUCTION

The computational approach to vision postulates that the understanding of the human perceptual system as an information–processing device can be divided into three levels (1). The first level is the computational theory of the task to be performed by the visual system. In this theory, the goals, assumptions and the general strategy of the computation must be stated and analyzed. The second level is the abstract implementation of the computational theory, and focuses on the representations and algorithmic transformations from the input to the output. Finally, the third level is the hardware implementation for the postulated representations and algorithms. In terms of the human visual system, this is a study of the neuronal and biophysical mechanisms underlying the proposed processes. In this paper, the first two levels will be emphasized.

One of the most important tasks of the human visual system is to derive an invariant shape description of the image. A crucial step in this direction is the derivation by the early vision processes, of the physical properties of visible surfaces and, in particular, their three–dimensional (3–D) geometry. The main problem for the recovery of the 3–D structure of objects is that the relevant information in the retinal image is two–dimensional (2–D), and therefore, underconstrained, which makes these problems ill–posed (2). The main approach to "solve" ill–posed problems is to restrict the class of admissible solutions by assumptions that reflect *a priori* knowledge about the physical world. The human visual system uses a wide variety of image cues and physical assumptions in order to recover 3–D shapes of objects.

An important source of 3–D information is given by the relative motions of elements in the changing 2–D image. The human visual system can recover structure from motion under both orthographic and perspective projection and in the absence of all other cues to 3–D structure. The relevant psychophysical studies included not only simple paradigms, but also complex patterns of motions (e.g. biological motion) encountered in everyday life (for a review of the psychophysical literature, see (3)).

Perceptual studies suggested that the rigidity of objects may play a role in the recovery of structure from motion in humans (3). Computational studies established that rigidity is a sufficiently powerful constraint to derive a unique interpretation of structure from the observation of image motions over a small extent in space and in time (for a review of the computational literature, using both optical flow and discrete methods, see (4)). These studies explored the fact that if an object is rigid, and if the different parts of its image change relative to one another, then a 3–D structure–invariant motion must be implied. This fixed geometry of the object may be computed from the changing image parameters.

Theoretical studies have directly given rise to algorithms for deriving the 3–D structure of rigid objects in motion. Experimentation with algorithms has revealed two important limitations. First, although it is possible in theory to recover structure from motion information that is integrated over a small extent in space and time, such a strategy may not be robust in practice. Small errors in image motion measurements can lead to very different solutions. This suggests that a robust algorithm for recovering structure should use motion information that is more extended in space or time. This conclusion is supported by computational studies that either integrate motion measurements over large areas of the image (or many image features) (5–15) or over long times (4, 14–21).

The second limitation that emerged from experimentation with these algorithms was that they derived a 3–D structure only when a strictly rigid interpretation was possible; otherwise they did not yield any interpretation of structure or yielded a solution that was incorrect or unstable. Regarding the human system, however, it is an everyday experience to perceive the structure and motion of deforming objects such as a flowing river, an expanding balloon, or a dancing ballerina. Perceptual studies show that some structure can be derived for a broad range of nonrigid motions, including stretching, bending and more complex deformations (22–27). Furthermore, displays of rigid objects in motion sometimes give rise to the perception of somewhat distorting objects (28–36).

With that in mind, algorithms that considered nonrigid motions were devised by some researchers (16, 18, 37–39). The computational part of this paper focuses on a method proposed by Ullman (18) that derives the structure of rigid and nonrigid objects in motion by integrating image motion information over an extended time. The algorithm, called the *incremental rigidity scheme*, maintains an internal model of the structure of a moving object which is continually updated as new positions of image elements are considered. Its postulated action maximizes (but does not impose) the rigidity in the transformation from the current model to the new positions, and can be applied to both rigid and nonrigid objects in motion. This scheme has two computational advantages: (1) by integrating information over an extended time, it provides a stable recovery of structure in the presence of error in the image measurements, and (2) it allows deviations from rigidity, while always maintaining some model of 3–D structure.

We developed Ullman's work further through analytical studies and computer simulations, reaching the following conclusions regarding the computation of structure from motion: (1) The computation of a stable long–term solution requires the use of views of a moving object that differ significantly. This implies the need for a recovery process with memory of past views. As a result, for rigid objects rotating about a fixed axis, the rate of convergence of the algorithm and the quality of the final solution decrease for smaller angular displacements. Furthermore, one can show that a continuous formulation of the algorithm, which may be devised in the limit of infinitesimally small angular displacements (i.e. when velocity instead of positional information is used), is unstable. (2) The structure of the initial internal model used by the algorithm affects its asymptotic behavior. In particular, we found that the algorithm converges to the correct solution if the initial model was flatter than the viewed object, but did not for stretched initial conditions. We will argue that some form of dependence on the initial models is to be expected for the incremental rigidity scheme following the early theoretical results on the recovery of structure from motion.

The conclusions from our computational studies suggested the two classes of psychophysical experiments that we also report in this paper. The first class of experiments followed conclusion (1) of the previous paragraph. We expected that a long temporal integration would be necessary for a reliable recovery of structure from motion. We measured this integration time and found that it can be as long as 1 sec and for our rotating objects this implied rotations of at least 45°. Although some integration time is expected for any realistic algorithm for structure from motion, we would like to stress the importance of our computational study in the choice of this class of experiments, as the concept of integration is fundamental in Ullman's incremental rigidity scheme.

The second class of experiments suggested by our computational studies addressed whether different performance in recovering structure from motion is observed if different initial models of the object geometry are given to the observer. We will argue in the computational part of this paper that this prediction is quite unique to the incremental rigidity scheme, because this algorithm cannot guarantee the correct solution in general as the minimization procedures may become trapped in solutions corresponding to local minima in the solution space and never reach the global minimum.

2 COMPUTATIONAL STUDIES: METHODS

Our studies used a slightly modified version of Ullman's incremental rigidity scheme, which appears in detail in (4). The incremental rigidity scheme maintains and updates an internal model $M(t)$ of the viewed object, which consists of a set of 3–D coordinates: $M(t) = (x_i(t), y_i(t), z_i(t))$. In this paper we assume orthographic projection (the case of perspective projection yields similar results and is analyzed in (40)) onto the $X - Y$ image plane, so that $(x_i(t), y_i(t))$ are the image coordinates of the i–th point, and $z_i(t)$ is the current estimate of the depth at the i–th point. (We assume a positive z axis pointing away from the observer.) In this paper all the initial configurations were flat, i.e. $z_i(t) = 0$, except for the cases where we specifically tested the effects of the initial conditions.

Given a current model $M(t)$ at time t, and the image of the moving points in a new frame at a later time t', the problem is to compute a new model $M(t')$ such that the transformation from $M(t)$ to $M(t')$ is as rigid as possible. Since $x_i(t')$ and $y_i(t')$ are known, this requires the computation of the unknown depth values $z_i(t')$. (It is assumed that the correspondence between points in the two successive frames is known.) The new depth values are computed as follows. Let $l_{ij}(t)$ denote the distance between points i and j at time t. To make the transformation as rigid as possible, the values $z_i(t')$ for the new model are chosen so as to make $l_{ij}(t)$ and $l_{ij}(t')$ as similar as possible. We defined a measure of the difference between $l_{ij}(t)$ and $l_{ij}(t')$ as:

$$d(l_{ij}(t), l_{ij}(t')) = (l_{ij}(t) - l_{ij}(t'))^2, \tag{2.1}$$

and formulated the recovery of structure as the computation of $z_i(t')$ that minimize the following overall deviation from rigidity:

$$D(t, t') = \sum_{i,j} d(l_{ij}(t), l_{ij}(t')). \tag{2.2}$$

After the $z_i(t')$ are determined using this minimization process, the new model $M(t') = (x_i(t'), y_i(t'), z_i(t'))$ becomes the current model. A new frame is then registered and the process repeats itself. In this way, the scheme maintains rigidity by keeping the total distances between points in the model as constant as possible. For orthographic projection, only relative depth values, $z_i(t) - z_j(t)$, can be recovered, rather than absolute depth values, and 3–D structure is determined only up to a reflection about the image plane, since the orthographic projection of a rotating object and its mirror image rotating in the opposite direction coincide. We point out that measures other than that in Eq. 2.1 may be used. For example, Ullman (18), used a weighting which reduced the importance of maintaining rigidity of distant features.

3 COMPUTATIONAL STUDIES: RESULTS

This section begins with an analysis of the behavior of the incremental rigidity scheme when applied to rigid objects in rotation, as a function of the angular displacement between frames. We illustrate this behavior through computer simulations (although elsewhere (4) we discuss the results of an analytical study where we look at the long–term stability of the solution). Our main conclusion is that for the particular class of motions considered, the algorithm always yields a 3–D solution that converges asymptotically to the correct solution, but the rate of convergence varies with the angular displacement. The rate of convergence increases with increasing angular displacement up to a maximum, and then decreases with further increases in this displacement. We also show that for the limit of infinitesimally small angular displacements the rate of convergence goes to 0, i.e. the solution becomes unstable.

Our computer simulations consider the rigid rotation of a set of discrete points about the vertical axis. In all of the examples presented here, the input consisted of five points in space. Figure 1 illustrates the behavior of the algorithm. Each figure shows a birds' eye view of the points (their projection onto the $X - Z$ plane) with filled circles representing the true positions of the points and open circles representing the structure computed by the algorithm. Figure 1a shows the true positions of the points and the initial model at time $t = 0$. For all of the simulations, the initial model was flat. Figures 1b and 1c show the true and computed configurations of points after 120° of rotation. This final position was reached by taking three steps of 40° (Fig. 1b), and 12 steps of 10° (Fig. 1c). Note that the use of a smaller number of more disparate views yields a 3–D model that is closer to the true solution.

Figure 2 shows a set of graphs that illustrate in a different way, the behavior of the algorithm, as a function of angular displacement. The set of points shown in Fig. 1 were rotated by four full revolutions. The graphs of Fig. 2 show the error between the true and computed structures, as a function of time, computed as follows:

$$\left[\sum_{i,j}(d_{ij} - l_{ij})^2\right]^{\frac{1}{2}} \tag{3.1}$$

where d_{ij} is the correct 3–D distance between points i and j in the object, and l_{ij} is the distance between points i and j in the computed model. The graphs shown in Fig. 2 correspond to rotations with angular displacements of 40°, 20°, 10°, 5° and 1°. It can be seen that the rate of convergence and quality of the solution improves with larger angular displacements. For a particular total angular extent, the error in the computed model decreases with increasing angular displacement between frames.

We also found a complete deterioration of the performance of the algorithm for rotations with infinitesimally small angular displacements. In that case a continuous formulation, which uses velocity information at discrete feature points, can be developed as a limiting case for the incremental rigidity scheme, and be shown to be only weakly stable, i.e. if the solution is perturbed, the algorithm never recover the correct structure but rather oscillates around it (4). (We will refer to the non-continuous formulation as "discrete" formulation)

We encountered another important characteristic of the incremental rigidity scheme. Computer simulations indicated that the algorithm sometimes behaves differently when started with

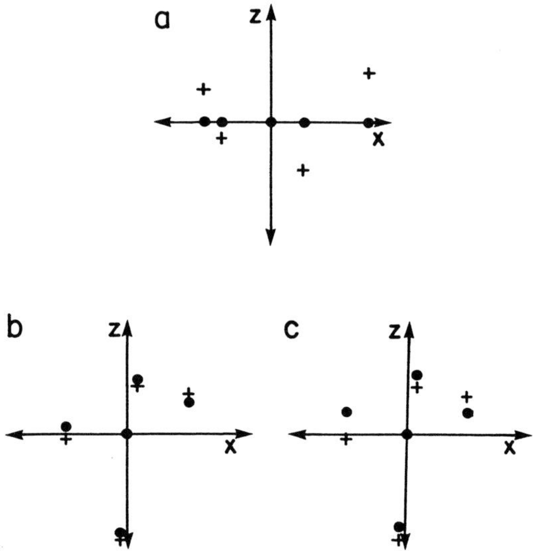

Figure 1. The behavior of the discrete formulation of the algorithm, under orthographic projection. (a) The true configuration of five points (plusses) is compared with the initial configuration of the points in the model (dots) at time $t = 0$. The projection is onto the $X - Z$ plane. (b) The comparison between the true and computed positions of the points after three steps of 40°. (c) The comparison between the true and computed positions of the points after 12 steps of 10°.

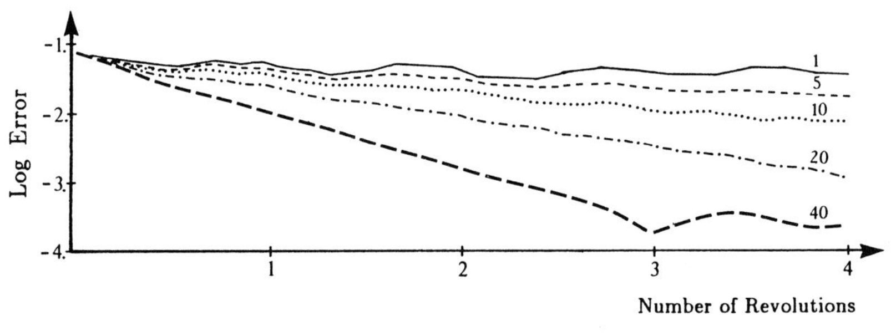

Figure 2. Graphs of the error in the internal distances between points in the computed 3–D model, as a function of time (in a log-linear plot). Simulations here used the discrete formulation and orthographic projection. The points are rotated 4 full revolutions, in steps of 40°, 20°, 10°, 5° and 1°.

different initial models of the 3–D structure of a set of points (Fig. 3). In general, if the algorithm begins with an initial model that is compressed relative to the true structure of the object, then the algorithm will converge toward the true structure. If on the other hand, the algorithm begins with a model that is very stretched (say, by a factor of two) relative to the true structure, then the algorithm often settles into a stretched and nonrigid structure, rather than converging toward the true, rigid structure.

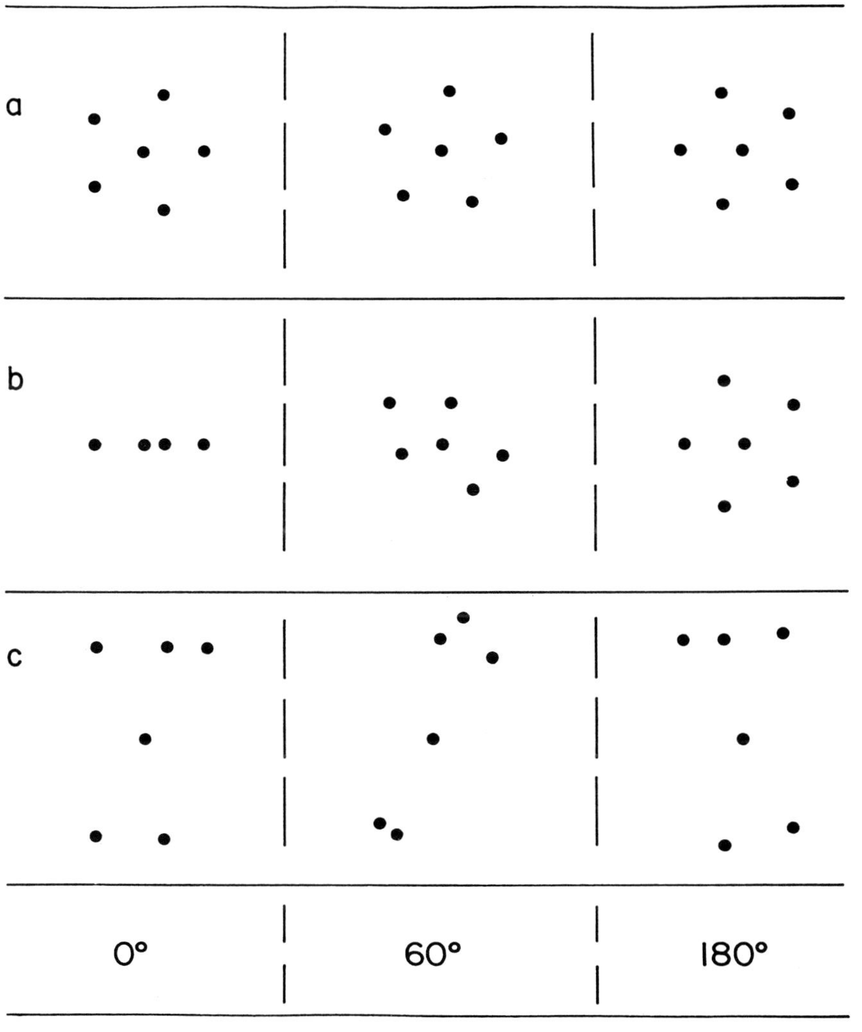

Figure 3. The effects of the initial internal model on the performance of the incremental rigidity scheme. The figure shows from left to right the initial structure, an intermediate structure (after 60°) and the final structure of a six points object which was rotated by 180° in increments of 30°. (a) Bird's eye view of the true structure of the object. (b) Bird's eye view of the internal model for the case where the initial model was flat. (c) Bird's eye view of the internal model for the case where the initial model was stretched.

4 COMPUTATIONAL STUDIES: DISCUSSION

In our explorations with the incremental rigidity scheme, we have studied the decrease in the rate of convergence of the internal model to the true structure of a moving object when the amount of change between consecutive frames decreases. In analogy to linear algebra, we found that the transformations from frame to frame of the internal model become ill–conditioned as the angular displacements decrease, i.e. they become sensitive to noise. Furthermore, we found that this reduction is such that in the limit of infinitesimally small displacements where only velocity information is used, the internal model does not have a full convergence to the true structure of the viewed object. This property is not unique to the present transformations, but occurs in other important situations such as numerical differentiation. Other numerical studies on structure from motion also suggest that temporally local schemes are inherently unstable (6, 8, 9, 11, 41–44).

Why does the rate of convergence fall with decreasing angular displacement between frames, eventually leading to instability for the case of the continuous formulation? If at a given instant the internal model does not coincide with the real structure of the object, the algorithm will try to correct this error. One expects poorer corrections for these perturbations if smaller displacements are used, because less resolution is available in the image data (the ratio between the mean distance traveled by the image features to the mean error from the internal model to the true structure of the object is reduced). Many more views, however, are seen in the case of small angular displacements which might trade off with the deterioration in the data resolution. A theoretical stability analysis of the algorithm (4) showed that the reason for the instability of the solutions for small angular displacements between frames is a fast deterioration of the algorithm's ability to recover from errors of the internal model with decreasing angular displacement (quadratic dependence) and a slow correction of the problem due to the increased number of frames (linear dependence). This deterioration can be shown to be equivalent to a decreased robustness of the algorithm against noise.

Our analysis shows, therefore, that an algorithm that overcomes temporal locality does not necessarily provide a robust solution to the structure–from–motion problem. It seems that in this case, a necessary condition for a stable solution is the use of views of the object that are significantly different. The use of pure velocity information allows only a rough solution, which is less stable over an extended time period.

Our results hint that even in a temporally extended algorithm, a memory of sufficiently distinct past views or internal models of the object is necessary for a robust recovery of its structure. The incremental rigidity scheme, as originally formulated, uses memory of only one past internal model. In principle, a memory of many past models could be used by the algorithm, but it is remarkable that such a minimal amount of past information is sufficient for a robust recovery of structure. It may still be the case that the use of multiple past views or internal models for establishing a new 3–D structure could increase the rate of convergence of the internal model.

In further support of the above conclusions, we note that in order to recover 3–D structure from measured image velocities, it is essential that these velocities at least approximate the true projected velocities of moving elements in space. In general, however, it is difficult to derive real projected velocities from the optical flow pattern on the eye or camera. Factors such as changing illumination, specularities, shadows, and rotation of an object surface relative to a light source, can create patterns of optical flow that do not correspond to the real movement of features on a physical surface (45). The difficulty of computing correct projected velocities provides an additional motivation for not basing the recovery of structure from motion directly on velocity information (although, of course, velocity information may help indirectly through the solution of the motion correspondence problem (50)).

We also found in our explorations with the algorithm that the recovery of structure from motion is dependent on the algorithms initial conditions. It performed better with flat initial internal models than with stretched ones.

This dependence on initial 3–D models of structure is quite unique to the incremental rigidity scheme and occurs for the following reason. Many of the early theoretical results on structure from motion established the set of minimal criteria to be met by a rigidly moving object such that its structure can be completely and correctly recovered from a noiseless image of its motion. For example, the first such result was due to Ullman (46) and stated that given *three* distinct orthographic projections of *four* non–coplanar points in a rigid configuration, the structure and the motion compatible with these three views are uniquely determined up to a reflection about the image plane. Furthermore, Ullman found that in the orthographic case, if less than three projections or less than four non–coplanar points are present in the image, a *unique* recovery of structure is not possible in general. The incremental rigidity scheme, however, can be shown to recover asymptotically the structure of objects with three points (4). Furthermore, the scheme works at any given instant with only two views of the object, (although the asymptotic convergence only happens after many frames). Due to this lack of minimal criteria met by the incremental rigidity scheme, it is not to be expected that this algorithm will always give the correct answer. Rather its minimization strategy will sometimes find the global minimum, but often will become trapped in a local minimum. The result reported above (e.g. Fig. 3), therefore, states that "luckily" if the algorithm begins with an initial model that is compressed relative to the true structure of the object, then it seems always to find the global minimum (at least for vertical axis rotation). This is not the case if the algorithm begins with a model that is stretched relative to the true structure. Under this condition the internal model remains stretched and nonrigid. It is worthwhile to note that displays of rigid objects in motion that give rise to nonrigid percepts were reported in the literature (28–36).

Finally, the orthographic projection is in general not physically valid, but was used here because it allows a deeper analysis of the phenomena studied. This analysis is further validated because these results are not limited to the case of orthographic projection, or to the case in which the object rotates. The incremental rigidity scheme can be generalized successfully to formulations that use perspective projection (4), and in this case, the structure of the viewed object is recoverable both for rotation and translation. Also, similar to the orthographic case, the convergence rate of the internal model to the true structure decreases, if smaller spatial changes between consecutive frames are used.

The first of our psychophysical experiments measures the integration time for structure from motion in humans and finds it to be relatively large. This long integration suggests the necessity for combining image frames that are quite disparate in space and time in the structure from motion process and supports our computational results.

The second of our experiments establishes the importance of the initial perceptual model of the object's shape in the determination of its structure from motion. This experiment shows once again that the human visual system seems to prefer a long term integration of the motion stimulus, including past knowledge of the object's structure, even if it sometimes yields the wrong result. The alternative of a temporally local analysis of the image would provide a correct recovery of structure, which, however, would not be robust against noise.

5 PSYCHOPHYSICAL STUDIES: METHODS

The first experiment in this section explored the accuracy of the 3–D structure computed by the human visual system and the time course of the buildup of perceived structure. The basic experimental design was guided by a number of considerations. First, the task relies on an objective judgment that does not require the observer to form an internal subjective scale of properties such as 3–D distance or amount of rigidity. Second, the subject is able to perform the task in a short time period so that we can measure the time course of the early buildup of 3–D structure. If, for example, it only requires one or two seconds for the visual system to compute an accurate model of 3–D structure, but at least several seconds of observation time is needed for the subject to make a judgment, then we could only assess the accuracy of the final computed 3–D model. Such a task would not allow us to explore the intermediate structures perceived in the first one or two seconds of viewing time.

A third consideration for the design of these experiments is that we wanted the task to rely on the derived 3–D structure of the moving elements, rather than their raw 2–D positions or velocities in the display. Because the recovery of 3–D structure from motion necessarily relies on properties of the changing 2–D projection, one cannot guarantee that observer's judgments are not based directly on 2–D information, but we chose an experimental design that we believe makes it very difficult for the observer to use strictly 2–D cues. Finally, we found in our initial explorations that it is hard to relate quantitative aspects of the 3–D structure derived from different visual sources. For example, it is difficult to relate depth sensations derived from relative movement to those derived from stereo or perspective. Consequently, the experiments make use of a quantitative judgment that requires only relative movement as the source of 3–D information.

A set of three points distributed in space was rotated smoothly around a central vertical axis and projected onto a 2–D computer display, using an orthographic projection. Figure 4a illustrates the basic setup in a suggestive way. To describe the stimuli in more detail, let us assume a coordinate system in which the X and Y axes are the horizontal and vertical axes in space and in the image plane, which are the same under orthographic projection, and the positive Z axis is directed perpendicular to the image plane, away from the viewer. When projected onto the $X - Z$ plane (bird's eye view), the positions of the three points always lie within an annulus, as indicated in Fig. 4b. The outer boundary of the annulus restricts the overall range of X and Z coordinates of the moving points. The reason for restricting the points to lie outside the inner boundary of the annulus is that points located near the center of this circular projection would move very little under rotation. A point that is moving by only a small amount could easily be identified as lying near the center of the cylindrical volume encompassing the set of three points. The use of an annulus as shown in Fig. 4b removes this potential cue to 3–D structure.

The particular $X - Z$ projections were chosen such that for any given rotation, the positions of the points were evenly spaced in depth in the final frame that was viewed. Let γ denote the displacements in depth for this final view, as shown in Fig. 4c. (The parameter, γ, will be defined in units of picture elements on the display. In the plane of the display, which is perpendicular to the observer's line of sight, one picture element corresponds to a visual angle of $2.5'$.) Figure 4d shows an example of the first and last frames of a sequence over which a set of points with $\gamma = 30$ was rotated by a total of $60°$. The Y coordinates of the three points (vertical positions in the image plane) were chosen randomly between specified bounds.

For each trial, the first frame of the movie appeared on the display, the subject pushed a button to indicate that he was ready and the movie was displayed. Viewing was monocular and in a dark room. The subjects were asked to order the points in depth; in particular, they were asked to specify which of the three points was located midway in depth between the other two. The accuracy of perceived structure can be assessed by measuring subjects' performance

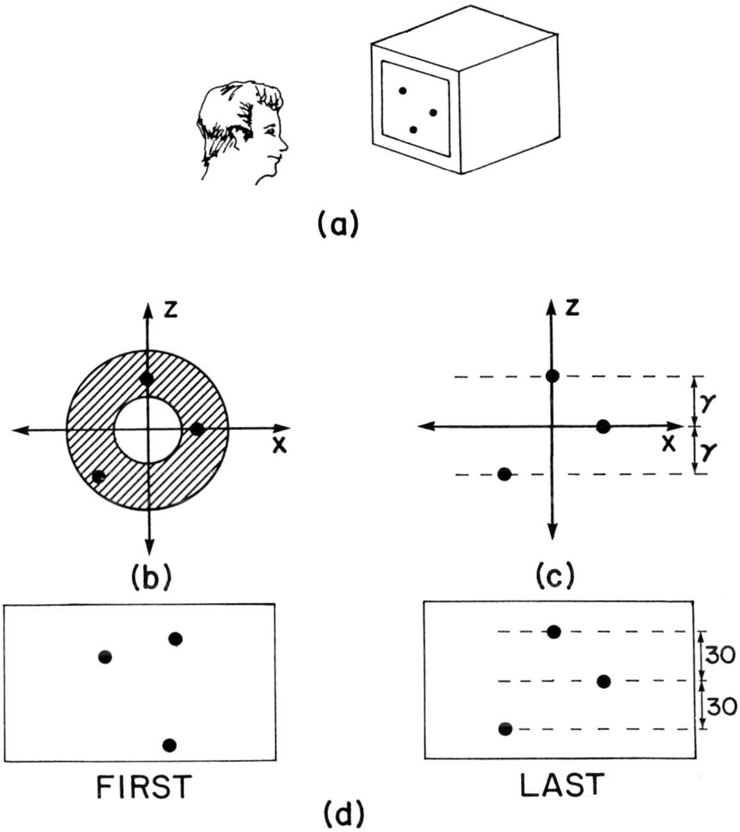

Figure 4. The ordinal experiment. (a) A suggestive drawing of the basic experimental setup. (b) Bird's eye view of the positions of the three points. The dotted lines indicate the annulus within which the points are located. (c) Bird's eye view of a final view of the three points, indicating their separation in depth, γ. (In the final view the center dot does not necessarily lie at the same depth as the center of rotation.) (d) Bird's eye views of the first and last frames of a sequence over which the three points were rotated 60° and $\gamma = 30$.

as the separation in depth between the points is decreased, and the buildup in accuracy over time can be assessed by measuring subjects' performance as the angular extent of rotation of the points is increased.

To specify their choice to the computer, subjects were given a box with three buttons oriented vertically. The subjects were told to associate the vertical ordering of the three buttons on the box with the vertical ordering of the positions of the points on the display. After each trial, the subject pushed the appropriate button depending on whether the top, middle or bottom point was perceived as being between the other two points in depth. The percentage of correct answers was then computed.

6 PSYCHOPHYSICAL STUDIES: INTEGRATION TIME FOR STRUCTURE FROM MOTION

Complete data were gathered for three subjects (ECH, VKI and NMG). Partial data were also obtained for a fourth subject, which confirmed the general trends seen in the data. Figure 5 shows the results of averaging the data obtained for the three subjects. The angular extents of rotation, as well as the associated temporal extents of motion, are indicated on the abscissa of the graphs shown in Fig. 5 and the percentage of correct responses appear on the ordinate. Data for different displacements γ are drawn on separate curves. Note that a chance level of performance corresponds to 33%. As a control, we verified that subjects were at a chance level of performance for 0° of rotation.

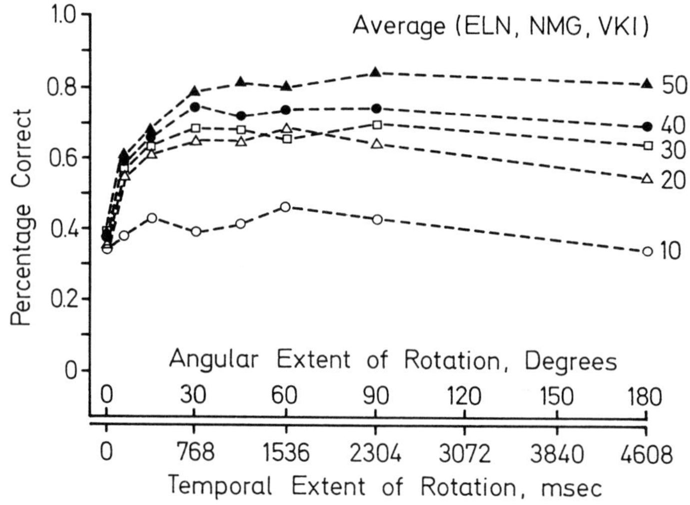

Figure 5. Results of the ordinal experiment: the percentage of correct responses against the total angular rotation. The data is the average for the three subjects. The total viewing time in *msec.* is indicated below the angular extents. The five different curves in each figure correspond to the five displacements in depth, $\gamma = 10, 20, 30, 40$ and 50.

The first observation that can be made is that subjects did exhibit a buildup in the accuracy of the perceived 3–D structure of the points. Particularly, there was a steady rise in performance level from chance at 0° of rotation towards a higher plateau at larger rotations. There was individual variability of the rotation extent necessary to reach plateau level (not shown in the figure). The plateau was reached roughly at 30° of rotation for subjects ECH and VKI and about 45° of rotation for NMG. After only 6° of rotation, subjects already reached a level of performance that was often within about 15% of the level at which performance reached a plateau. The level of performance sometimes significantly dropped for very large angular extents such as 180°. Some of the subjects found that it was difficult to maintain the perception of a rigid structure over such a long viewing period; the points sometimes appeared to separate and move independently of one another in a nonrigid manner. It should be emphasized that the total viewing time and total angular extent of viewed motion are directly coupled in this experiment. The rotation of 30°, for example, corresponds to a total viewing time of roughly 768 *msec*. The times reported here for the buildup of an accurate 3–D model are in rough agreement with those expected from the reaction times for detecting 3–D structure observed in monkeys (47).

The overall level of performance at which the plateau was reached increased with larger displacements in depth. (This level of performance also decreased very little with the introduction of high–frequency noise in the stimulus (3). This is probably due to the long temporal integration of the structure from motion process, which is able to low–pass filter the noise.) Performance was generally worse for smaller γ, as one would expect. For the largest value of γ tested, $\gamma = 50$, performance for some subjects reached the 90% level of performance. For the smallest displacement, $\gamma = 10$, subjects were not significantly above chance, while for $\gamma = 20$, subjects were well above chance for all angular extents of rotation. This suggests that the threshold for discrimination of relative depth from motion probably lies somewhere in this range of displacements. For the case where $\gamma = 20$, the relative depth between the points is 1.45% of the distance between the observer and computer display (i.e., the threshold was roughly 0.6 *cm* from a viewing distance of 40 *cm*). Measurements need to be made at additional viewing distances, however, to derive an appropriate measure of structure–from–motion 'acuity.'

In the above experiment, the temporal extent of viewed motion was coupled with the angular extent of rotation of the points. From two other experiments (3) we obtained evidence that both the angular extent of rotation and to a lesser extent its duration, influenced the quality of recovery of structure from motion. In one experiment, objects were oscillated indefinitely until subjects felt they could determine its structure. We found that the quality of the recovery of structure increased with the angular extent of oscillation. In another experiment, objects were oscillated with fixed angular extents. It was found that the subjects performance increased (weakly but significantly) with oscillation time.

7 PSYCHOPHYSICAL STUDIES: EFFECTS OF PAST MODELS OF STRUCTURE ON HUMAN PERFORMANCE

The motion of the three–point configurations viewed in the previous experiment was now immediately preceded by the motion of a different configuration of three points. The last view of the preceding configuration was arranged to coincide with the first view of the configuration on which the observer was tested, so that there was no abrupt transition between the motion of the first and second configurations. The experiment examined three different types of motion preceding the movement of the test configurations:

(1) In the first condition, the preceding configurations were both stretched relative to the test configurations and the ordering of the points in depth was randomized. The ratio between the stretched and true depths was, on average, between 4:1 and 5:1. Both the stretched and test configurations were rotated around a central vertical axis.

(2) In the second condition, the preceding configuration was such that at the transition point at which the motions of the two configurations join, the preceding configuration was flat and in the plane of the computer display. Both the planar and test configurations of points were rotated around a central vertical axis.

(3) In the third condition, the initial configuration was identical to the test configuration, but was rotated around the line of sight. (This rotation should not convey any information

about the real 3–D structure of the points.) After being rotated around the line of sight, the configuration was then rotated around the vertical axis as before.

Based on observations from computer simulations, we expect that if the incremental rigidity scheme is an appropriate model for the human recovery of structure from motion, then the first manipulation above should lead to a substantial degradation in the subjects' ability to judge accurately the structure of the test configuration. This is because we are initially 'priming' the subject to see a stretched 3–D configuration in which the ordering of the points may be different from that of the test configuration. Suppose that the initial stretched configuration is used by the visual system as an explicit source of constraints on the subsequent recovery of the 3–D structure of the test pattern. The test pattern might then be forced to look somewhat stretched and with an incorrect ordering of the points in depth relative to the true structure. If this were the case, one would expect a degradation in the quality of the 3–D structure attributed to the test pattern.

The second and third manipulations, on the other hand, should not lead to substantial degradation in performance. This expectation is based on the observation that in general, the incremental rigidity scheme will converge quickly to the true structure of a rotating object, if it begins with an initial model that is compressed in depth, relative to the true structure. The algorithm assumes that in the absence of other 3–D cues, the initial model is flat and parallel to the image plane. The second and third manipulations described above only serve to strengthen the perception of a flat initial 3–D structure, prior to the motion of the test configuration.

For comparison, experimental sessions were run with and without the previous configuration. Figure 6 shows the average data of three subjects for the first condition, in which the preceding configuration is stretched relative to the true structure of the points. The stretched pattern was then immediately followed by the rotating test configurations, which were constructed as described in the first experiment. The test configurations were rotated only for small total angular extents.

The equivalent total times of rotation are the same as those shown in Fig. 5. We show both the data for the control experiment, in which only the test configuration appeared, and the data for the case in which the stretched configuration appeared first. *The drop in performance at low rotation extents was statistically significant.* Once again there was individual variability for rotations extents in which this effect held (not shown). For the subjects ECH and VKI, there was a significant drop in performance for 6° and 15° of rotation of the test pattern when the stretched configuration occurred first. For the subject NMG, the mean performance was lower when the stretched configuration occurred first, for all angular extents tested, showing a significant overall effect. Note that 45° of rotation corresponds to about 1200 *msec.* of total viewing time. The presence of the initial stretched configuration clearly influences the subsequent computation of the structure of the test configuration (although with large subject variability). For the subjects ECH and VKI, this influence extended roughly 500 – 600 *msec.*, while for the subject NMG, an influence could still be seen after about 1200 *msec.*

Figure 7 show the average data for the subjects ECH and NMG, for the first control experiment in which the preceding configuration was flat and rotated around the vertical axis. Figure 8 show the average data (for the same two subjects) for the second control experiment, in which the preceding configuration was rotated about the line of sight. In contrast to the initially stretched case there was a lack of a significant effect for the flat and line–of–sight configurations.

Some of the degradation in performance seen in this experiment can be attributed to the size of the temporal integration period that is used to measure retinal image motion. It appears, however, that this integration period could only account for the degradation seen for

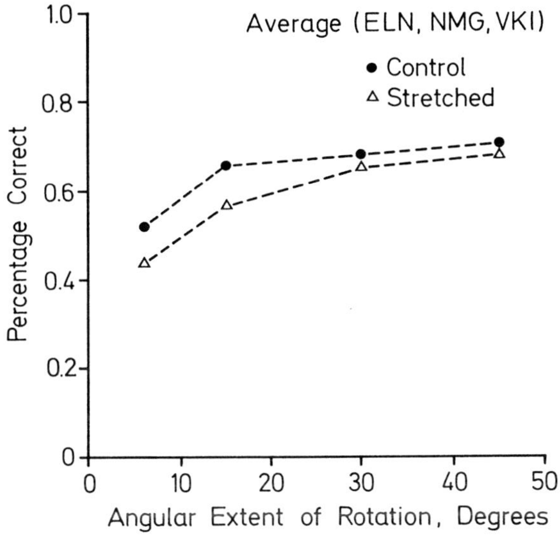

Figure 6. Psychophysical data for the experiment with an initial stretched structure. Average data for three subjects.

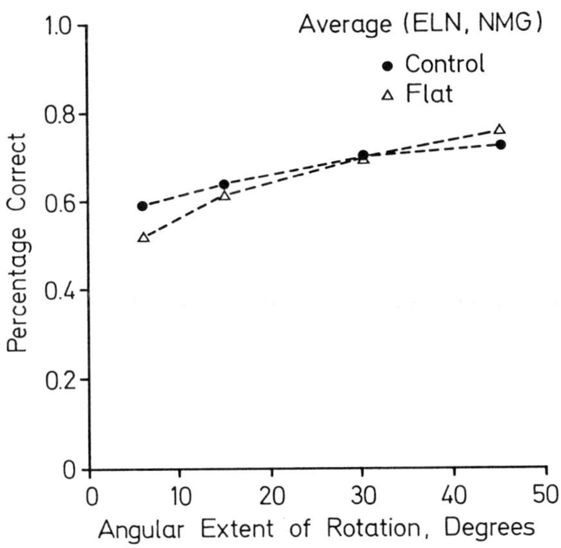

Figure 7. Average psychophysical data for subjects ECH and NMG for the experiment in which an initial planar configuration is presented.

the smallest angular extent of rotation, 6°, which extended over a viewing time of roughly 150 *msec.* Psychophysical studies indicate that the size of the temporal integration window for measuring image velocities ranges between 80 *msec.* for high velocities to about 200 *msec.* for medium range velocities (48). Integration times for measuring very slow velocities may be longer (S. McKee, personal communication), but such low velocities rarely occur in the visual stimuli used here. We therefore do not believe that the extended influence seen here (from half a second up to a second or so) can be accounted for on the basis of the mechanisms by which retinal image motion is first measured. In further support of these conclusions, the second and third manipulations used to generate the preceding configurations (data shown in Figs. 7 and 8) should also influence the performance of the motion measurement stage, but did not lead to an extended influence on the quality of the perceived 3–D structure.

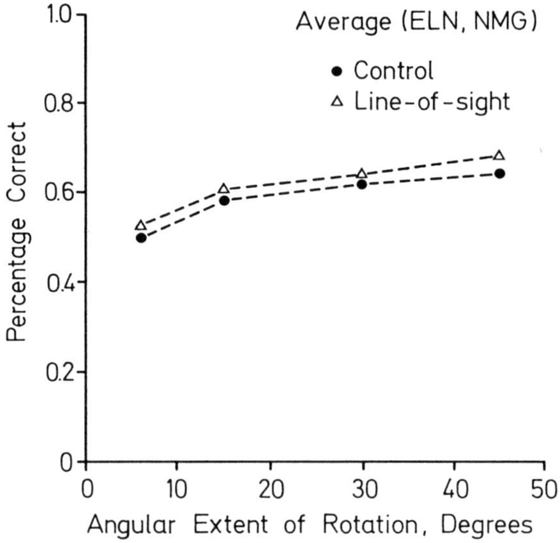

Figure 8. Average psychophysical data for subjects ECH and NMG for the experiment in which the test configuration is initially rotated around the line of sight.

We conclude that the computations underlying the recovery of 3–D structure from motion integrate motion information over an extended time period. For the case of the incremental rigidity scheme, the structure computed at a particular moment appears to depend on previous 3–D models. Based on computer simulations with this model, we expected that if observers initially viewed a stretched configuration, then there would be a significant degradation in the quality of the structure perceived at later times. On the other hand, if a flat configuration were viewed initially, we expected no degradation at later times (except for the shortest angular extent of rotation). Our experimental data showed these expectations to hold true for a limited time frame of a second or so. When the initial configuration was stretched, however, the perceived 3–D structure eventually 'collapsed' to the true, more compact structure, while the algorithm typically remains in the stretched 3–D interpretation indefinitely. Thus our experiments lend support to the notion that previous perceptions of 3–D structure constrain future models, but suggest that some modification is required to the incremental rigidity scheme to fully account for human behavior. We should note that it may be possible to design a model to account for this data that only integrates the visual input over an extended time period, without regard for previous 3–D models. Further computational and experimental studies are needed to resolve this question.

8 PSYCHOPHYSICAL STUDIES: DISCUSSION

Our everyday experience tells us that we are quite adept at deriving the 3–D structure of the environment from the relative motion of features in the changing visual image. Few laboratory studies, however, have examined the quantitative aspects of how we perform this analysis. This paper presented two experiments that assess the nature and accuracy of the structure buildup over time, and its dependency on the spatial extent of viewed motion. In our search for a possible experimental paradigm to study these issues, we first considered a variety of complex and elaborate visual displays that were interesting to explore, but which unfortunately were either too rich in additional cues to 3–D shape, or involved tasks that required long viewing times. Our early frustrations led us to consider the simple paradigms presented here, in which subjects performed the simple task of ordering the depths of a set of three points rotating in space. These displays are far removed from the richness of our typical visual experience, but allow a careful study of some of the quantitative aspects of perceived 3–D structure from motion.

Our main conclusions are the following: First, the human visual system can derive quite an accurate model of the relative depths of isolated moving points even in the presence of noise in their image positions. Second, the accuracy of the 3–D model improves with time, eventually reaching a plateau beyond which there is no further improvement. Third, there is evidence that the 3–D structure currently perceived depends on previous 3–D models. Finally, the accuracy of perceived 3–D structure depends on both the temporal, but mainly the spatial extent of the viewed motion.

The questions addressed were motivated by computational models of the recovery of structure from motion. A limitation of many computational models has been an extreme sensitivity to noise in the visual input. Some models have attempted to overcome this sensitivity by integrating motion measurements over large spatial areas, while others overcome this problem through integration of motion measurements over time. The results of our experiments with added noise (3) in the visual stimulus suggests that the human system can derive a rough estimate of structure in the presence of large amounts of noise, even when viewing only three points in motion. We may integrate motion information over large spatial regions for some tasks, such as the recovery of observer motion, but an extensive spatial integration by itself cannot account for our experimental observations. This suggests that the integration of motion measurements over time (which may be coupled with viewing the motion over larger spatial extents) may be a more important factor in reducing sensitivity to noise. These noise experiments also suggest that the human visual system may not rely on precise measurements of the velocities and accelerations of image features, but rather may require only rough estimates of the motion of image features, perhaps over an extended time period.

The issues of the time course of the buildup of accuracy of perceived 3–D structure and the possible dependence of the currently perceived structure on past 3–D models were specifically motivated by Ullman's incremental rigidity scheme. (Note, however, that in a strict quantitative sense the performances of humans and that of the incremental rigidity scheme do not seem to agree. A comparison between Figs. 2 and 5 shows that humans require a smaller extent of rotation than the scheme in order to reach their performance plateau. We are currently investigating modifications of the scheme which seem to correct this discrepancy.) It is expected that there will be some temporal buildup in accuracy, due to the extended temporal window over which image motion is first measured. (Indeed, although we argued that velocity information is not directly used in the structure from motion process, this information may be crucial in the establishment of correspondence between features in the different frames (49,50) and in the extrapolation of their positions (51–53)). Psychophysical studies indicate, however, that the extent of this temporal window is typically on the order of 80 − 100 *msec.* (48). Ullman's model proposes that the recovery of structure itself takes place incrementally over a longer time frame. This possibility is supported by our experiments. Subjects showed a buildup in accuracy of perceived structure over a second or so, with some variation between subjects.

The results of our experiments are consistent with those of Andersen and Siegel (47), which indicate that for the monkey the perception of 2–D properties of the pattern of image motion occurs more quickly than the perception of 3–D structure. In a further experiment (54), they used multiunit recordings to map the retinotopic representation of the middle temporal area (MT) of the extrastriate cortex (an area which predominantly contains neurons that respond to visual motion) and then lesioned a selected part of this region with ibotenic acid. Following the lesion, the animal performed a structure from motion task at chance at a location within the lesioned MT field, although it performed relatively well in another 2–D task. However, he could still correctly do the 3–D task at a second location not in the lesioned field. Their demonstration of the monkey's ability to perceive structure from motion, the dependence of this function on the intactness of MT, and the consistency of its performances to that found by us for humans, raises hopes that soon the computational and neuronal principles underlying structure from motion will be studied at the cortical level.

9 ACKNOWLEDGMENTS

We thank Tomaso Poggio, Shimon Ullman and Hanspeter Mallot for valuable comments on this paper. This report describes research done within the Artificial Intelligence Laboratory and the Center for Biological Information Processing (Department of Brain and Cognitive Sciences) at the Massachusetts Institute of Technology. Support for the A.I. Laboratory's artificial intelligence research is provided in part by the Advanced Research Projects Agency of the Department of Defense under Office of Naval Research contract N00014–80–C–0505. Support for this research is also provided by a grant from the Office of Naval Research, Engineering Psychology Division. NMG is supported by a Fairchild Fellowship.

10 REFERENCES

(1) D. Marr, Vision, Freeman, New York 1985.

(2) T. Poggio, V. Torre, C. Koch, Nature 317 (1985) 314–319.

(3) E. C. Hildreth, V. K. Inada, N. M. Grzywacz, E. H. Adelson, MIT Artif. Intell. Lab. Memo 920, 1986.

(4) N. M. Grzywacz, E. C. Hildreth, J. Opt. Soc. Am. (In press).

(5) S. Negahdaripour, B. K. P. Horn, MIT Artif. Intell. Lab. Memo 821, 1985.

(6) A. M. Waxman, K. Wohn, Int. J. Robotics Res. 4 (1985) 95–108.

(7) K. Wohn, A contour–based approach to image flow, PhD. Thesis, Dept. Computer Science, Univ. Maryland 1984.

(8) G. Adiv, IEEE Patt. Anal. Machine Intell., PAMI–7 (1985) 384–401.

(9) G. Adiv, Proc. IEEE Conf. Computer Vision Pattern Recognition, San Francisco, CA, (1985) 70–77.

(10) S. Bharwani, E. Riseman, A. Hanson, Proc. IEEE Workshop on Motion: Representation and Analysis, Charleston, SC, (1986) 73–80.

(11) G. L. Scott, Local and global interpretation of moving images, PhD Thesis, Cognitive Studies Program, Univ. Sussex 1986.

(12) A. Bruss, B. K. P. Horn, Comput. Vision Graph. Image Proc. 21 (1983) 3–20.

(13) K. Wohn, A. M. Waxman, Center for Automation Tech. Rep. 134, Univ. Maryland 1985.

(14) Y. Yasumoto, G. Medioni, Proc. IEEE Conf. Computer Vision Pattern Recognition, San Francisco, CA, (1985) 89–94.

(15) D. T. Lawton, Comput. Vision Graph. Image Proc. 22 (1983) 116–144.

(16) M. Subbarao, Proc. IEEE Workshop on Motion: Representation and Analysis, Charleston, SC, (1986) 157–165.

(17) T. J. Broida, R. Chellappa, Proc. IEEE Workshop on Motion: Representation and Analysis, Charleston, SC, (1986) 95–100.

(18) S. Ullman, Percept. 13 (1984) 255–274.

(19) R. C. Bolles, H. H. Baker, Proc. Third IEEE Workshop on Computer Vision: Representation and Control, Bellaire, MI, (1985) 168–178.

(20) M. S. Landy, Mathematical Studies in Perception and Cognition 86–2, Department of Psychology, New York Univ. 1986.

(21) H. Shariat, K. E. Price, Proc. IEEE Workshop on Motion: Representation and Analysis, Charleston, SC, (1986) 119–124.

(22) G. Johansson, Scand. J. Psychol. 5 (1964) 181–208.

(23) G. Johansson, in R. Held, H. W. Leibowitz, H.–L. Teuber (eds.), Handbook of Sensory Physiology, Springer–Verlag, Berlin 1978.

(24) G. Jansson, G. Johansson, Perception 2 (1973) 321–326.

(25) J. T. Todd, J. Exp. Psychol. 8 (1982) 238–252.

(26) J. T. Todd, Percept. Psychophys. 36 (1984) 97–103.

(27) J. T. Todd, J. Exp. Psych.: Human Perc. Perf. 11 (1985) 689–710.

(28) H. Wallach, A. Weisz, P. A. Adams, Am. J. Psych. 69 (1956) 48–59.

(29) B. W. White, G. E. Mueser, J. Exp. Psychol. 60 (1960) 1–11.

(30) B. F. Green, J. Exp. Psychol. 62 (1961) 272–282.

(31) M. L. Braunstein, J. Exp. Psychol. 6 (1962) 41–420.

(32) G. Sperling, M. Pavel, Y. Cohen, M. S. Landy, B. J. Schwartz, in O. J. Braddick, A. C. Sleigh (eds.) Physical and Biological Processing of Images, Springer–Verlag, Berlin 1983.

(33) B. J. Schwartz, G. Sperling, Invest. Ophthal. Visual Sci. Suppl. 24 (1983), 239.

(34) M. L. Braunstein, G. J. Andersen, Perception 13 (1984) 213–217.

(35) E. C. Hildreth, The Measurement of Visual Motion, MIT Press, Cambridge 1984.

(36) E. H. Adelson, Invest. Ophthalmol. Vision Sci. Suppl. 26 (1985) 56.

(37) D. D. Hoffman, B. E. Flinchbaugh, Biol. Cybern. 42 (1982) 195–204.

(38) B. M. Bennett, D. D. Hoffman, Biol. Cybern. 51 (1985) 293–300.

(39) J. J. Koenderink, A. J. Van Doorn, J. Opt. Soc. Amer. A 3 (1986) 242–249.

(40) N. M. Grzywacz, E. C. Hildreth, MIT Artif. Intell. Lab. Memo 845, 1985.

(41) D. D. Hoffman, J. Opt. Soc. Amer. 72 (1982) 888–892.

(42) K. Prazdny, Biol. Cybern. 36 (1980) 87–102.

(43) S. Ullman, MIT Artif. Intell. Lab. Memo 870, 1985.

(44) S. Ullman, in J. Beck, B. Hope, A. Rosenfeld (eds.), Human and Machine Vision, Academic Press, New York 1983.

(45) A. Verri, T. Poggio, MIT Artif. Intell. Lab. Memo 917, 1986.

(46) S. Ullman, The Interpretation of Visual Motion, MIT Press, Cambridge 1979.

(47) R. A. Andersen, R. M. Siegel, Soc. Neurosci. Abstr. 12, Part 2 (1986) 1183.

(48) S. P. McKee, L. Welch, J. Opt. Soc. Am. 2 (1985) 243–251.

(49) V. S. Ramachandran, S. M. Anstis, Vision Res. 23 (1983) 83–85.

(50) N. M. Grzywacz, Technical Digest of the Opt. Soc. Am. Meeting (1986) 90.

(51) K. Nakayama, C. W. Tyler, Vision Res. 21 (1981) 427–433.

(52) M. Biederman-Thorson, J. Thorson, G.D. Lange, Vision Res. 11 (1971) 889–903.

(53) S. P. Mckee, K. Nakayama, Vision Res. 24 (1984) 25–32.

(54) R. M. Siegel, R. A. Andersen, Soc. Neurosci. Abstr. 12, Part 2 (1986) 1183.

The Temporal Integration of 3-D Structure From Motion: A Computational and Psychophysical Study

Discussion

edited by *Graeme Mitchison, Gerard Toulouse* and *Christian Wehrhahn*

Question: To what extent can problems in visual motion perception in general, and structure from motion in particular, be reduced to a simple optimization problem?

Answer: The recovery of structure from motion is an example of what is called in computational terms an early vision problem, because it is a process that extracts a physical property (the 3-D structure) of the viewed object. This extraction is, in the case of the structure from motion problem, an ill-posed problem, because the retinal image is two-dimensional and, therefore, underconstrained to retrieve a three-dimensional geometry. An a priori assumption, therefore, about the physical properties of the object (e.g. rigidity) must be made by any computational theory of structure from motion. These assumptions are often expressed in terms of minimization procedures (mainly from calculus of variations) by the computational vision community. These procedures are used for two reasons: the existence of general mathematical theories underlying these methods, and the easy implementation of these problems in terms of parallel analog networks (1).

Questions: Is it justifiable to pose the solution for the structure from motion problem in terms of small number of features? What about continuous surfaces?

Answer: It is not necessary to postulate a solution of the structure from motion problem in terms of isolated features. In fact, optical flow approaches to the problem have been suggested (2-7). There are reasons, however, to consider feature-based schemes. The main reason is that the optical flow field (a 2-D field that can be associated with the variation of the image brightness pattern) and the 2-D motion field (the projection on the image plane of the 3-D velocity field of a moving scene), almost never coincide (8). The problem stems from the tact that image brightness patterns and their changes do not correspond directly to physical entities and their motion (9). Not surprisingly, however, it turns out that the optical flow and motion field nearly coincide at brightness edges and thus at the most elementary type of features.

Question: Can the solutions proposed for the structure from motion problem be implemented by parallel networks of simple elements? Is there any reason to choose a particular implementation over another? Should the dynamics be deterministic or stochastic?

Answer: Although we did not focus on the problem of building a parallel network to simulate our solution to the structure from motion problem, it does not seem difficult to imagine how to build one. There are four considerations to take into account for such a task. 1. The function that we minimize in our algorithm is not quadratic and therefore does not accept a simple linear resistive network of the type suggested by Poggio, Koch and Torre (1). Rather, nonlinear elements must be used. 2. The equilibrium values of the variables should be continuous as it is the depth of features that the algorithm must find. Equilibrium binary values, as in the case of the traveling salesman problem or the motion correspondence problem, (10, 11), are not enough. 3. The algorithm in the form we presented (chosen for mathematical convenience) is not spatially local, i.e. long and short distances between features were treated with equal weight. Ullman (12) pointed out that this treatment is not necessary and the interactions in the algorithm may be modified such that they decrease with distance. A network that implements our version of the algorithm will force interactions between all the features available, whereas an implementation of Ullman's version requires only interactions of close features. 4. We concluded in the paper that the equilibrium solution of the algorithm may depend on its initial condition. This conclusion could be considered a reason to choose a stochastic network to implement it. The noise of the network may pull the algorithm out of local minima. On the other hand stochastic algorithms seem to have slower convergence rates. We believe that the rate-of-convergence problem is more serious than the local-minima problem and therefore favor a deterministic network implementation. This preference is because our analysis indicated that compressed initial conditions seem always to yield the correct answer, and if chosen will eliminate the local-minima problem.

Question: Is the algorithm not assuming a specific type of motion, i.e. rotation around an axis perpendicular to the viewing axis, and hence in a sense prejudicing the solution?

Answer: There are many papers in the literature studying the recovery of structure under specific types of motion. The studies cover motions such as planar surfaces in motion, pure translatory motion of the observer, planar of fixed axis rotation, and translation perpendicular to the rotation axis (for a review of the literature see (13)). These types of algorithms are important because it is to expected that knowledge of the type of motion being performed could be used to enhance the quality of recovery of the object's structure. It is important to note, however, that the only assumption used by our algorithm is that the rate of change in the object's structure is as small as possible. In fact it can be applied to arbitrary types of motion (including nonrigid motions). The algorithm, however, does not always give the physically correct answer. For example, one can show that if the rotation axis is sufficiently slanted in depth, an incorrect answer will be obtained. This phenomenon is to be expected to some extent, because in the limiting case of the rotation axis being parallel to the viewing axis, no depth information is available. Lack of physical correctness by the algorithm does not imply, however, incorrect simulation of human perception. Indeed in informal experiments we performed, subjects reported a loss of ability to recover structure from motion when the rotation axis was slanted, which is in rough agreement with the algorithm's behavior.

Question: A small change in image position can correspond to a large depth change. Can one reduce the error in the z-direction by imposing a constraint that the distance between two points must always exceed the input value seen in the image coordinates?

Answer: Our analysis of rigidly moving objects seems to indicate that such a constraint is automatically used by the algorithm. The only case for which our equations could be fully solved analytically was that of the two-feature object. In that case one can show that if the initial condition is compressed relative to the true structure of the object, then the internal model expands, such that the distance between two points exceeds (or is equal to) the input value in the image coordinates. The expansion happens until the model of the object reaches the maximal dimensions that can be seen in the image and then the shape of the perceived object becomes invariant. If, on the other hand, the initial condition is stretched, the structure of the object's model remains stretched throughout the entire extent of the motion. This result was confirmed by our computer simulations for objects with higher numbers of features.

REFERENCES

(1) T. Poggio, V. Torre, C. Koch, Nature 317 (1985) 314-319.

(2) K. Prazdny, Biol. Cybern. 36 (1980) 87-102.

(3) H.C. Longuet-Higgins, K. Prazdny, Proc. Royal Soc. London B. 208 (1981) 385-397.

(4) D.D. Hoffmann, J. Opt. Soc. Amer. 72 (1982) 888-892.

(5) M. Waxman, S. Ullman, Internat. J. Robotics Res. 4 (1985) 72-94.

(6) A.M. Waxman, B. Kamgar-Parsi, M. Subbarao, Center for Automation Research Technical Report CAR-TR-190, Univ. Maryland 1986.

(7) M. Subbarao, Proc. IEEE Workshop on Motion: Representation and Analysis, Charleston, SC, 157-165, 1986.

(8) A. Verri, T. Poggio, MIT Artif. Intell. Lab. Memo 917, 1986.

(9) S. Ullman, The Interpretation of Visual Motion, MIT Press, Cambridge 1979.

(10) J.J. Hopfield, D.W. Tank, Biol. Cyber. 52, (1985) 141-152.

(11) N.M. Grzywacz, A. Yuille, AIP Conference Proceedings 151, Meeting on Neural Networks for Computing, Snowbird, UT, 200-205, 1986.

(12) S. Ullman, Percept. 13 (1984) 255-274.

(13) N.M. Grzywacz, E.C. Hildreth, J. Opt. Soc. Am. (In press).

Review:

Some Basic Organizational Features of the Somatosensory Nervous System

by *Terry Allard* and *Michael M. Merzenich*

SUMMARY

In this chapter, we review the organizational features of the dorsal column-medial lemniscal pathway from the peripheral receptors to their ultimate cortical targets. Peripheral receptors can be subdivided into many classes on the basis of the type of physical stimuli most effective in driving them. Primary cutaneous afferents associated with superficial or deep receptors comprise the principal inputs to the lemniscal system. Neurons with similar sensory attributes form parallel perceptual channels both functionally and anatomically within the dorsal column nuclei and the ventroposterior nuclei of the thalamus. The somatosensory nuclei of the thalamus project upon multiple somatotopic representations of the periphery in several fields in primary somatosensory cortex. Contiguous thalamic nuclei appear to project to contiguous cortical fields in parallel. These functionally segregated projection systems interact at the cortical level in a topographic fashion: neurons from different cortical fields representing the same body parts are linked. Units at different locations in cortex share a common vertical organization. The intrinsic and extrinsic anatomical features underlying this vertical organization are discussed.

1 INTRODUCTION

The somatosensory nervous system is organized into functionally distinct anatomical pathways conveying different aspects of mechanoreceptive, kinesthetic, thermal, and nociceptive stimulation of the body to multiple somatotopic representations in different cortical fields (1). In this chapter, we selectively review the organizational features of one major somatosensory pathway, the tactile "lemniscal system" subcomponent that conveys inputs from the skin to the primary somatosensory cortical fields 3b and 1.

The tactile lemniscal system constitutes the principal pathway conveying information about discriminative touch (2). Among other features, it is characterized by: a) relatively restricted contralateral receptive fields (RFs);

b) submodality response specificity, i.e., neurons at subcortical levels respond as if they receive their inputs from a single type of specialized peripheral receptor; c) large myelinated fibers that rapidly transmit precise temporal and spatial features of a peripheral stimulus; d) orderly topographic representations of the skin surface within the major nuclei of the system (the dorsal column nuclei = DCN, and the ventroposterior nucleus = VP) as well as within the somatosensory cortical fields fed directly by this subsystem (areas 3b and 1 in primates); e) a regional functional segregation or clustering of inputs from different receptor classes across all system levels; f) topographically ordered corticofugal projections; g) widespread convergence and divergence of anatomical connections across the projection axis; h) RFs derived from neural interactions between excitatory and inhibitory elements; i) dynamically-maintained response properties that depend on specific temporal and spatial patterns of inputs throughout life.

2 AFFERENT PATHWAYS

Segregation of function is the rule rather than the exception in the overall organization of the somatosensory system, from the peripheral receptor to different fields in the cortex. At the periphery, complex tactile stimuli are represented by specific classes of receptors responding selectively to different aspects of mechanical stimulation of the skin. These receptors and their associated neural elements constitute distinct sets of elementary sensory channels. Neurons with similar sensory attributes form parallel channels within the ascending lemniscal pathway in the central nervous system and are ordered topographically.

2.1 Primary Afferents

Somatic stimuli are transduced in the periphery by morphologically-distinct receptor types encoding different aspects of somesthetic sensory information. Cutaneous receptors selectively signal contact, pressure, flutter, vibration, cold, warm, and noxious stimulation of the skin (3). Thermoreceptors are exquisitely sensitive to changes in temperature but respond poorly to mechanical deformation. Nociceptors respond to stimuli that damage tissue; these receptors are characterized by free nerve endings that respond to intense mechanical, thermal, or chemical stimulation. Mechanoreceptors are activated by innocuous mechanical stimulation and provide graded responses to different parameters such as the intensity and velocity of stimulation. A subset of mechanoreceptors (including Pacinian corpuscles, pallisade hair follicle endings, Merkel cell endings innervating glabrous skin and facial hair, Ruffini endings, Meissner's corpuscles) are marked by highly specialized epithelial cell structures that determine the special mechanical selectivity of each receptor type. These organized tactile receptors all convey information to the central nervous system *via* large-diameter afferent fibers. They provide the principal inputs to the tactile dorsal column-lemniscal system that is the focus of this review.

Primary sensory afferents conveying information from different receptor types can be characterized by their adaptation characteristics (4). Rapidly-adapting (RA) afferents such as those innervating the lamellated Meissner and Pacinian corpuscles respond best to transient events: a burst of action potentials signals the onset or offset of a mechanical indentation, while afferents are quiet during periods of steady skin indentation. Meissner corpuscles are located in the dermal ridges of the glabrous or non-hairy skin of the primate hand. Afferents innervating

Meissner receptors have restricted skin fields and probably encode skin contact, displacement velocity, the sense of low-frequency vibration or "flutter," and some aspects of surface textures encoded temporally. They outnumber Pacinian afferents in the glabrous skin of the hand by an order of magnitude, approximately 150 afferents/cm^2 vs. 10 afferents/cm^2, respectively (5, 6). Afferents innervating Pacinian corpuscles are found in deeper layers at the junction of the dermis and subcutaneous plexus. They are exquisitely sensitive to vibrations transmitted mechanically through skin and bone, and commonly have large receptive fields. Pacinian afferents mediate vibration sense and make an important contribution to the detection of different textures perceived with active touch *via* temporal cues. Pacinian corpuscles are selectively responsive to skin acceleration; they commonly phase-lock to stimuli at frequencies well above 300 Hz.

Slowly-adapting (SA) afferents innervating Merkel discs are associated with the basement membrane of the epidermis and have small receptive fields with distinct borders. They respond in a sustained fashion to steady skin displacement and appear to code the magnitude of skin displacement in a graded fashion. According to Phillips and Johnson and their colleagues, SA afferents innervating Merkel discs and Ruffini endings account for the high spatial acuity found in the glabrous skin of the hand of primates including man (7).

Electrical stimulation of individual physiologically-identified cutaneous afferents in human subjects produces specific sensory perceptions associated with the different classes of primary afferents (8). Stimulation of RA afferents putatively innervating Meissner corpuscles in the glabrous surface of the hand evokes a sense of tapping or vibration at frequencies of less than 100 Hz; increasing the frequency of stimulation beyond this point produces a perceived increase in stimulus intensity. Stimulation of Pacinian corpuscles at frequencies up to 300 Hz evokes a frequency-dependent vibratory sensation. In contrast, stimulation of slowly-adapting afferents induces a sensation of pressure at all frequencies tested. Such results indicate that cutaneous receptors in the skin selectively extract features that are associated with segregated perceptual channels in the central nervous system. The perception of complex tactile stimluli is a product of the simultaneous, parallel operation of these several perceptual channels.

Input from these specialized mechanoreceptors is conveyed into the spinal cord *via* large-diameter myelinated fibers of the dorsal root ganglion cells. Entering primary afferents bifurcate near the root entry zone, and may also give off collaterals into the spinal grey matter at many levels (9, 10). Primary afferents from a restricted skin surface may enter the spinal cord over any of several dorsal roots. The skin field of a given dorsal root is called a "dermatome" and there is a high degree of overlap across adjacent dermatomes. Similarly, there may be a considerable degree of overlap between the skin fields of individual cutaneous nerves (e.g., 11).

A primary collateral of each large-fiber afferent of the appropriate receptor class projects rostrally in the ipsilateral dorsal columns, the gracile and cuneate fasciculi of the spinal cord, to the dorsal column nuclei (DCN) of the medulla. The second-order neurons of the DCN project to the contralateral ventroposterior (VP) nuclei of the thalamus by way of the medial lemniscus. This prominent "dorsal column-medial lemniscal" pathway is dominated by inputs from specialized cutaneous mechanoreceptors. Inputs from "deep" receptors comprise a minor part of this projection and are conveyed forward to separate regions of the DCN and the thalamus.

2.2 Central Ascending Pathways

The dorsal column nuclei of the medulla include the cuneate, gracile, and external cuneate nuclei. These cell regions show a high degree of submodality specificity as if individual neurons receive input from a single class of primary afferents (12). In monkey, fibers from lower spinal segments representing the lower trunk and extremities, project topographically upon the gracile nucleus (13), while fibers from the upper spinal segments representing the upper extremities and upper trunk, terminate laterally in the cuneate nucleus (14). Within each submodality region of the DCN in cat, the body is represented in fine detail as a continuum of overlapping RFs gradually shifting across the periphery(15); this smooth progression of RFs is broken only at the abrupt borders of each submodality-specific region. The primary projection of these second-order DCN neurons crosses the midline *via* the medial lemniscus to project to the contralateral ventroposterolateral nucleus (VPL) and posterior group of the thalamus, hence the name, dorsal column-*lemniscal* pathway.

The ventroposterior nuclei in the thalamus are characterized by different submodality properties and segregated connections with their respective cortical targets (see below and figure 2). VPL receives cutaneous input that is organized in a strict topographic fashion (16, 17). The nucleus can be divided into 4 subdivisions on anatomical and physiological grounds corresponding to different portions of the contralateral body surface such as the tail, the foot, the hand, and the trunk and proximal extremities which together form a continuous 3-dimensional representation of the 2-dimensional body surface. The ventroposteromedial nucleus (VPM) receives the equivalent of the lemniscal projection *via* a special trigeminal pathway and forms a cutaneous representation of the contralateral face and both the ipsilateral and contralateral mouth (18). The VPL and VPM provide the predominant source of input to the lemniscal pathway representation in areas 3b and 1 of the anterior parietal cortex of primates.

In the DCN and VPL, the two-dimensional surface of the skin is represented within a 3-dimensional structure. A given patch of skin is represented by a column or slab of neurons that may be several hundred microns in length containing thousands of neurons with similar RFs and submodality characteristics. Jones and his collaborators have proposed that submodality segregation from the level of the medial lemniscus through VPL underlies the formation of described submodality-specific columns in somatosensory cortex in monkey (19) and in cat (20).

3 SOMATOSENSORY CORTEX

The existence of a topographic representation of the contralateral body in cortex was first surmised by John Hughlings Jackson on the basis of clinical data from patients with epileptic seizures (21). He noted a spread of motor symptoms from the finger to the hand, leg, and trunk before convulsions developed into a generalized epileptic episode. However, it was not until the electrical stimulation work of Cushing in 1909 in awake patients (22), that somatosensory cortex was first clearly identified as separate from motor cortex. These patients reported somatosensory sensations during direct electrical stimulation of the postcentral gyrus. Using similar clinical techniques, Foerster observed that the entire contralateral body surface was represented along the central sulcus in a crude topography, the dorsal cortex representing the lower extremities and the lateral cortex representing the speech musculature (23). Later studies by Penfield and Rasmussen clearly demonstrated the existence of a perceptual map of the contralateral body surface on the postcentral parietal cortex of humans (24).

In the 1930's, a new laboratory technique was developed by Bard and his collaborators (25, 26) for use with experimental animals demonstrating that natural and electrical stimulation of a restricted skin surface could evoke a local electrical response at the cortical surface. This evoked-potential recording technique was exploited by Woolsey and his colleagues(e.g., 26) who reconstructed an "homuncular" representation of the body surface, similar to Penfield's stimulation maps, in a wide variety of mammalian species. The advent of intra-cortical microelectrode recording techniques in the 1950's allowed the exploration of this somatosensory representation in the postcentral parietal cortex at the single neuron level.

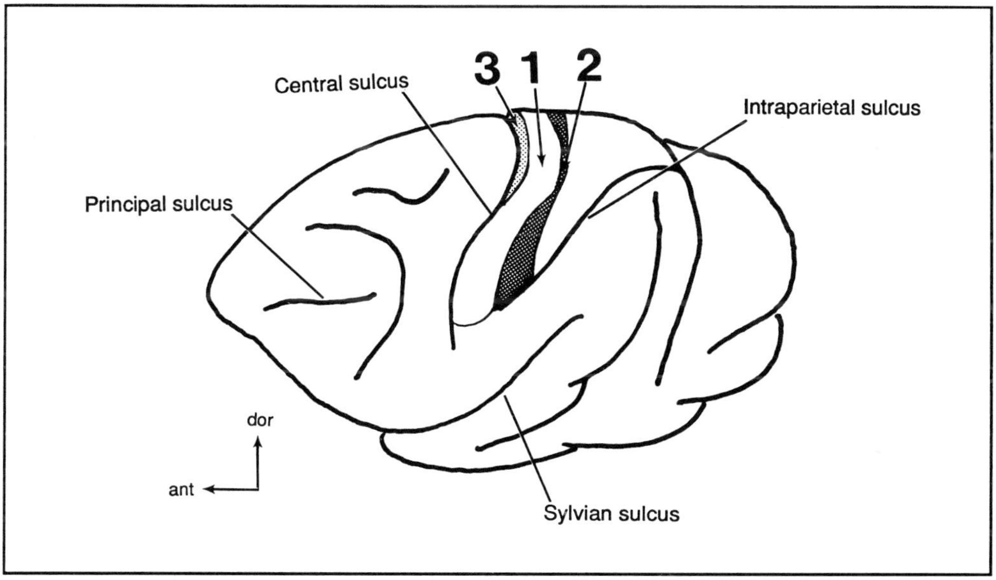

Figure 1. Schematic view of lateral surface of macaque brain showing location of primary somatosensory cortex.

3.1 Cortical Fields

The anterior postcentral parietal cortex in primates including man can be subdivided into four cytoarchitectonic fields, areas 3a, 3b, 1, and 2 (27, 28; Figure 1). Area 3a is found in the depths of the central sulcus on its posterior bank. It is fed principally by inputs from muscles and joints (29) and appears to be a transition zone between motor and sensory cortices. Area 3b is found on the posterior bank of the central sulcus. Its characteristic granular appearance is caused by the presence of many densely packed stellate cells in the middle cortical laminae, layers III-VI. This feature is associated with a heavy and early-myelinated thalamic input , a pattern also found in the primary visual cortex, Brodmann's area 17, and primary auditory cortex, Brodmann's area 41 (30). Posterior to area 3b, areas

1 and 2 show clearer laminar definition with an increasing number of pyramidal cells and less granularization of the middle cortical layers.

Intra-cortical microelectrode recording techniques made possible two major contributions by the Mountcastle group in the late 1950's (31,32). First, they demonstrated that individual neurons responded maximally to peripheral stimulation of separate classes of peripheral receptors. Thus neurons were submodality-specific, and could be classified as either cutaneous or deep depending on their apparent sources of peripheral input. Second, in a given microelectrode penetration vertical to the pial surface, neurons shared common "static" properties including largely overlapping RF's and similar modality characteristics.

For example, neurons in area 3a, like those in motor cortex, commonly respond to stimulation of muscle afferents and are probably important in coordinating movements. The large majority of neurons of areas 3b and 1 respond to cutaneous stimulation (33, 34). Again, these fields constitute the principal destination of thalamic inputs from the cutaneous portion of the dorsal column-lemniscal system. The more posterior Area 2 receives inputs from both cutaneous afferents and joint and muscle afferents.

Among neurons responding to cutaneous inputs across these cytoarchitectonic divisions, systematic receptive field differences exist. Units in area 3b have small, homogenous receptive fields without strong direction preferences or inhibitory surrounds. In contrast, cutaneous receptive fields in areas 1 and 2 are larger than in area 3b; they often have complex response properties including inhibitory areas and sensitivity to direction and movement (35, 36).

This segregation of modality in primates is associated with separate topographical representations or "maps" of the body surface at least in the middle cortical layers. Merzenich, Kaas, and their collaborators report separate, mirror-image maps of skin surfaces in areas 3b and 1 running in dorsolateral bands across the postcentral parietal cortex of anesthetized monkeys (37, 38, 39). There is some evidence that these multiple representations extend to areas 3a and 2 as well (for discussion of this point, see 40). This type of parallel organization allows the mapping of different functional attributes onto a precise spatial representation of the peripheral receptor sheet.

Segregation of function is supported by ablation-behavior studies showing specific behavioral deficits associated with lesions of different cytoarchitectonic areas. Destruction of the entire postcentral parietal cortex (41) or area 3b alone (42) in macaque monkeys caused profound deficits in the detection and discrimination of a wide variety of somesthetic cues including the judgement of vibration, texture, and form. However, Randolph and Semmes also showed that restricted ablations of areas 1 and 2 produced a double dissociation of deficits for texture and shape discrimination (42) . This combination of deficits is consistent with the physiologically-based notion that area 1 is important in mediating vibration sense through the convergence of cutaneous and deep inputs while area 2 is important in form recognition through deep inputs signalling joint position particularly in the hands.

3.2 Input-Output Relations

Recent evidence using modern anatomical tracers in monkey depicts a complex pattern of thalamo-cortical connectivity (Figure 2). Cortical regions representing the same body surface in area 3b and 1 receive a cutaneous input in the middle cortical layers from a "slab" of neurons in the VP thalamic nucleus; this thalamo-cortical connection is reciprocal (43, 44). Cortical fields responding primarily to deep inputs from muscle afferents and joint receptors, areas 3a and 2, receive inputs from outside of VPL proper (45): area 3a receives muscle spindle afferents from a rostral "transition" zone between VPL and the ventrolateral nucleus which projects to motor cortex; area 2 receives input from a region dorsal to VPL proper and projects rostrally to motor cortex. Thus, contiguous thalamic nuclei appear to project to contiguous cortical fields. Unlike other parts of somatosensory cortex, area 2 also receives inputs from the anterior pulvinar (46). This recent finding distinguishes area 2 from other "primary" sensory areas of the cerebral cortex.

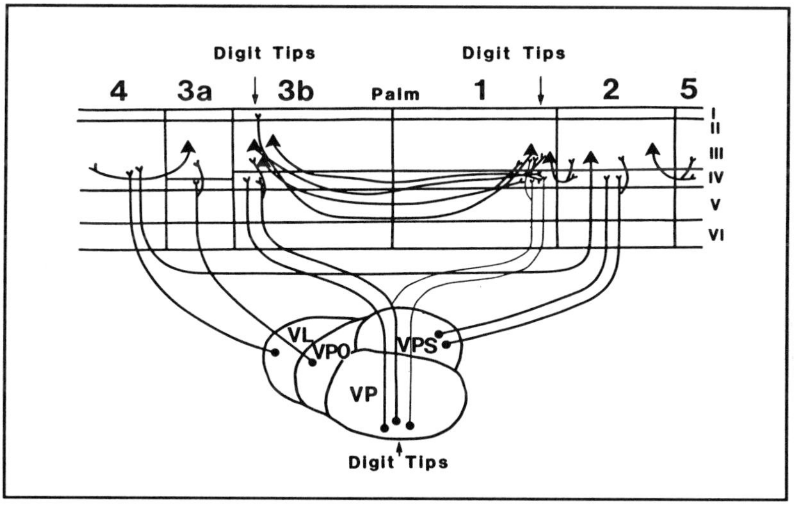

Figure 2. Schematic of thalamo-cortical projections proposed by Jon Kaas (40). Areas 3b and 1 receive driving cutaneous and deep inputs from the ventroposterior nucleus proper (VP). Area 2 receives deep inputs from the dorsal caudal shell region, the ventroposterior superior nucleus (VPS) while area 3a receives a probable muscle-spindle projection from the ventroposterior oral nucleus (VPO).

Neurons in VPL that share modality and place characteristics appear to be organized in clusters in the thalamus of the cat (20). Microlesions in VPL that may destroy one or several clusters produce anterograde terminal degeneration in a pattern of patches in the middle laminae of somatosensory cortex of cats (47). Similarly, small injections of HRP in area 3b result in the labeling of small clusters of neurons in VPL. These data are consistent with the notion that neuronal aggregates with common functional qualities are anatomically segregated within the nervous system even within distinct cytoarchitectonic regions.

These functionally segregated projection systems appear to interact at the cortical level (Figure 3). The cytoarchitectonic divisions of monkey somatosensory cortex are interconnected in a somatotopic fashion so that neurons with different functional properties representing the same body part are linked (48, 49, 50, 51). Area 3b receives a dense myelinated thalamic projection within its middle layers. This type of "feed-forward" input from the thalamus to the middle cortical laminae characterizes the projection of area 3b upon its anterior and posterior neighbors, areas 3a, 1, 2, and 5. Similarly, area 1 projects to the middle cortical layers of area 3a anteriorly but not 3b and posteriorly to areas 2 and 5. Area 2, in turn, also sends a "feed-forward" input anteriorly to 3a but not areas 3b and 1 and posteriorly to area 5. In contrast to this "feed-forward" pattern of connections, area 3b receives "feed-back" inputs in supragranular and infragranular laminae from its posterior neighbors, areas 1 and 2, while area 1 receives "feed-back" input from area 2. The efferent connections of areas 3a and 5 have not yet been determined.

The sequence of projections from anterior cortical fields to posterior fields may reflect a hierarchy of information processing (e.g., 35). Alternatively, it has been argued that individual cytoarchitectonic fields receive separate thalamic inputs allowing the parallel processing of somatosensory information (1). Under this scheme, the "feed-forward" and "feed-back" connections among the somatosensory cortical fields indicate a high degree of integration without *a priori* assumptions regarding the direction of information flow within the system.

Somatosensory cortical fields are reciprocally interconnected with topographically related zones within and across hemispheres (51). In addition, cortical efferents send topographic projections to the basal ganglia, the midbrain, pontine nuclei, and to sources of somatosensory inputs in specific thalamic projection nuclei, the dorsal column nuclei and spinal cord (see 52, for review). These efferent connections may provide an anatomical basis for feedback filtering of incoming afferent inputs.

3.3 Vertical Organization

One major contribution of the Mountcastle group in the 1950's was the demonstration that individual cortical neurons encountered in a single vertical penetration shared common functional characteristics of place and modality (31, 32). This observation led to the proposal that cortical neurons are organized in vertical aggregates or mini-columns. Mountcastle (32, 53) has theorized that the phylogenetic trend toward encephalization or increased neocortical surface area has been achieved through the propagation of mini-columns with similar intrinsic processing mechanisms. These mini-columns are organized into clusters, cylindrical columns, or slabs that have similar organizational features and are classified as belonging to a given system by merit of a common set of inputs and outputs. According to this view, cortical fields are distinguished from one another only by their patterns of extrinsic connections. Histological differences between cytoarchitectonic areas are responses to differences in these patterns.

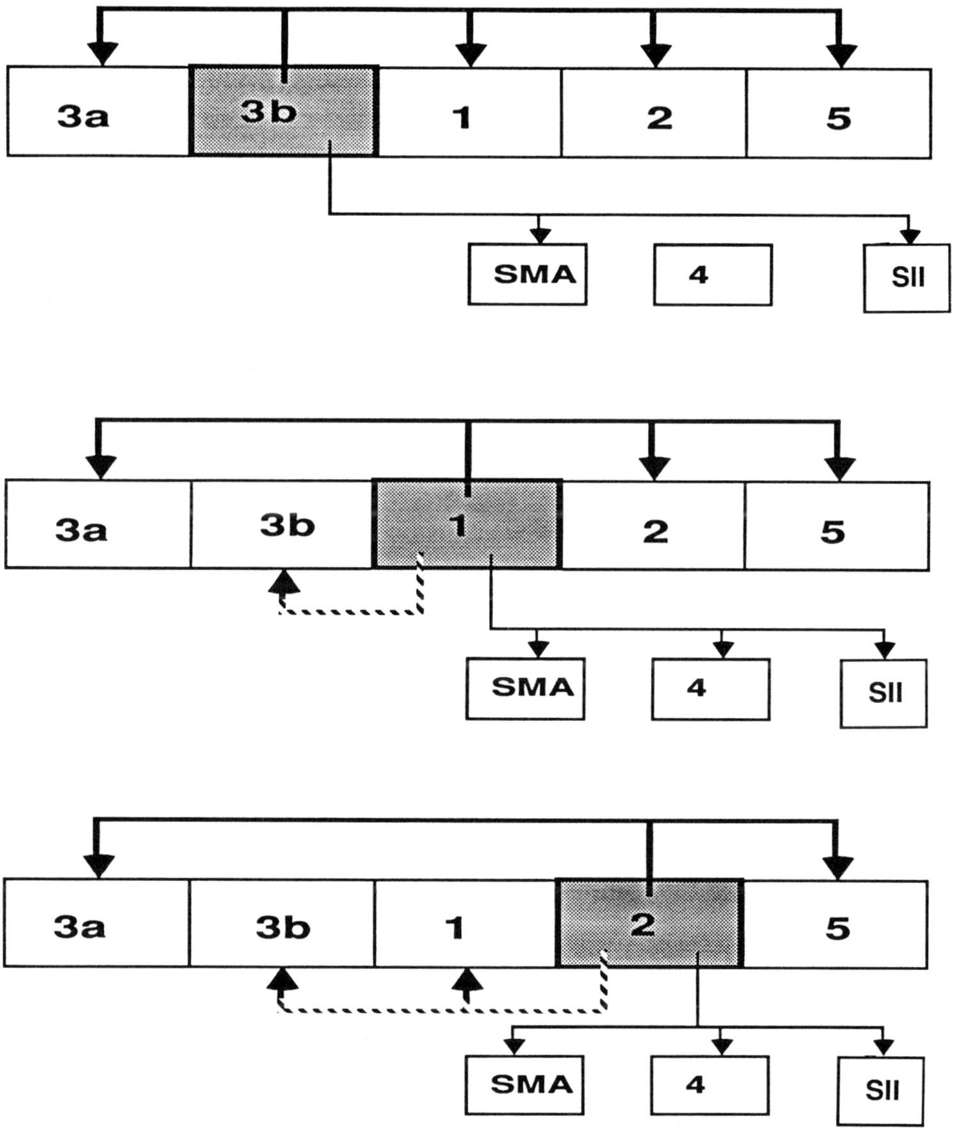

Figure 3. Efferent connections of areas 3b, 1, and 2 of monkey somatosensory cortex. Heavy solid lines indicate "feed-forward" inputs to the middle cortical layers. Dashed lines indicate "feed-back" projections primarliy to supragranular and infragranular laminae. SMA = supplementary motor cortex of area 6. S II = secondary somatosensory cortex. 4 = primary motor cortex. (Modified from 48-51)

According to Mountcastle (53), cortical columns can be characterized by three "static" properties: 1) RF location, 2) modality, and 3) adaptation characteristics. These static properties are determined by the segregated projection of small sets of first-order afferents with closely overlapping RF's. Modality and adaptation characteristics are closely linked as specific classes of primary afferents have well-defined adaptation characteristics. The "static" nature of these properties are determined by the transducer properties of primary receptors and the synaptic linkages between first-order afferents and the neurons upon which they project in the central nervous system. These properties are relatively fixed and change little with anesthesia or behavioral state.

Recent work by Merzenich, Kaas, and their colleagues indicates that the synaptic linkages determining the spatial location of specific cortical RF's can be drastically altered by changes in the pattern of afferent input (see review by Merzenich, this volume). Restricted peripheral deaffentation causes dramatic shifts in cortical responses to skin surfaces adjacent to the denervated skin. Rearrangement of peripheral skin fields causes major changes in the representation of the periphery in the cortical map. Similarly, differential use of a skin surface expands the representation of that surface in cortical fields. These results indicate that the specification of cortical receptive fields results from dynamic processes within the central nervous system rather than from any fixed anatomical convergence from the periphery.

Some workers have claimed recently that receptive field size changes systematically with cortical depth (54, 55). RF's in the primary thalamic recipient zones of the middle cortical layers are significantly smaller than in supragranular or infragranular laminae. Receptive fields in supragranular and infragranular layers must be the result of active intra-cortical processes and do not reflect the passive relay of information from thalamic afferents.

Adaptation characteristics of neurons representing the glabrous digits of macaques and owl monkeys may be organized into bands in area 3b (56) similar to orientation and ocular dominance bands in visual cortex. Sur and his colleagues have reported discrete regions within somatosensory cortex that have slowly-adapting (SA) or rapidly-adapting (RA) response properties in ketamine-anesthetized monkeys. In the RA bands, RA neurons were found in all cortical layers. However, in the so-called SA bands, SA units were found in only the middle cortical layers while RA units were found in supra-granular and infra-granular layers. It appears from these results that RA characteristics of some neurons in area 3b may be a dynamic function of intra-cortical mechanisms and not a static property of the column. These RA neurons cannot be part of the segrated RA channel from VP to areas 3b and 1 that has been proposed by Sur and his colleagues (57). Furthermore, connectivity arguments that RA units in cortex must receive projections from RA units in thalamus are negated by the demonstration that RA properties can be derived from SA inputs.

3.4 Physiological Tuning

There is now strong evidence that RFs defined at given cortical locations reveal only a subset of the actual array of inputs potentially available to any specific cortical unit. Recent anatomical, physiological, and behavioral studies have demonstrated that the neural substrate underlying the detailed physiological topography of the cutaneous representation in area 3b is itself quite diffusely organized. Cortical RFs and the detailed overall topography associated with them reflect the physiological tuning of a "degenerate" anatomical system and are not "hard-wired" in any traditional sense (e.g., 58).

Inhibitory mechanisms play an important role in the formation of cortical receptive fields. Iontophoretic application of bicuculline methiodide, an antagonist to the inhibitory neurotransmitter gamma-amino butyric acid (GABA), greatly enlarges receptive fields of RA units in cat somatosensory cortex (e.g., 59). Similarly, direct electrical stimulation of the large fiber afferents of a single cutaneous nerve in cat causes a rapid, several-fold increase in cortical RF area (60). Such data indicate that there is a greater degree of anatomical convergence on single units in the cortex than is apparent through standard recording techniques.

Consistent with these physiological data, Landry and Deschênes have shown that thalamocortical axons have broad terminal arborizations in primary somatosensory cortex of the cat (61). Physiologically-identified axons from the ventrobasal nuclei in the cat had bushy terminal arborizations in the middle cortical layers of area 3b that extended 1400 μ in the mediolateral direction and 817 μ in the anterior-posterior direction on average. Unfortunately, the extent of overlap or segregation of adjacent VB afferents in the cortex can only be inferred with this particular technique.

Recent cat experiments by Snow and his collaborators suggest that the terminal arborizations of individual thalamic afferents do not respect functionally defined somatic boundaries in the cortical map of the body surface (62). In these experiments, the cortical representations of adjacent digits were mapped in cat somatosensory cortex including the abrupt boundary between digits. Receptive fields of individual thalamic neurons were defined on one or the other adjacent digits. These "single-digit" thalamic neurons projected to cortical regions representing at least two digits, extending up to 800 μ into the "somatotopically inappropriate" cortical zone. The mechanisms underlying the selection of specific driving inputs from a far greater set of different anatomical possibilities is the subject of another chapter in this volume. It seems unlikely that the anatomical segregation of inputs can account for such a result.

Sensory convergence from the periphery upon cortical neurons imposes an anatomical limit on the range of possible driving inputs. In an extreme example, cortical neurons with RFs on the hindlimb have no anatomical connections with thalamic units with forelimb RFs. Therefore, units in hindlimb cortex should not be able to assume RFs on the forelimb under any circumstance in mature adult animals.

Sur and colleagues (63) have proposed that an analogous distance limit exists in somatosensory cortex. According to these authors, the horizontal distance between cortical neurons is linearly related to the percent overlap of their corresponding receptive fields; cortical RF's do not overlap when they are separated by an average of about 600 μ. It is probable that the physiological distance limit observed by Sur and his colleagues reflects some limiting aspect of peripheral convergence upon cortical neurons.

3.5 Intrinsic Elements

It is generally believed that some essential aspects of intrinsic neocortical processes are common to all areas of neocortex with idiosyncratic variations related to the pattern of extrinsic connections. With this assumption in mind, the following discussion of intrinsic structure and functions of the neocortex extends beyond the limits of somatosensory cortex *per se*.

The majority of synapses within the cortex are from intrinsic or local circuit neurons (LCN's) with short or no axonal ramifications (64). These cells have increased in both absolute and relative number with the progressive encephalization seen in phylogenetic ascent suggesting their importance in higher-order cortical functions (65). This increase in number of LCN's has been accompanied by an increase in the complexity of axonal and dendritic ramifications. Rakic has noted that the rhesus monkey has a full complement of neocortical neurons by fetal day 100 yet at this age the brain looks lissencephalic; the enormous amount of subsequent gyrification of the cortex is almost entirely due to the elaboration of neuropil (66). This enrichment of local interactions with ascending phylogeny led Cajal (67, p. 459) to speculate that "intellectual power ... (does) not depend on the size or number of cerebral neurons, but on the richness of the connective processes" (cited by Rakic, 68).

The current state of knowledge regarding intrinsic cortical circuits is extremely sketchy. It is apparent, however, that the traditional concept of sequential hierarchical processes resulting in progressive elaboration of receptive field properties is a gross over-simplification of neocortical functions. For example, recent studies have shown that cells with complex RF properties in visual cortex undoubtedly receive direct thalamic inputs as well as intrinsic inputs from the middle cortical layers (e.g., 69). In an excellent critical review of intrinsic cortical circuit elements, Jones argues that a more realistic model must include "multiple potential circuits" where "some elements may, under one set of circumstances, form part of one circuit and, under another, part of a second" circuit (70, p.20). The reader is referred to this article for details of current morphological-functional data.

Jones and Powell have found a bilaminar distribution of thalamic afferents and incoming ipsilateral and commissural cortico-cortical fibers in somatosensory cortex (e.g., 71). For the most part, extrinsic afferents penetrate the cortex from the white matter and arborize within the cortex in vertically oriented columns. All extrinsic inputs appear to be excitatory in nature with asymmetrical synapses and spherical vessicles. Specific thalamocortical afferents project to layer IV and lower layer III while generalized thalamocortical and brain stem afferents terminate in all layers especially in Layer I. Both ipsilateral and commissural cortico-cortical fibers terminate in supragranular layer III. The majority of these input fibers are oriented vertically in the cortex giving rise to a vertical flow of information.

The output cells of somatosensory cortex are the pyramidal cells which are found in the supragranular and infragranular layers. These cells are characterized by a conical somata, spiny basal dendrites, long spiny apical dendrites that project vertically to layer I, and extrinsically-projecting axons with many horizontal collaterals extending up to several millimeters. They are excitatory in nature, making asymmetrical synapses with other pyramidal cells locally as well as to targets outside their original cytoarchitectonic area. Recurrent collaterals can branch in the local neighborhood of the cell soma and dendrites or project at some horizontal distance in any cortical layer. In motor cortex, these cells emit recurrent collaterals that have a profound inhibitory effect on the subsequent firing of a given cell (72, 73). Because pyramidal cells are themselves excitatory, this inhibitory influence must be mediated by some set of inhibitory interneurons. Mountcastle has used these data to propose a form of peri-columnar inhibition whenever there is intense, focal activity in a restricted number of pyramidal cells.

Corticothalamic fibers originate in Layer IV while corticocortical fibers originate in layer III. Other cortico-fugal fibers to the brain stem, basal ganglia, and spinal cord originate in Layer V. Pyramidal cells receive excitatory inputs from thalamic and cortico-cortical afferents as well as from neighboring pyramidal cells (74) in addition to strong inhibitory inputs from certain classes of intrinsic neurons such as the large basket cells (see below).

The large majority of thalamo-cortical synapses are onto the stellate-shaped interneurons of the middle cortical layers. These interneurons dominate layers III, IV, and V giving area 3b its granular appearance with cell-body stains. Many of these neurons are inhibitory in nature and use GABA as a neurotransmitter. Jones has classified these layer IV cells into four major categories (70): 1) Small spiny cells have restricted dendritic fields with vertically oriented axons that terminate on the spiny apical dendrites of pyramidal cells in layers II-IV. These cells are not GABA-ergic and have asymmetrical synapses indicating excitatory function. 2) The neurogliaform or spiderweb cells have spherical dendritic and axonal arborizations within layer IV and probably make local synapses on spiny cell somata; the type of synapse is unknown but it is likely that they are inhibitory in nature. 3) Basket cells have been studied extensively with Golgi preparations in human motor and visual cortex (75). They are GABA-ergic and make symmetric synapses on pyramidal cell somata forming pericellular nests around pyramidal cells in layers II, III, and V. Large basket cells have broad axonal arborizations with horizontal ramifications extending 1-2 mm in the saggital plane. Each basket cell may arborize around several pyramidal cells and, conversely, one pyramidal cell may be surrounded by axonal processes from several basket cells. 4) Aspiny stellate cells are also found in layers II and III as well as in layer IV. For this reason, many aspiny cells probably do not receive direct thalamic inputs. They are probably GABA-ergic and make symmetric synapses upon the somata and basal or apical dendrites of pyramidal cells in layers III and V (76).

Other intrinsic neurons are found largely in layers II and III. Chandelier cells are GABA-ergic and make inhibitory synapses on the initial segments of pyramidal cell axons. Double-bouquet cells have slender vertically-oriented axonal ramifications to symmetric synapses on pyramidal cell spines in layers II-V. Bipolar cells have long aspiny dendrites and vertically-oriented axons with excitatory synapses.

It is clear from the foregoing information that much more work remains to be done before a schematic model of specific cortical circuitry can be drawn. Many opportunities exist for neuronal interactions among all classes of neurons. The time-consuming delineation of connections among different morphological classes of neurons is merely an initial step in understanding neocortical function. Detailed examination of different neurochemical systems and physiological functions, ultimately with behaving animals, will be necessary to uncover the dynamic interactions of these cell classes.

Any neural model of information transformation must include neural processes based on excitatory and inhibitory interactions. Excitatory inputs from thalamic and cortico-cortical afferents make synapses on pyramidal cells directly and indirectly through the spiny stellate cells of layer IV. If unchecked, these pyramidal cells could spread the excitatory influence to other neighboring pyramidal cells through intrinsic recurrent collaterals. However, concurrent with direct thalamic activation of pyramidal cells of all layers, intrinsic inhibitory interneurons are also excited by thalamic afferents and probably by recurrent collaterals of pyramidal cells themselves. These powerful inhibitory agents act to contain the spread of pyramidal cell activation through vertical axonal projections possibly mediating a form of peri-columnar inhibition (58). Without specific anatomical segregation, this peri-columnar inhibition would necessarily be dynamic in nature and the functional columns that result would pass across regions of somatosensory neocortex as a function of the specific focus of afferent excitation. Whether there exists an anatomical basis for discrete, non-overlapping cylindrical columns in the somatosensory cortex remains to be proven.

4 RELATIONSHIP TO OTHER SYSTEMS

Due of space limitations, we have not attempted to describe the relation of the somatosensory nervous system to other systems within the brain as a whole. It may be an error to consider this system as separate in function from the motor system to which it is closely allied both anatomically and functionally. The representation of external space in the tactile system and the interpretation of sensory input is inextricably linked to the perception of somatic movement and there exist many opportunities for activity in what has been called the motor system to influence the pattern of incoming somatosensory information. An initial step in unraveling this mysterious alliance between sensory and motor systems has been made by Nelson (77) through the study of sensory responses to well-defined stimuli during movements. This type of investigation will undoubtedly reveal functional aspects of sensorimotor integration and interdependence heretofore unimagined.

Generally speaking, the study of somatosensory responses in awake behaving animals will greatly increase our understanding of the function of these somatosensory circuits. Ablation-behavior techniques uncover specific behavioral deficits that suggest global functions for the systems in question. Physiological studies of response plasticity as a function of behavioral state have already challenged the orthodox manner in which we conceptualize information transformation in the auditory and visual systems, especially in what has been called association areas of cerebral cortex (e.g., 78). Analogous studies remain to be pursued in the somatosensory system.

6 ACKNOWLEDGEMENTS

The authors were supported by NIH grants NS07363 (T.A.) and NS10414 (M.M.M.). We are grateful to G. Recanzone for his critical review of the manuscript.

7 REFERENCES

(1) M.M. Merzenich, J.H. Kaas, Prog. in Psychobiology and Physiol. Psych., 9 (1980) 1-42.

(2) G.F. Poggio, V.B. Mountcastle, J. Neurophysiol. 26 (1963) 775-806.

(3) A. Iggo, Prog. in Br. Res. 43 (1976) 15-31.

(4) R.S. Johansson, Sensory Functions of the Skin of Humans, Plenum Press, New York 1979, pp. 129-145.

(5) R.S. Johansson, A.B. Vallbo, J. Physiol. 286 (1979) 283-300.

(6) I. Darian-Smith, P. Kenins, J. Physiol. 309 (1980) 147-155.

(7) J.R. Phillips, K.O. Johnson, J. Acoust. Soc. Am. 77 (1985) 220-224.

(8) A.B. Vallbo, K.A. Olsson, W.G. Westberg, F.J. Clark, Brain 107 (1984) 727-749.

(9) W.D. Willis, R.E. Coggeshall, Sensory Mechanisms of the Spinal Cord, Plenum Press, New York 1978.

(10) K.W. Horch, P.R. Burgess, D. Whitehorn, Br. Res. 117 (1976) 1-17.

(11) R.L. Kitchell, D.D. Canton, R.D. Johnson, S.A. Maxwell, J. Comp. Neurol. 210 (1982) 400-410.

(12) P.R. Douglas, D.G. Ferrington, M. Rowe, J. Physiol. 285 (1978) 493-513.

(13) M.B. Carpenter, B.M. Stein, J.E. Shriver, Am. J. Anat. 123 (1968) 75-118.

(14) J.E. Shriver, B.M. Stein, M.B. Carpenter, Am. J. Anat. 123 (1968) 27-74.

(15) R.W. Dykes, D.D. Rasmusson, D. Sretavan, N.B. Rehman, J. Neurophysiol. 47 (1982) 389-416.

(16) J.H. Kaas, R.J. Nelson, M. Sur, R.W. Dykes, M.M. Merzenich, J. Comp. Neurol. 226 (1984) 111-140.

(17) C.S. Lin, M.M. Merzenich, M. Sur, J.H. Kaas, J. Comp. Neurol. 185 (1979) 355-372.

(18) E.G. Jones, H.D. Schwark, P.A.Callahan, Exp. Br. Res. 63 (1986) 310-320.

(19) E.G.Jones, D.P. Friedman, S.H.C. Hendry, J. Neurophys. 48 (1982) 545-568.

(20) C.-T. Yen, M. Conley, E.G. Jones, J. Neurosci. 5 (1985) 1316-1338.

(21) J.H. Jackson, Selected Writings of John Hughlings Jackson, Vol 1, Staples Press, London 1958.

(22) H. Cushing, Brain 32 (1909) 44-53.

(23) O. Foerster, Lancet 221, II (1931) 309-312.

(24) W. Penfield, T. Rasmussen, The Cerebral Cortex of Man, Macmillan , New York 1950.

(25) R.W. Gerard, W.H. Marshall, L.J. Saul, Arch. Neurol. Psychiat. 36 (1936) 675-738.

(26) C.N. Woolsey, W.H. Marshall, P. Bard, Bull. Johns Hopk. Hosp. 70 (1942) 399-441.

(27) K. Brodmann, Vergleichende Lokalisationslehre der Grosshirnrinde, Barth, Leipsig, East Germany 1909.

(28) C. Vogt, O. Vogt, J. Psychol. Neurol. Leipsig 25 (1919) 279-462.

(29) E.G. Jones, R. Porter, Br. Res. Rev. 2 (1980) 1-43.

(30) F. Sanides, in C.R. Noback & W. Montagna (eds.), The Primate Brain, Appleton-Century-Crofts, New York 1970, pp. 137-208.

(31) V.B. Mountcastle, J. Neurophys. 20 (1957) 408-434.

(32) T.P.S. Powell, V.B. Mountcastle, Bull. Johns Hopkins Hosp. 105 (1959) 133-162.

(33) D.A. Dreyer, P.R. Loe, C.B. Metz, B.L. Whitsel, J. Neurophys. 38 (1975) 714-733.

(34) J. Tanji, S.P. Wise, J. Neurophys. 45 (1981) 467-481.

(35) J. Hyvarinen, A. Poranen, J. Physiol. 283 (1978) 523-537, 539-556.

(36) M. Sur, Br. Res. 198 (1980) 465-471.

(37) R.L. Paul, M.M. Merzenich, H. Goodman, Br. Res. 36 (1972) 229-249.

(38) M.M. Merzenich, J.H. Kaas, M. Sur, C.S. Lin, J. Comp. Neurol. 181 (1978) 41-74.

(39) M. Sur, R.J. Nelson, J.H. Kaas, J. Comp. Neurol. 211 (1982) 177-192.

(40) J.H. Kaas, Physiol. Rev. 63 (1983) 206-231.

(41) R.H. LaMotte, V.B. Mountcastle, J. Neurophys. 42 (1979) 400-419.

(42) M. Randolph, J. Semmes, Br. Res. 70 (1974) 55-70.

(43) W.E. Le Gros Clark, T.P.S. Powell, Proc. R. Soc. London 141 (1953) 567-487.

(44) C.-S. Lin, M.M. Merzenich, M. Sur, J.H. Kaas, J. Comp. Neurol. 185 (1979) 355-372.

(45) D.P. Friedman, E.G. Jones, J. Neurophys. 45 (1981) 59-85.

(46) T.P.Pons, J.H. Kaas, J. Comp. Neurol. 240 (1985) 16-36.

(47) E. Kosar, P.J. Hand, J. Comp. Neurol. 198 (1981) 515-539.

(48) E.F. Jones, T.P.S. Powell, Brain 92 (1969) 477-502.

(49) M.F. Shanks, A.J. Rockel, T.P.S. Powell, Br. Res. 98 (1975) 166-171.

(50) B.A. Vogt, D.N. Pandya, J. Comp. Neurol. 177 (1977) 179-191.

(51) M.F. Shanks, R.C.A. Pearson, T.P.S. Powell, Br. Res. Rev. 9 (1985) 67-88.

(52) A.L. Towe, in A. Iggo (ed.), Handbook of Sensory Physiology, Vol II, Springer, New York (1973), pp. 701-718.

(53) V.B. Mountcastle, in The Mindful Brain, MIT Press, Cambridge, MA 1978, pp. 7-50.

(54) Y. Iwamura, M. Tanaka, M. Sakamoto, O. Hikosaka, Exp. Br. Res. 58 (1985) 400-411.

(55) M. Sur, P.E. Garraghty, C.J. Bruce, Br. Res. 342 (1985) 391-395.

(56) M. Sur, J.T. Wall, J.H. Kaas, J. Neurophys. 51 (1984) 724-744.

(57) M. Sur, J.T. Wall, J.H. Kaas, Science 212 (1981) 1059-1061.

(58) G.M. Edelman, L.H. Finkel, in G.M. Edelman, W.E. Gall, W.M. Cowan (eds.), Dynamic Aspects of Neocortical Function, John Wiley & Sons, New York 1984, pp. 653-695.

(59) R.W. Dykes, P. Landry, R. Metherate, T.P. Hicks, J. Neurophys. 52 (1984) 1066-1093.

(60) G.H. Recanzone, T. Allard, W.M. Jenkins, M.M. Merzenich, Neurosci. Abst., (1986).

(61) P. Landry, M. Deschenes, J. Comp. Neurol. 199 (1981) 345-371.

(62) P.J. Snow, R.J. Nudo, W. Rivers, W.M. Jenkins, M.M. Merzenich, Australian Winter Conference on Br. Res., Queenstown, New Zealand, Abs. (1986).

(63) M. Sur, M.M. Merzenich, J.H. Kaas, J. Neurophys. 44 (1980) 295-311.

(64) E.G. Jones, T.P.S. Powell, Philos. Trans. R. Soc. Lond. 257 (1970) 45-62.

(65) N.L. Mitra, J. Anat. 89 (1955) 467-483.

(66) P. Rakic, in N.A. Buchwald & M.A.B. Brazier (eds.), Brain Mechanisms in Mental Retardation, Academic Press, New York 1975, pp. 3-40.

(67) S. Ramon y Cajal, Recollections of My Life, MIT Press, Cambridge, Mass. 1937.

(68) P. Rakic, Local Circuit Neurons, MIT Press, Cambridge, Mass. 1976.

(69) G.H. Henry, A.R. Harvey, J.S. Lund, J. Comp. Neurol. 187 (1979) 725-744.

(70) E.G. Jones, in G.M. Edelman, W.E. Gall, W.M. Cowan (eds.), Dynamic Aspects of Neocortical Function, John Wiley & Sons, New York 1984, pp. 7-40.

(71) E.G. Jones, J. Comp. Neurol. 160 (1975) 167-204.

(72) C. Stephanis, H. Jasper, J. Neurophys. 27 (1964) 855-877.

(73) H. Asanuma, I. Rosen, Exp. Br. Res. 16 (1973) 507-520.

(74) D.A. Winfield, R.N.L. Brooke, J.J. Sloper, T.P.S. Powell, Neurosci. 6 (1981) 1217-1230.

(75) M. Marin-Padilla, Z. Anat. Entwicklungsgesch. 144 (1974) 123-135.

(76) A. Peters, A. Fairen, J. Comp. Neurol. 181 (1978) 129-172.

(77) R.J. Nelson, Exp. Br. Res. Suppl. 10 (1985) 59-76.

(78) D.M. Diamond, N.M. Weinberger, Br. Res. 372 (1986) 357-360.

Self-Organizing Processes in Adult Neo-Cortex

by *Michael M. Merzenich, Terry Allard, William M. Jenkins* and *Gregg Recanzone*

SUMMARY

A dominant theme of contemporary neuroscience has been the interpretation of response properties of neurons in the central nervous system as a product of anatomical and functional processes that are fixed through an early critical period of nervous system maturation. By a commonplace if extreme interpretation of this view, the functional roles of neurons are determined by their anatomical maturation and by the maturation of their connections early in life. By that view, the fixed, lifelong roles of neurons could be defined at any point in their lifetime, an assumption made countless times in the predominant neurophysiological experimental paradigm.

Recent experiments conducted principally in the somatosensory system challenge this strictly static view of the origins of forebrain responses. They have shown that the response selectivity of cortical neurons can be substantially altered throughout the lives of rodents, carnivores, and primates. In this brief review, we summarize some of the evidence. Specific aspects of these results relevant to understanding the organization and rules of operation of the underlying networks of neocortical neurons will be briefly described. Some wider implications of these findings will be noted.

1 SELF-ORGANIZATION OF SOMATOSENSORY CORTEX IN ADULT MAMMALS

Recent experiments have shown that there is a substantial reorganization of the topographic representations of the skin surfaces within somatosensory cortical fields of adult rats, raccoons, cats and monkeys following: a) peripheral nerve transection (1 - 4); b) dorsal root section (5); c) digit amputation (6 - 8); d) peripheral nerve transection and reconnection (9); e) restricted cortical lesions (10); f) behavioral training in which a restricted skin surface is heavily, differentially stimulated (10 - 12); g) transfer of an island of skin and its intact innervation to a new peripheral skin location (10, 13); and h) surgical fusion of adjacent digits (10, 14). Some basic features of this representational plasticity include the following:

a) *Neurons across broad cortical regions are driven by "new" previously ineffective inputs*. Thus, for example: 1) The representation of skin surfaces surrounding a denervated skin zone expands to be represented in finer grain following peripheral nerve transection or digit amputation (2 - 4, 6). With such reorganization, cortical neurons which had previously represented the now-denervated skin, as well as neurons across a broad bordering zone, acquire newly effective driving inputs. 2) A completely novel - but still orderly - cortical topography is recorded after the transection and reconnection of a large peripheral cutaneous nerve (9). This topography cannot be accounted for by peripheral reinnervation patterns, which randomize peripheral to central connections in regeneration; central input selection processes must be involved (9, 12). 3) Skin surfaces represented exclusively within the zone of a cortical lesion emerge in the undamaged cortex surrounding the infarct, with an often substantial displacement of the representation of the skin surfaces formerly represented in these cortical sectors (10). Again, with this representational reorganization, all neurons over a cortical region several millimeters across can acquire new driving inputs. 4) Selective stimulation of and attention to restricted skin surfaces during a behavioral task greatly enlarge cortical representations of those surfaces (10 - 12; see Figure 1). These distorted representations return to a more normal skin surface representation after the behavioral task is halted (10, 11 and unpublished observations). 5) Effective driving inputs are dramatically altered all across the cortical zone representing the hand as a consequence of the movement of an island of skin with its innervation intact from one digit to another (12, 13; see Figure 2). In these experiments, care is taken to conserve the anatomical relationships between the skin and central neurons, i.e., these representational changes occur with little (if any) alteration of skin-to-central nervous system connections.

Again, in all of these experimental models, effective driving inputs are altered for neurons at all locations across extensive cortical sectors. In most of these preparations, many of the newly emergent receptive fields do not overlap with receptive fields defined for neurons at those same locations prior to the experimental manipulation.

It should be emphasized that spatial aspects of receptive field organization constitute only one manifestation of input selection. These studies should be considered not only with respect to the formation and maintenance of topographically ordered cortical representations but, more generally, for their implications for mechanisms underlying the selection of effective excitatory inputs throughout the forebrain.

b) *Despite substantial distributed changes in cortical input selection, topographic order within the cortical representation is usually maintained.* In general, certain intrinsic features of cortical topography are retained despite the substantial remodeling of representations obtained in these experiments (2 - 4, 9, 10, 12 - 14). For example, a continuously shifting overlap in peripheral receptive fields is commonly recorded in microelectrode penetration sequences across the horizontal dimensions of the cortex through all recorded stages of reorganization.

c) *As cortical magnification changes over time, mean receptive field sizes vary inversely; a percentage overlap vs cortical distance "rule" is roughly maintained.* In normal somatosensory representations, receptive fields of neurons at nearby locations commonly overlap within cortical distances less than about 500 - 600 microns (15). On average, the percentage overlap of receptive fields is a linear function of distance within this "hypercolumn" domain. Following peripheral nerve transections, digital amputations, restricted cortical lesions or differential stimulation of limited skin surfaces in a behavioral task, specific skin surfaces come to be represented over only a fraction of their original representational territory in one extreme (e.g., after cortical lesions), or over a cortical zone up to about 10X the normal size in another extreme (e.g., after heavy differential skin use). With these substantial changes in magnification, roughly inverse changes in receptive field sizes are recorded. That is, the percentage overlap vs distance "rule" observed in normal animals is roughly maintained. As the cortical representation of specific skin surfaces enlarges or contracts, those skin surfaces are represented in finer or coarser grain, respectively.

We have noted one important exception to this "percentage-overlap rule". Under some special conditions, the cortical zone over which a given, specific cutaneous receptive field is represented can be greatly extended. This basic phenomenon is described in a later section.

d) *There is a distance limit for reorganization.* There is a limit to the horizontal distance that the representation of a given skin surface can be moved across the cortex. Within the koniocortical field of somatosensory cortex, area 3b, sites of representation do not move farther than about 600 - 700 microns in most cases, although translocations up to about 2 mm have been recorded in a few experiments. Representational translocations can be substantially greater in cortical area 1 than in area 3b.

2 UNDERLYING MECHANISMS

We have hypothesized (3, 6, 12, and 16) that:

1. This representational plasticity involves alterations in the effectiveness of existing inputs.
2. These input selection processes involve closely coupled excitatory and inhibitory mechanisms.
3. Input selection is by cooperating groups of neurons
4. These neuronal groups are dynamic entities; they move; their neuronal constituencies are subject to continual remodeling.
5. In detail, cortical representations map probabilities of coincident inputs.
6. The effectiveness of individual inputs in creating long-lasting representational changes are modulated as a function of behavioral state.
7. Dynamic input-selection processes are general in the neocortex. They probably constitute a major component of the adaptive processes underlying higher brain function.

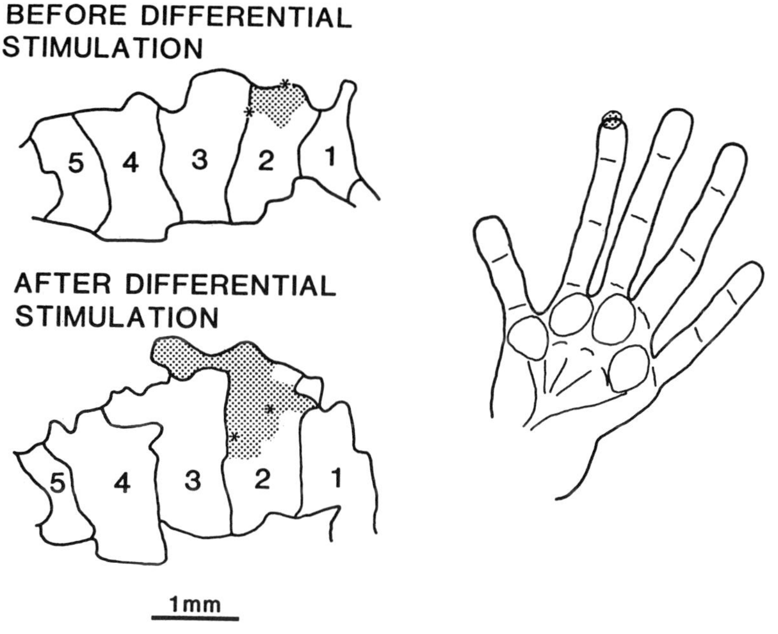

Figure 1. Outline of territories representing the surfaces of the digits within cortical field 3b of a normal adult owl monkey. These microelectrode maps were derived before and 3 months after the initiation of a special differential-stimulation behavior. During this time, this monkey had performed a behavioral task involving heavy, differential stimulation of the tips of the distal phalanges of digits 2, 3, and occasionally 4 for about 1 hour/day. These representational "maps" were drawn from definition of cutaneous receptive fields defined in several hundred parallel microelectrode penetrations introduced into the middle cortical layers in area 3b at both experimental stages. Outlined territories are zones over which receptive fields were centered on digits 1 to 5. The small, differentially-stimulated skin surface at the tip of digit 2 is dotted in the hand drawing at the right. The cortical zones over which this skin surface was represented before and after the behavioral task are dotted in both drawings. In these zones, all receptive fields included this heavily stimulated surface. Note that after a period of differential digit tip stimulation in this and in 5 other monkeys studied with this paradigm, there was a several-fold enlargement of its territory of representation. In this monkey, most of the gain in territory was across the area 3b-3a border as defined functionally prior to the initiation of this digital stimulation behavior. On average, about half of the new representation in these monkeys is gained internally from within area 3b and about half from across its functionally-defined rostral border. Redrawn from Jenkins and Merzenich (14).

A number of studies have shown that the repertoire of excitatory inputs that are anatomically available to a given set of neurons is far larger than ordinarily revealed in physiological studies of cortical neurons driven by natural stimuli. In the area 3b representation of the adult monkey hand, neurons at any given location are commonly excited by stimulation of a small fraction of the surface of a palmar pad or digit. However, neurons at any given location have additional "masked" excitatory inputs from a third or more of the surfaces of the hand - and likely from surfaces of the face and wrist as well. This "degenerate" (17) divergent-convergent input organization has been revealed in a number of experimental studies. Individually filled thalamic terminals in cats and monkeys spread from a few hundreds of microns to as much as 2 mm across the koniocortical field, area 3b of cats and monkeys (18, 19). Sectors of area 3b labelled by 2-deoxyglucose following stimulation of a restricted skin surface are far larger than the zone in the cortical map over which stimulation of that skin surface effectively excites cortical neurons (20). Microstimulation maps of the zones of termination of thalamic axons within the neocortex conducted in this laboratory reveal that single ventroposterior neurons can project to cortical regions up to about 2 mm across. Most of the field of termination of the projecting arbor of a single ventroposterior neuron can be outside of the cortical territory over which that skin surface is functionally represented (Snow, Nudo, Rivers, Jenkins, and Merzenich, unpublished studies). EPSPs are commonly generated in neurons at a given neocortical location by stimulation far outside of their cutaneous receptive field (21, 22). Excitatory receptive fields are greatly enlarged by local bicuculline injection; enlarged fields are very uniformly driven (23, 24). Receptive fields are similarly enlarged by the electrical stimulation of the large-fiber afferents of a peripheral cutaneous nerve, which engages an opioid-based modulatory control system (25).

Taken together, these studies reveal that the anatomical 'maps' of excitatory inputs within even the koniocortical somatosensory field, area 3b, are crude when compared with functional maps defined physiologically at any given time in the life of the adult cat or monkey. This degenerate anatomical projection provides an extended input repertoire that is selected from, by use, in the generation and continual behavioral remodeling of functional maps. There are countless possible versions of those distributed behaviorally-selected representations and countless possible receptive fields could be recorded for neurons at any given cortical location, at different times over a lifetime.

If neurons at any given cortical location can respond to a variety of different inputs located over a relatively large skin area at different times in life, by what criteria are specific inputs selected? What mechanism accounts for the conservation of map topography as representations change? It is increasingly evident that, considered from the perspective of afferent inputs, input coincidence underlies the selection of effective inputs by use, and accounts for local map order (12, 20, 26). Cortical representations are not static anatomical constructs; considered in detail, they map probabilities of coincident or nearly coincident responses.

Thus, for example, the abrupt representational discontinuities between adjacent digits in the normal cortical map of an owl monkey hand are broken down by connecting the skin of adjacent digits (14, 12). In this surgical fusion or syndactyly, there is no significant change in the innervation of the digits. Nonetheless, the cortical representation of the syndactyly changes greatly. Over a broad cortical zone, most neurons come to be excited by stimulation of receptive fields overlapping the two digits. In normal monkeys, such two-digit receptive fields are rarely encountered. We conclude that the substantially non-simultaneous excitation of the surfaces of adjacent digits accounts for their discontinuous representation in the normal map; when the probability of input coactivation across two fingers is increased through artificial syndactyly, the normal representational discontinuity is broken down.

The role of coincident inputs for shaping cortical receptive fields is also demonstrated in experiments in which skin is transferred with its innervation intact from one digit to another (13, 12; see Figure 2). In these experiments, the relationships between skin locations and central nervous system locations are minimally disturbed. At the same time, the fully innervated island of skin is surrounded by new neighboring skin. If topographical cortical representations were strict anatomical constructs, they should not be altered by such manipulations. To the contrary, they reorganize dramatically after such a transfer. Many receptive fields emerge that have components extending from the island skin onto the surrounding recipient skin. These skin surfaces are normally represented in the cortex at locations separated by hundreds of microns; these skin surfaces are never incorporated into the same receptive fields in area 3b in normal monkeys.

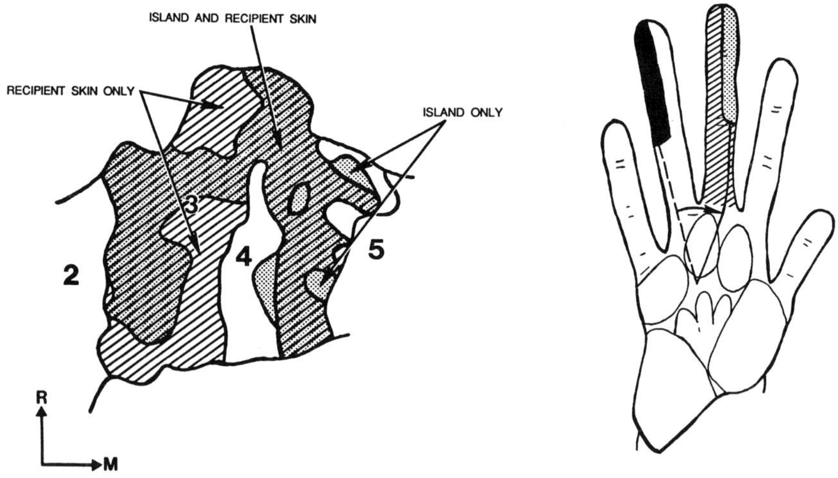

Figure 2. One of the class of experiments revealing that cortical input selection is based upon input coincidence. This cortical map was derived 6 months after transfer of an innervated, vascularized island pedicle from the ulnar aspect of digit 4 to the radial aspect of digit 3 in an adult owl monkey. A denervated graft derived from digit 3 was transferred back to close the defect on digit 4. With such an island pedicle transfer (indicated in the hand drawing at the right) in which the skin is cut along the boundaries of distribution of peripheral cutaneous nerves, there is no significant alteration of skin-to-central nervous system connections. However, the island skin (from digit 4) is now associated with different surrounding skin increasing the probability of coincident activity from the newly juxtaposed skin surfaces (on digit 3).

The cortical map on the left was derived by definition of receptive fields at several hundred microelectrode penetration sites. The cortical zone with receptive fields including skin from the recipient finger, digit 3, is marked by diagonal striping; the area with receptive fields extending onto the island are stippled. Note the large zone of overlap where receptive fields are partly on the island skin and partly on the skin of the recipient finger. These skin surfaces are normally represented in cortical zones separated by more than 500 microns from each other in adult owl monkeys. Control experiments confirm that this dramatic representational reorganization is due to central, and not peripheral effects: when the digital nerves innervating the island or ulnar recipient skin are anesthetized or cut, receptive field boundaries invariably terminate along the scar line.

Considering receptive field topography in detail, there is a complex topographic order in these representations, by which the island skin is represented in sequence with the adjacent skin of the finger receiving the island.

Features of reorganization following nerve transection, reconnection and regeneration can also be explained by the hypothesis that map order, in detail, is coincidence-based (see 12).

It is almost certain that the modulation of input effectiveness as a function of behavioral state is a basic feature of these self-organizing processes (12, 16). The most convincing demonstrations of this modulatory control come from studies conducted in the motor cortex and auditory cortex, described below. We have limited evidence indicating that alterations of effective inputs within the somatosensory koniocortex, area 3b, are under the same cognitive control. Acetylcholine-, norepinephrine-, and opioid-based systems are conjointly implicated in this control (25, 27).

There is also growing evidence that dynamic input selection is the product of neuronal groups (26) with significant horizontal extents. Groups are dynamic, functional (not fixed, anatomical) units; sizes of neuronal groups can be manipulated experimentally.

In the somatosensory koniocortex, we believe those groups are relatively sharply bounded, and have significant (10s of microns) horizontal extents (12). They have been unequivocally demonstrated in several special classes of experiments. First, following peripheral transection of a large cutaneous nerve, neurons at given cortical loci have multiple receptive fields located at several different locations across the skin field of distribution of the nerve (9, 12, 28). Specific unique combinations of receptive fields are recorded for all neurons that respond to tactile stimuli recorded in vertical penetrations across the depths of cortex, and for all neurons across a significant horizontal distance. These special mapping studies reveal discretely bounded cortical neuronal groups.

Second, under a variety of experimental circumstances, greatly enlarged neuronal groups have been generated. These "supergroups" often extend hundreds of microns across the cortex. Within them, all member neurons have identical or nearly identical response characteristics. These enlarged groups have been recorded, for example, in animals wearing chronic digital casts, in adult cats studied after stimulation of peripheral cutaneous nerves, in two cats in which differential stimulation of small hair patches was facilitated over a several day period. In these instances, all neurons over cortical sectors up to nearly 1.5 mm across had virtually identical receptive fields.

It is clear that the dimensions of input-selecting neuronal groups are dynamic. They move. They cannot be fixed anatomically-static entities. It is highly likely that they are generated, sustained, and compete for domination of neurons along their mutual borders under control of the magnitudes of synchronized afferent inputs (12, 26).

3 SELF-ORGANIZING PROCESSES ARE GENERAL IN THE NEOCORTEX

Dynamic representational changes have been recorded in the auditory cortex with classical conditioning (29, 30). Changes in auditory tuning curves for frequency - and necessarily, in representational territories - result from behavioral conditioning. Representations revert to near the pre-experimental form with behavioral extinction.

Similarly, conditioned movements in classically conditioned animals are represented at a far greater number of motor cortex sites than prior to conditioning; again, the extended representation shrinks with extinction (31). Microstimulation maps within motor cortex have long been known to be highly alterable by surface stimulation (32, 33); brief episodes of intracortical microstimulation also result in a substantial remodeling of movement representations (34). We have initiated behavioral studies in area 4 in adult monkeys, deriving detailed microstimulation maps before and after motor training. Preliminary results indicate that motor 'maps' are also highly alterable by use.

A series of representational plasticity studies has been conducted in another cortical field, area 1 (see 12). Changes in area 1 representation are of as great or greater magnitude as are those in area 3b.

It appears likely that these self-organizing processes are general in the neocortex. The magnitudes of possible changes are presumably dependent, field by field, upon the extents of convergent-divergent projectional anatomy. If that is the case, the map alterability by use in cortical area 3b, the most anatomically constrained somatosensory representation, is probably modest in comparison with the potential remodeling by use of other cortical areas that have substantially more complex input combinations and substantially greater divergence and convergence of projections.

4 SOME IMPLICATIONS

These intrinsic self-organizing processes probably constitute _the_ principal neocortical operation. It is important to note that the predominant experimental paradigms now employed for studying the cortex are inappropriate for studying these processes. Stimulus-driven neuronal responses recorded in an adult mammal at any given window of time presumably reflect behaviorally significant input histories (12, 16, 35, 36). These behaviorally derived representations will be idiosyncratic in detail, in every adult individual. To be effectively studied, this machinery must be investigated _in operation_.

The strategy of studying awake animals after they are trained to asymptotic performance levels is also an inappropriate one. Cortical representations will obviously be shaped by this training, and distributed cortical responses like those recorded in these preparations would almost certainly not be recorded in a naive animal. The critical period for study is over the time of _acquisition_ of skill - as that acquisition is presumably attributable to the experience-induced representational changes. It would also appear likely that studies designed to explain the bases of input response specificity by cataloging anatomical details are doomed to fail.

These adaptive, self-organizing processes presumably constitute a major component of the adaptive processes underlying higher brain functions.

5 REFERENCES

(1) J. Kalaska, B. Pomeranz, J. Neurophysiol. _42_ (1979) 618-633.

(2) M. M. Merzenich, J. H. Kaas, J. T. Wall, R. J. Nelson, M. Sur, D. J. Felleman, Neurosci. _8_ (1983) 3-55.

(3) M. M. Merzenich, J. H. Kaas, J. T. Wall, M. Sur, R. J. Nelson, D. J. Felleman, Neurosci. _20_ (1983) 639-665.

(4) J. T. Wall, J. Neurosci. _4_ (1984) 1499-1515.

(5) J. L. Franck, Brain Res. _186_ (1980) 458-462.

(6) M. M. Merzenich, R. J. Nelson, M. P. Stryker, M. S. Cynader, A. Schoppmann, J. M. Zook, J. Comp. Neurol. _224_ (1984) 591-605.

(7) D. Rasmusson, J. Comp. Neurol. 205 (1982) 313-326.

(8) A. M. Kelahan, G. S. Doetsch, Somatosensory Res. 2 (1984) 49-81.

(9) J. T. Wall, J. H. Kaas, M. Sur, R. J. Nelson, D. J. Felleman, M. M. Merzenich, J. Neurosci. 6 (1985) 218-233.

(10) W. M. Jenkins, M. M. Merzenich, Prog. in Br. Res. 41 (1987) 249-266.

(11) W. M. Jenkins, M. M. Merzenich, M. T. Ochs, Soc. Neurosci. Abstr. 10 (1984) 665.

(12) M. M. Merzenich, The Neural and Molecular Bases of Learning, Dahlem Konferenzen, Springer-Verlag, Berlin 1987, in press.

(13) S. A. Clark, T. Allard, W. M. Jenkins, M. M. Merzenich, Soc. Neurosci. Abstr. 12 (1986) 391.

(14) T. Allard, S. A. Clark, W. M. Jenkins, M. M. Merzenich, Soc. Neurosci. Abstr. 11 (1985) 965.

(15) M. Sur, M. M. Merzenich, J. H. Kaas, J. Comp. Neurol. 44 (1980) 295-311.

(16) M. M. Merzenich, W. M. Jenkins, J. C. Middlebrooks, Dynamic Aspects of Neocortical Function, J. Wiley and Sons, New York 1984, pp. 397-424.

(17) G. M. Edelman, The Mindful Brain. MIT Press, Cambridge 1978, p. 51.

(18) P. E. Garraghty, M. Sur, Soc. Neurosci. Abstr. 13 (1987).

(19) P. Landry, M. Deschenes, J. Comp. Neurol. 199 (1981) 345-371.

(20) J. L. Juliano, B. L. Whitsel, Brain Res. 342 (1985) 242-251.

(21) P. Zarzecki, P. S. Blum, D. A. Bakker, D. Herman, Exp. Brain Res. 48 (1983) 28-42.

(22) P. Zarzecki, D. M. Wiggin, Exp. Brain Res. 48 (1982) 28-42.

(23) R. W. Dykes, P. Landry, R. Metherate, T. P. Hicks, J. Neurophysiol. 52 (1984) 1066-1093.

(24) T. P. Hicks, R. W. Dykes, Brain Res. 274 (1983) 160-164.

(25) G. H. Recanzone, T. Allard, W. M. Jenkins, M. M. Merzenich, Soc. Neurosci. Abstr. 12 (1986) 1436.

(26) G. M. Edelman, L. H. Finkel, Dynamic Aspects of Neocortical Function, J. Wiley and Sons, New York 1984, pp. 653-695.

(27) M. F. Bear, W. Singer, Nature 320 (1986) 172-176.

(28) R. L. Paul, H. Goodman, M. M. Merzenich, Brain Res. 39 (1972) 1-19.

(29) N. M. Weinberger, D. M. Diamond, Auditory Function, J. Wiley and Sons, New York 1987.

(30) D. Diamond, Physiological Plasticity of Single Neurons in Auditory Cortex of the Cat During Learning, Ph. D. Thesis, University of California at Irvine 1985.

(31) C. D. Woody, J. Engel, Jr., J. Neurophysiol. 35 (1972) 230-241.

(32) T. G. Brown, C. S. Sherrington, Proc. Roy. Soc. B 85 (1912) 250-277.

(33) A. S. F. Layton, C. S. Sherrington, Quart. J. Exp. Physiol. 11 (1917) 135-222.

(34) R. J. Nudo, M. M. Merzenich, Soc. Neurosci. Abstr. 13 (1987).

(35) M. M. Merzenich, R. J. Nelson, J. H. Kaas, M. P. Stryker, W. M. Jenkins, J. M. Zook, M. S. Cynader, A. Schoppmann, J. Comp. Neurol. 258 (1987) 281-297.

(36) M. M. Merzenich, Comparative Neurobiology: Modes of Communication in the Nervous System, J. Wiley and Sons, New York 1985, pp. 138-157.

Self-Organizing Processes in Adult Neo-Cortex

Discussion
edited by *Klaus-Peter Hoffmann*

Question: What is the meaning of the increase or shift in topographical representation? Is there a gain in acuity or better performance in extraction of functional parameters?

Answer: A change in topographic representation also means a change in the output relation. It is unclear whether there is also a change in the temporal domain.

Question: Are shifting connections ruled out by this process? What about long term processes?

Answer: It is only clear that changes can be generated whithin a few minutes. No one has tested and adaequately looked for the idea of shifting connections that could account for long term changes but it is still a possibility.

Question: Is there only the one governing principle synchrony of input setting up the map. What about the role of gamma-amino-butyric acid and local inhibition?

Answer: The problem is how tightly the excitatory and inhibitory processes are linked. There is clearly an inhibitory map.

Comment: The type of plasticity shown here for the sensory map also holds for the motor side. Sherrington already had carried out experiments to show this. But where do we have to see limitations on the extent of adaptation? Is it within the radius of projection of the afferents?

Answer: A child up to the age of 8 with a transection of a nerve in the hand can have complete recovery. If the skin field denervated is small and the lesion confined to one finger there will be excellent recovery. Even in the adult it is difficult to distinguish after some time a lesioned finger from the corresponding one in the other hand. If the skin lesion is large it is unavoidable that there is temporal overlap of input from different regions of this skin patch after recovery. This is consistent with false localisation within this area of skin even after very long time after the lesion. A patient with a skin transplant from the fifth digit to the thumb located stimulation of the thumb as on the fifth digit, which is the region from where the skin came. Within 6 weeks, however, the skin is perceptually intergrated into the thumb.

Question: Do we have to give up the notion of the fixed maps?

Answer: There is definitely an anatomically defined map.

Question: What did you mean by saying the cortex has an "input time structure representation" instead of an "anatomical representation"?

Answer: The anatomical map is relatively crude but the inputs driving the neurons are clearly selected from a large repertoire and are clearly related to each other topographically on the basis of their time relation. We can drive the map into thousands of different forms by manipulating temporal relationships. The anatomical projection establishes the overall border. There are no special command neurons allowing the input to make connections but it happens by coincidence. This connection formation is almost certainly modulated by the behavioural state of the animal.

Question: In similar nerve cut experiments on the motor side you never regain the detailed order of recruitment. Is the output end exceedingly rigid in contrast to the plastic somatosensory input side?

Answer: There are a number of examples in which there is substantial recovery of motor control involving new populations of motor neurons.

Question: Do you expect changes in subcortical structures or different changes in different cortical layers?

Answer: The shift in restricted receptive fields is consistent across all cortical layers. The refined maps in the cortex must be a product of self organization of the cortex. But it seems likely that there is a cascade of changes that also involves subcortical nuclei.

Question: Is the visual system more complex and complicated so that it needs more hardware?

Answer: If the appropriate experiments were done the same sort of plasticity on a limited scale might be seen. There is also the possibility that some cortical fields are more rigid than others. Area 17 might be relatively rigid but you could perhaps find substantial plasticity if you looked for changes within anatomical boundaries like the ocular dominance bands.

Comment: Cutting out the influence of eye muscle proprioceptors will create changes in binocularity in area 17 neurons even in the adult cat.

Question: Could receptive fields already have extended under normal conditions to the region where you found them later to have shifted? Was plotting of receptive field borders sensitive enough?

Answer: The observed shifts from the representation of one finger to the next would require shifts in the normal map of 600μm.

Comment: The scatter of receptive fields found in visual cortex would correspond to a shift of 1mm in a perfect map.

Question: Do the receptive fields of neurons change instantaneously in relation to the task the finger is involved in?

Answer: We believe that this is very likely but we have not looked at the system in operation.

Final comment: The shifting and reorganization of the cortical receptive fields is clearly of great importance. Rarely does one encounter phenomena in the study of the cortex which are so strikingly paradoxical as are the movements of the discontinuous boundaries of the receptive fields separating the digits. An important question is how fast can these reorganizations take place? Further, how long do the changes last in different situations? In addition to the large spatial changes in the receptive fields of these neurons, it should be of great interest to look for more subtle temporal changes in their post stimulus time histograms. The full understanding of the boundary shifts and the possible temporal changes should be subject to a major theorectical effort.

Review:

Aspects of Human Memory and Amnesia Relevant to Neurobiological and Computational Models of Memory

by *Arthur P. Shimamura* and *Larry R. Squire*

SUMMARY

A complete explanation of neural plasticity must consider and ultimately account for the diverse and complex memory capabilities observed in human behavior. In the past three decades, a large body of information about human memory has emerged from psychological studies, and numerous theories have been proposed to explain the behavioral data. Whereas most approaches to neural plasticity focus on individual neurons and the synapses between them, much of the work with human memory proceeds at a more global level of analysis--generally focusing on whole behavior. Perhaps as a result, neuroscientists and psychologists often have difficulty communicating pertinent information to each other. One approach that attempts to bridge these two disciplines is rooted in the tradition of neuropsychology, which concerns itself with brain substrates of behavior and cognition at the level of neural systems and their connections. One fruitful strategy for neuropsychological work has been to study how behavior is disrupted by brain injury. We focus here on findings from the neuropsychological approach to human memory that seem particularly relevant to biological models. We outline some critical facts and data, suggest a descriptive framework for understanding the organization of memory, and discuss how a parallel distributed approach to model building might help connect the disciplines of neuroscience and psychology.

1 NEUROPSYCHOLOGICAL STUDIES OF HUMAN MEMORY

Ever since the clinical observations of Korsakoff (1) and Ribot (2)--both published in the 1880s--it has been known that a relatively circumscribed memory deficit can occur as a result of brain injury or disease. That is, memory impairment can occur with little or no impairment of other intellectual abilities or social skills. The hallmark feature of this "amnesic syndrome" is anterograde amnesia--a loss of the ability to learn and remember information presented after the onset of memory impairment. For example, amnesic patients have difficulty either recalling or recognizing a list of words, even when testing is conducted only minutes after list presentation. The impairment affects information arriving from all sense modalities, and it affects memory for both verbal and nonverbal material (for reviews see 3-8). In addition to anterograde amnesia, amnesic patients also frequently exhibit retrograde amnesia, the loss of memory for events that occurred prior to the onset of amnesia (9,10).

Our understanding of the anatomical basis of amnesia has been improved enormously by investigations of the noted amnesic patient H.M., who in 1953 sustained a bilateral medial temporal lobe resection in an attempt to relieve severe epilepsy (11-13). The removal included bilateral excision of amygdala, uncus, hippocampal gyrus, and the anterior two thirds of the hippocampus. Following surgery H.M.'s seizure activity was reduced, but he exhibited a severe and enduring anterograde amnesia. Presently, he has very little memory for personal and public events that have occurred since his surgery in 1953. The hippocampus and surrounding cortex were considered to be critical for producing this amnesia, because amnesia was not observed in other patients whose surgical resection included the amygdala and uncus but did not extend posteriorly enough to include the hippocampus and hippocampal gyrus.

A recent anatomical study involving a different patient has shown that damage to the hippocampus itself can cause memory impairment (14). This patient became amnesic following an ischemic episode. Until his death five years later, he exhibited anterograde amnesia, little if any retrograde amnesia, and no signs of cognitive impairment other than amnesia. Thorough histological examination of the brain revealed a circumscribed bilateral lesion involving the entire CA1 field of the hippocampus, but no other pathology that could reasonably explain the memory impairment.

The diencephalic midline of the brain has also been implicated in amnesia. The best-studied example of diencephalic amnesia is Korsakoff's syndrome (see 5,6,15 for review), which can develop after years of chronic alcohol abuse and is characterized especially by bilateral lesions along the walls of the third and fourth ventricles. Neuropathological studies have linked the memory deficits in patients with Korsakoff's syndrome to damage in the mammillary bodies, the dorsomedial nucleus of the thalamus, and possibly other medial thalamic structures (16,17). Recently established animal models of amnesia in the monkey should be able to determine which structures must be damaged, either alone or in combination, to produce amnesia (18-20).

The memory impairment observed in patients with medial temporal damage and in patients with diencephalic damage are similar. One should expect to find some differences, because it is unlikely that two different regions of the brain have identical functions. Nevertheless, the similarities between the effects of medial temporal and diencephalic damage suggest that the two brain areas are part of one neural system, or perhaps comprise two (or more) closely interconnected systems. For the purposes of the present discussion, these two brain areas will be referred to as belonging to one functional system.

Because intellectual and language functions are spared in amnesic patients, the neural system that is damaged appears to be specialized for memory function. The system is not a site of memory storage, because amnesic patients are able to recall remote events from their past. Indeed, amnesic patients often have quite efficient recollection of premorbid events, despite their severe anterograde amnesia (10,14,21). The deficit in amnesia appears to be, at least in part, a deficit in the ability to store new information in an enduring way. Because patients can often retrieve information acquired before the onset of amnesia, the deficit is not easily explained solely as an impairment in memory retrieval.

The data from retrograde amnesia provide additional clues about the function of the damaged neural system. For example, in the case of the amnesia associated with head trauma, the extent of retrograde amnesia can be predicted by the severity of anterograde amnesia (22). Traumatic amnesia is reported to be temporally graded, affecting recent memories more severely than remote memories. A similar pattern can be seen in the memory impairment associated with electroconvulsive therapy (ECT). Following ECT, retrograde amnesia can be temporally graded, affecting especially the 1 to 3 years prior to the onset of amnesia (23,24). As the severity of anterograde amnesia decreases during the months after treatment, the severity of retrograde amnesia also decreases (25). Theories of memory must account for the finding that anterograde amnesia often occurs together with retrograde amnesia for recent events. Events that occurred long before the onset of amnesia are relatively resistant to memory impairment.

Some amnesic patients (e.g., patients with Korsakoff's syndrome) can exhibit extensive retrograde amnesia, which affects memory for events that occurred throughout adult life (9,10). Such extensive and severe remote memory impairment may be indicative of more widespread brain damage than is needed to cause anterograde amnesia and may not be directly related to the severity of anterograde amnesia (26). Theories of memory must also consider these facts about retrograde amnesia.

The new learning deficit in amnesia is pervasive, exhibiting itself whenever amnesic patients are asked to recall or recognize information learned since the onset of amnesia. Patients express neither familiarity toward nor conscious awareness of recent experiences. There are, however, aspects of memory that are preserved, even in severely amnesic patients. First, amnesic patients exhibit intact performance on tests of immediate or short-term memory, such as the digit span test (27,28). In this test, subjects must repeat numbers immediately after presentation. Amnesic patients like normal subjects, can report about seven digits without error.

There are two additional domains of preserved memory function that have implications for neurobiological models of memory. Amnesic patients can learn and retain certain skills, and they exhibit intact priming effects. Early work had shown that even severely amnesic patients can learn perceptual-motor skills (29-31), such as the ability to improve performance on a rotary pursuit tracking task. More recently, it has been shown that certain cognitive skills can also be learned in a normal fashion. In one study (32), amnesic patients learned the skill of reading mirror-reversed words and then exhibited normal retention of the skill after a three-month retention interval. Yet the same patients, unlike control subjects, could not recognize the words that they had read during the training session, and often they could not recall the training experience at all. Preserved skill learning performance has also been observed on tests in which amnesic patients are asked to solve jigsaw puzzles (29), learn a mathematical rule (33), or learn the solution to a puzzle (34). These findings suggest that skill learning can occur despite lack of familiarity for the training session. One important feature of all of these skill tasks is that memory is tested in an implicit manner. That is, performance is based on proficiency or speed and does not depend on any explicit reference to previously presented information or on explicit recall of any stored information.

A second domain of preserved memory function is priming. Priming occurs when performance is facilitated or biased by recently encountered information (for review see 35). One example of priming can be illustrated with the word completion task (36,37). Subjects are shown a list of words (e.g., MOTEL,

ABSTAIN) and then are asked to complete three-letter word beginnings (e.g. MOT, ABS) with the first word that comes to mind. Amnesic patients exhibit a normal tendency to use previously presented words. This facilitation (or priming effect) disappears in both amnesic patients and control subjects after a two-hour retention interval (36,38). For amnesic patients, the words appear to "pop" into mind, without conscious awareness that the words were recently presented. Similar preserved priming effects can be found when amnesic patients are shown words and then asked to identify the same words when they are briefly presented at near-threshold exposure times (39). Also, amnesic patients exhibit priming when they are first shown words (e.g., BABY) and later asked to "free associate" to semantically related words (e.g. CHILD) (40).

The finding of preserved priming effects in amnesia suggests that preexisting representations can be transiently activated and that activation occurs independently of the brain structures damaged in amnesia. Patients with mild to moderate Alzheimer's disease exhibit gross deficits in word completion priming (41). This finding is pertinent, because the neuropathology of Alzheimer's disease includes widespread neocortical damage (42,43). Priming may depend on the integrity of cortical ensembles that contain representations of lexical information.

Some basic facts about amnesia are summarized below:

A. Amnesia is a relatively circumscribed deficit that can occur in the absence of noticeable deficits in intelligence, language, or personality.

B. Damage to the medial temporal region--including the hippocampus--or damage to the diencephalic midline can cause a permanent amnesic disorder.

C. The medial temporal and diencephalic regions that are damaged in amnesia are not themselves the permanent sites of memory reprsentations, because damage to these areas does not result in a loss of all past memories.

D. A temporally-limited retrograde amnesia--covering one or two years just prior to the onset of amnesia--is observed in some cases of amnesia. The severity of retrograde amnesia is often correlated with the severity of anterograde amnesia (e.g. in patients with head injury).

E. An extensive remote memory impairment covering the majority of adult life is observed in some forms of amnesia.

F. The following memory functions are spared in amnesia:
   Short-term or immediate memory
   Perceptual-motor and cognitive skills
   Priming

2 A FRAMEWORK FOR UNDERSTANDING THE ORGANIZATION OF HUMAN MEMORY

This section outlines a theoretical framework that attempts to accommodate the findings from neuropsychological studies of memory. The first principle in this framework is that the neural system damaged in amnesia facilitates or enables the storage of memories but does not affect other information processing systems. The impairment can be strikingly circumscribed, affecting memory disproportionately more than other higher cortical functions. The selectiveness of memory impairment in amnesia is only one example among many where a particular aspect of cognitive function is found to be impaired selectively as a result of discrete brain injury (see 44-46). This fact suggests that brain organization is modular and can be characterized as a collection of separable information processing systems (47-50).

Amnesia is viewed as a deficit primarily in memory storage and in the consolidation of stored memory over time, because the global and permanent amnesia that occurs as a result of medial temporal or diencephalic damage disproportionately affects new learning capacity. Memory for events that occurred before the onset of amnesia can also be affected, and this retrograde effect appears to be temporally graded. Very remote memories are usually not affected. The finding that damage to medial temporal or diencephalic regions causes amnesia for information derived from all the sense modalities suggests that these regions affect supramodal information processing. Indeed, neuroanatomical studies show that the hippocampus receives projections from supramodal neocortical areas via the entorhinal cortex (see 51). In addition, both medial temporal and diencephalic midline regions (particularly midline thalamic nuclei) send projections to neocortical association areas (51). Thus, the memory system affected in amnesia may interact with cortical areas that actually store information. By this view, the medial temporal and diencephalic regions belong to a functional system that enables, strengthens, or consolidates memory storage. The memory storage sites themselves are presumably distributed widely, especially in various cortical association areas.

The view that medial temporal and diencephalic brain areas are involved in establishing durable memory representations is consistent with the finding that <u>remote</u> memories are generally preserved in amnesia. However, this view must be expanded to account for the fact that amnesia does affect memory for recent events that occurred just prior to the onset of amnesia. That is, retrograde amnesia is often temporally-graded or temporally-limited. One way to account for temporally-limited retrograde amnesia is to suppose that the affected neural system is involved not only in establishing memory storage but in maintaining it and/or retrieving from it for a period of time after learning. By this view, memory storage does not involve a rapid or discrete process that can establish new long-lasting memories immediately after the time of learning. Instead, long-lasting memories require redintegration over time, wherein some neural elements are lost through forgetting and others are strengthened. This process, termed consolidation, requires extended organization and reorganization with the passage of time after learning (52). This idea is similar to principles of neural competition (53,54).

It is hypothesized that the neural system damaged in amnesia is either necessary for the process of efficient strengthening or consolidation of newly formed memories over time, or necessary for retrieving from memory storage while consolidation occurs. This process occurs over a period of weeks, months, and perhaps years. This idea does not deny that some information can become established at the moment of learning, or that some information can be acquired even when it is encountered on only one occasion. However, with the passage of time, memories become better and better organized and more resistant to disruption. The time needed for consolidation of memory storage may well be a necessary consequence of the idea that human memory requires the participation of widely distributed neural ensembles. The neural system damaged in amnesia may ordinarily participate in reorganizing and restructuring these neural ensembles. This view can explain temporally-limited retrograde amnesia. Damage to the medial temporal/diencephalic neural system will affect the organization of recently established memories that have not had enough time to be fully consolidated (see 5,52). In summary, disruption of a consolidation process which ordinarily takes considerable time to establish and organize long-lasting memories can explain both anterograde amnesia and temporally-limited retrograde amnesia for events that occurred prior to the onset of amnesia.

Any description of memory organization must account for the finding of spared memory function in amnesic patients. The critical evidence is that amnesic patients can exhibit entirely intact performance on some tests (e.g., skill learning, priming). By definition, memory functions that are completely spared in amnesia cannot depend on the brain regions that cause amnesia. One important feature of spared memory performance by amnesic patients on tests such as skill learning and priming is that memory can be assessed without subjects even knowing that their memory is being tested. Subject are asked simply to read words, perform a perceptual-motor task, work at puzzles, or complete word stems. By contrast, amnesic patients exhibit impairment when they are asked explicitly to recall (or recognize) a particular piece of newly learned information (e.g., "Can you recall the words that you just read?" or "Do you remember in what country I just told you Angel Falls can be found?").

There is a growing consensus among cognitive psychologists and neuropsychologists that the memory system damaged in amnesia is responsible for conscious or explicit recollection. An often-cited distinction between this memory function and the memory functions that are spared in amnesia is the distinction between "declarative" and "procedural" memory (see 4,5,32). Declarative memory is available to conscious awareness and includes the facts, episodes, and images of everyday experiences. Amnesia impairs the establishment or consolidation of declarative memory, which depends on the integrity of the medial temporal and diencephalic structures damaged in amnesia. Procedural memory is implicit and is available only by engaging the operations in which the memory is embedded. Procedural memory is created by tuning, biasing, or combining preexisting elements; and it is expressed through performance. Procedural mmeory can be acquired and stored independently of the neural system damaged in amnesia. This conceptual framework accounts for much of the neurobehavioral data. It predicts that amnesic patients will exhibit preserved learning on tests of memory in which 1) memory can be expressed implicitly and 2) preexisting memory representations can be tuned or combined. Similar distinctions between two memory processes or systems have been suggested by others (55-59).

The declarative/procedural distinction can be explored further and tested from both a biological and psychological perspective. One important finding is that this distinction is found in neurobehavioral studies of non-human primates. Monkeys with bilateral medial temporal lesions exhibit deficits on memory tests such as delayed nonmatching to sample and delayed response but are normal on tests of skill learning (60; for a similar dissociation, see 58). Thus, a distinction between kinds of memory seems to be prominently reflected in the organization of brain systems, and the distinction has some generalizability beyond human memory. From a psychological perspective, however, the distinction remains somewhat vague. What is still needed is a formal model at the neural systems level that can account for the neurobehavioral data. The following section considers a connectionistic approach that could eventually provide such a model.

3 CONNECTIONISM: PARALLEL AND DISTRIBUTED PROCESSING MODELS

One recent advance in neuroscience has been the development of computational models of neural systems (see this volume, 61,62). These neural models have benefitted greatly from formal, computational models of behavior (see 63-67). One reason for the convergence in modeling style by neuroscientists and cognitive scientists is that both groups have adopted principles of neuronal behavior as basic building blocks for modeling. Thus, computations are based on the pattern of activation across a matrix of neuron-like units, with the level of activation in each unit dependent on the degree of excitation and inhibition received from input units. Key factors in this kind of modeling are that units are activated in parallel and that information flow and storage are distributed across a matrix of units. Models of this type have been called connectionistic or parallel-distributed processing models.

In models of neural systems, computational units are typically meant to represent actual or prototypical neurons in an identifiable neural circuit (e.g., cerebellum). In such models, the architecture for the neural circuitry is based on neuroanatomical data. In connectionistic models of behavior, however, units are used which could represent single neurons or populations of neurons, and the models vary in the extent to which they incorporate the neural circuitry of any known brain system. In behavioral models, the pattern of activation in a matrix of units is used to represent cognitive features, such as perceptual dimensions in models of pattern recognition (e.g., 64) or semantic concepts in models of memory (68). The best models would seem to be ones where the model builder tries to adopt real neural circuitry so that the model has some biological plausibility based on present knowledge about the brain. At the same time, even models that are relatively removed from biological reality can be useful, insofar as they can rigorously show what computations are needed to perform particular operations. Such information can point the way to searching for neural mechanisms that perform these functions (see 69).

Connectionistic models of behavior may ultimately provide a useful link between behavior and the neural systems that subserve behavior. In fact, an attempt has been made recently to model neuropsychological findings from amnesia (70). In this model, a hypothetical mechanism was employed in order to try to account for anterograde amnesia, temporally-limited retrograde amnesia, and the relative sparing of some forms of skill learning. The mechanism incorporated the basic characteristics of the consolidation process described above. That is, a mechanism was formalized that strengthened new memory traces gradually across time. The interesting characteristic of the model was that both anterograde amnesia and temporally-limited retrograde amnesia were exhibited by introducing a single "deficit" in the hypothetical (consolidation) mechanism. No additional deficits in a retrieval mechanism were required.

However, this particular model did not successfully mimic all the features of amnesia. For example, although learning by repetition (skill learning in the model) was possible when the consolidation mechanism was impaired, it was not normal. Second, this particular model does not clearly predict whether retrograde amnesia develops immediately following the onset of anterograde amnesia or develops only gradually. In human amnesia, retrograde amnesia appears abruptly at the moment of the disruptive event. A more general problem with this kind of model is that it is not based on known neural circuitry nor on known or even plausible biochemistry. However, the model has stimulated interactions between cognitive scientists and neuroscientists. Further developments in this and other models of amnesia could improve our understanding of memory at the neural systems level by making hypotheses more precise and by making explicit what is and is not implied. In this way, model building can generate predictions and lead to experimental tests that will be missed when ideas are stated only in descriptive, general language.

4 CONCLUSIONS

We have outlined basic facts about human memory that may help constrain neural models of memory. We have also suggested a conceptual framework that accounts for these facts. In addition, we have suggested that connectionistic modeling may provide a useful bridge for linking behavior and neural events. An interactive approach involving neurobiology, neuropsychology, and cognitive science is needed to develop quantitative, complete theories of brain mechanisms.

5 ACKNOWLEDGEMENTS

Supported by the Medical Research Service of the Veterans Administration, NIMH Grant MH24600, and the Office of Naval Research.

6 REFERENCES

(1) S. S. Korsakoff, Vestnik Psichiatrii, Vol. 4 (1887).

(2) T. A. Ribot, Les Maladies de la Memoire, Germer Baillere, Paris 1881.

(3) N. Butters, L. S. Cermak, Alcoholic Korsakoff's Syndrome: An Information Processing Approach, Academic Press, New York 1980.

(4) L. R. Squire, Science, 232 (1986) 1612-1619.

(5) L. R. Squire, Memory and Brain, Oxford University Press, New York 1987 in press.

(6) G. A. Talland, Deranged Memory, Academic Press, New York 1965.

(7) E. K. Warrington, L. Weiskrantz, Neuropsychologia, 20 (1982) 233-248.

(8) C. W. M. Whitty, O. L. Zangwill, Amnesia, Butterworths Press, London 1977.

(9) M. S. Albert, N. Butters, J. Levin, Arch. Neurol., 36 (1979) 211-216.

(10) N. J. Cohen, L. R. Squire, Neuropsychologia, 19 (1981) 337-356.

(11) W. B. Scoville, B. Milner, J. Neurol. Neurosurg. Psychiat., 20 (1957) 11-21.

(12) B. Milner, S. Corkin, H-L. Teuber, Neuropsychologia, 6 (1968) 215-234.

(13) S. Corkin, Sem. Neurol., 4 (1984) 249-259.

(14) S. Zola-Morgan, L. R. Squire, D. G. Amaral, J. Neurosci., 6 (1986) 2950-2967.

(15) N. Butters, Sem. Neurol., 4 (1984) 226-244.

(16) W. G. P. Mair, E. K. Warrington, L. Weiskrantz, Brain, 102 (1979) 749-783.

(17) M. Victor, R. D. Adams, G. H. Collins, The Wernicke-Korsakoff Syndrome, Davis Press, Philadelphia 1971.

(18) M. Mishkin, Phil. Trans. Roy. Soc., 298 (1982) 85-95.

(19) L. R. Squire, S. Zola-Morgan, in J. A. Deutsch (ed.), The Physiological Basis of Memory (2nd ed.), Academic Press, New York 1983, pp. 199-268.

(20) H. Mahut, M. Moss, in L. Squire, N. Butters (eds.), The Neuropsychology of Memory, Guilford Press, New York 1984, pp. 297-315.

(21) W. D. Marslen-Wilson, H-L. Teuber, Neuropsychologia, 13 (1975) 353-364.

(22) W. R. Russell, P. W. Nathan, Brain, 69 (1946) 280-300.

(23) L. R. Squire, P. C. Slater, P. M. Chace, Science, 187 (1975) 77-79.

(24) L. R. Squire in W. B. Essman, R. Abrams (eds.), Electroconvulsive Therapies: Biological Foundations and Clinical Applications, Spectrum, Jamaica, NY 1982, pp. 169-186.

(25) L. R. Squire, P. C. Slater, P. L. Miller, Arch. Gen. Psychiat., 38 (1981) 89-95.

(26) A. P. Shimamura, L. R. Squire, Behav. Neurosci., 100 (1986) 165-100.

(27) A. D. Baddeley, E. K. Warrington, J. Verb. Learn. Verb. Behav., 9 (1970) 176-189.

(28) D. A. Drachman, J. Arbit, Arch. Neurol., 15 (1966) 52-61.

(29) D. N. Brooks, A. D. Baddeley, Neuropsychologia, 14 (1976) 111-122.

(30) S. Corkin, S. Neuropsychologia, 6 (1968) 225-265.

(31) L. S. Cermak, R. Lewis, N. Butters, H. Goodglass, Percept. Mot. Skills, 37 (1973) 259-263.

(32) N. J. Cohen, L. R. Squire, Science, 210 (1980) 207-209.

(33) F. Wood, V. Ebert, M. Kinsbourne, in L. S. Cermak (ed.), Human Memory and Amnesia, Erlbaum Press, Hillsdale, NJ, pp. 167-193.

(34) N. J. Cohen, N. J., in L. Squire, N. Butters (eds.), The Neuropsychology of Memory, Guilford Press, New York 1984, pp. 83-103.

(35) A. P. Shimamura, Quart. J. Exper. Psychol., Sec. A, 38A (1986) 619-644.

(36) P. Graf, L. R. Squire, G. Mandler, J. Exp. Psychol.: Learn., Mem., Cog., 10 (1984) 164-178.

(37) E. K. Warrington, L. Weiskrantz, Nature, 228 (1970) 628-630.

(38) L. R. Squire, A. P. Shimamura, P. Graf, Neuropsychologia, 1987 in press.

(39) L. S. Cermak, N. Talbot, K. Chandler, L. R. Wolbarst, Neuropsychologia, 23 (1985) 615-622.

(40) A. P. Shimamura, L. R. Squire, J. Exp. Psychol.: Gen., 113 (1984) 556-570.

(41) A. P. Shimamura, D. Salmon, L. R. Squire, N. Butters, Behav. Neurosci., 1987 in press.

(42) B. T. Hyman, G. W. Van Hoesen, A. R. Damasio, C. L. Barnes, Science, 225 (1984) 1168-1169.

(43) R. C. A. Pearson, M. M. Esiri, R. W. Hiorns, G. K. Wilcox, T. P. S. Powell, T. P. S., Proc. Nat. Acad. Sci., 82 (1985) 4531-4534.

(44) D. E. Broadbent, L. Weiskrantz, L. (eds.), The Neuropsychology of Cognitive Function, The Royal Society, London 1982.

(45) K. M. Heilman, E. Valenstein, (eds.), Clinical Neuropsychology, Oxford University Press, New York 1979.

(46) A. R. Luria, Higher Cortical Functions in Man, Basic Books, New York 1980.

(47) H. Gardner, Frames of Mind, Basic Books, New York 1983.

(48) A. R. Luria, The Working Brain, Basic Books, New York 1973.

(49) M. Mishkin, in D. E. Broadbent, L. Weiskrantz, (eds.), The Neuropsychology of Cognitive Function, The Royal Society, London 1982, pp. 85-95.

(50) T. Shallice, in L-G. Nillson (ed.), Perspectives in Memory Research, Erlbaum, Hillsdale 1979, pp. 257-277.

(51) D. J. Amaral, in J. M. Brookhart, V. B. Mountcastle (eds.), Handbook of Physiology: Higher Functions of the Nervous System, American Psysiological Society, Bethesda 1987 in press.

(52) L. R. Squire, N. Cohen, L. Nadel, in H. Weingartner, E. Parker (eds.), Memory Consolidation, Earlbaum, Hillsdale 1984, pp. 185-210.

(53) D. Purves, J. W. Lictman, Science, 210 (1980) 153-157.

(54) M. M. Merzenich, J. H. Kaas, Prog. Psychobiol. Physiol. Psychol., 9 (1980) 1-42.

(55) A. D. Baddeley, A. D., Psychol. Rev., 89 (1982) 708-729.

(56) J. O'Keefe, L. Nadel, The hippocampus as a Cognitive Map, Oxford University Press, London 1978.

(57) G. J. Thomas, in L.R. Squire, N. Butters (eds.), Neuropsychology of Memory, Guilford Press, New York 1984, pp. 374-384.

(58) M. Mishkin, B. Malamut, J. Bachevalier, in J. L. McGaugh, G. Lynch, N. Weinberger (eds.), The Neurobiology of Learning and Memory, Guilford Press, New York 1984, pp. 65-77.

(59) G. Mandler, G., in L-G. Nilsson (ed.), Perspectives in Memory Research, Erlbaum, Hillsdale 1979, pp. 293-327.

(60) S. Zola Morgan, L. R. Squire, in N. Weinberger, J. McGaugh, G. Lynch (eds.), Memory Systems of the Brain: Animal and Human Cognitive Processes, Guilford Press, New York 1985, pp. 463-477.

(61) J. J. Hopfield, D. W. Tank, Science, 233 (1986) 625-633.

(62) A. Pellionisz, R. Llinas, Neurosci., 4 (1979) 323-348.

(63) D. Marr, Vision, Freeman, San Francisco 1982.

(64) G. M. Edelman, G. N. Reeke, Jr., Proc. Natl. Acad. Sci. USA, 79 (1982) 2091-2095.

(65) G. E. Hinton, J. A. Anderson (eds.), Parallel Models of Associative Memory, Erlbaum, Hillsdale 1981.

(66) J. L. McClelland, D. E. Rumelhart, Parallel Distributed Processing, MIT Press, Cambridge 1986.

(67) D. H. Ballard, Behav. Brain Sci., 9 (1986) 67-120.

(68) J. L. McClelland, D. E. Rumelhart, J. Exp. Psychol.: Gen., 114 (1985) 159-188.

(69) T J. Sejnowski, in J. L. McClelland, D.E. Rumelhart (eds.), Parallel Distributed Processing, MIT Press, Cambridge 1986, pp. 372-389.

(70) J. L. McClelland, D. E. Rumelhart, in J. L. McClelland, D. E. Rumelhart (eds.), Parallel Distributed Processing, MIT Press, Cambridge 1986, pp. 503-527.

Categorization in a Neural Network

James A. Anderson

SUMMARY

Parallel systems composed of many interconnected elements are both simple brain models and possible novel computer architectures. Potential advantages of such systems are massive parallelism with resulting speedup of computation as well as general ability to compute with noisy, corrupted, or missing data. Parallel, distributed, associative models have pronounced psychologies. Some ways of handling information are natural for them, and some things that we might want them to do are unnatural and quite difficult to do easily. Since these models virtually all had their beginnings as 'brain' models, a question of considerable interest is whether their capabilities are features of human psychology. We will discuss here the ability of a simple model of this type to form categories and concepts.

1. Human Concepts. Concepts are of great importance to many intelligent behaviors: memory, pattern recognition, problem solving and understanding among them. They have become a central issue of cognitive psychology. Psychologists investigate how concepts are learned, how they are represented in the mind, how they are used to categorize the world, and how they are used to answer questions and solve problems. Although the human conceptual system is by no means the only way to represent knowledge, any system that captures its properties will be well on the way to intelligent behavior. The remainder of this chapter will be devoted to illustrating how a parallel memory model can capture a few of the main properties of the human conceptual system.

At a general level, it is easy to explain what a concept is: a concept is a mental representation or rule that picks out a class of objects or events. However, this description may lead to an oversimplifed notion of what concepts are like. Specifically, one might have the idea that a concept is like a definition, that it gives the criteria by which all and only the members of the concepts are identified. However, one important fact about human concepts is that they are fuzzy. They do not provide all-or-none criteria for defining classes of objects: in many cases, there probably are no such criteria, and when the criteria do exist, people often do not know what they are.

The fuzziness of concepts has two specific components. First, it includes the phenomenon of unclear examples--objects which do not fall clearly into or out of a category. For example, a tomato is not clearly either a fruit or a vegetable. But if concepts were like definitions, every object would be clearly in or out of a concept. Unclear cases do not demonstrate a fault of our concepts but rather unavoidable variation and uncertainty in the world.

The second component of fuzziness is graded category membership, often called typicality. Even among objects that can definitely be categorized, some objects seem to be 'better' category members than others: a robin is a better bird than a chicken is. Judgments of how typical objects are of some concept are highly predictable, and typical objects can be classified more quickly and with greater reliability than atypical objects (Rips, Shoben & Smith, 1973).

One way to explain concept fuzziness is to propose that people learn a prototype -- the most typical example -- in order to learn the concept. The usual prototype of bird is a small, flying animal with wings that eats worms and builds a nest.

The first part of this paper reviews material published elsewhere. See particularly Knapp and Anderson (1984), Anderson and Murphy (1986) and Anderson, Golden and Murphy (1986). Sections 4 to 7 are new.

Fuzziness results from the continuous gradation of similarity of objects to the prototype. For example, most robins are similar to this prototype but most penguins are dissimilar to it, suggesting why subjects are faster at identifying robins as birds. Rosch (1978) and Smith and Medin (1981) provide good reviews of such theories.

Concepts seem to be organized in a nested fashion: robin-bird-animal-living thing, for example. This structure allows information to be represented about increasingly general categories (Collins & Quillian, 1969). The facts known about birds apply to robins, larks, crows, etc., and the facts known about animals apply to birds, mammals, reptiles, etc. These hierarchies allow us to draw inferences about objects. For example, if Tweety is a bird, we can deduce that Tweety is an animal. Such inferences are important, because we can now infer properties true of all animals to be true of Tweety as well. This system is economical in that information about all animals is not repeated with each animal, but is stored as true about the animal concept in general. The details of representation of such hierarchies is still a matter of argument in psychology, but it is clear that hierarchies are important means of organizing concepts and storing information.

1.1. Stimulus Coding and Representation. Our fundamental modelling assumption is that information is carried by the pattern or set of activities of potentially very many neurons in a group of neurons. This set of activities carries the meaning of whatever the nervous system is doing. We represent these sets of activities as state vectors. Percepts, or mental activity of any kind, are similar if their state vectors are similar. Our basic approach is to consider the state vectors as the primitive entities and see how state vectors can lawfully interact, grow and decay. The component values in the state vectors might correspond to the activities of moderately selective neurons.

For reviews of such models and their applications to engineering and psychology, see a book edited by Hinton and Anderson (1981), and two books by Kohonen (1977, 1984). A recent two volume collection edited by Rumelhart and McClelland (Volume 1, 1986) and McClelland and Rumelhart (Volume 2, 1986) is an essential reference for psychological and cognitive applications.

1.2. The Linear Associator. We use an 'outer product' associator, also called the 'linear associator', as a starting point. This simple distributed memory system uses Hebb synapses to make a simple associative system. (Kohonen, 1984; Anderson, 1970). The two primary quantitative assumptions are: First, the neuron acts as a linear summer of its inputs. That is, the ith neuron in the second set of neurons will display activity $g(i)$ when a pattern f is presented to the first set of neurons according to the rule,

$$g(i) = \sum_j A(i,j)\, f(j).$$

where A(i,j) are the connections between the <u>ith</u> neuron in the second set of neurons and the <u>jth</u> neuron in the first set. We can then write g as the simple matrix multiplication, g = A f. Second, we assume that matrix elements are modifiable according to a generalized Hebb rule, that is, the change in an element of A, $\delta A(i,j)$, is given by

$$\delta A(i,j) \propto f(j)\, g(i).$$

We can then write the matrix A as a sum of outer products,

$$A = \eta \sum_{i=1}^{n} g_i f_i^T$$

where η is a learning constant.

<u>1.3.</u> <u>Linear Concept Formation.</u> Let us make the reasonable but powerful coding assumption that the activity patterns representing similar stimuli are themselves similar, that is, their state vectors are correlated. This means the inner product between two similar patterns is large. Now assume the matrix has satisfactorily formed the association $f \rightarrow g$. With an input pattern f′ then

$$\text{(output pattern)} = g\, [f, f']$$

If f and f′ are not similar, their inner product [f, f′] is small. If f is similar to f′ then the inner product will be large. The model responds to input patterns based on similarity to f.

Consider a situation in which a category contains many similar items. Here, a set of similar activity patterns (representing the category members) becomes associated with the same response, for example, the category name. It is convenient to discuss such a set of vectors with respect to their mean. Let us assume the mean is taken over all potential members of the category. Consider a set of correlated vectors, {f}, with mean p. Each individual vector in the set can be written as the sum of the mean vector and an additional noise vector, d, representing the deviation from the mean, that is,

$$f_i = p + d_i .$$

If there are n different patterns learned and all are associated with the same response, the final connectivity matrix will be

$$A = \sum_{i=1}^{n} g f_i^T$$

$$= n g p^T + \sum_{i=1}^{n} d_i$$

Suppose that the term containing the sum of the noise vector is relatively small, as could happen if the system learned many randomly chosen members of the category (so the d's cancel on the average and their sum is small) or if d is not large. In that case, the connectivity matrix is approximated by

$$A = n g p^T .$$

The system behaves as if it had repeatedly learned only one pattern, p, essentially n times the mean of the set of vectors it was exposed to. In this respect the distributed memory model behaves like a psychological 'prototype' model, because the most powerful response will be to the pattern p, which it may never have seen.

However if the sum of the d's is not relatively small, as might happen if the system only sees a few patterns from the set or if d is large, the response of the model will depend on the similarities between the novel input and each of the learned patterns, that is, the system behaves like a psychological 'exemplar' model.

1.4. Dot Prototypes. Knapp and Anderson (1984) applied this model to the formation of simple 'concepts' composed of nine randomly placed dots. These stimuli were used in a classic experiment on concept formation by Posner and Keele (1968). Many variants of the initial experiments have been performed, with many different kinds of stimuli, but the general pattern of results is consistent. For a complete literature review, see Knapp and Anderson (1984) and McClelland and Rumelhart (1985).

A prototype experiment can be very simple. Described in terms of the Posner and Keele stimulus set, the experimenter starts with the generation of several (usually three or four) patterns of dots, randomly placed on a computer screen. These initial patterns are called 'prototypes'. Then distortions of the prototypes are generated by the experimenter, for example moving the individual dots in a random direction a given distance. These distortions are called 'exemplars.' If the distance moved is small, the exemplars are visually similar to the prototypes; if the distance moved is large, an exemplar may appear to be a completely new pattern.

The experiment is divided into two parts, a learning phase and a testing phase. During the learning phase, subjects are presented with examplars. They are taught with feedback after a response is made to group all the exemplars derived from a particular prototype with the same response. Subjects are required to reach a criterion number of correct responses or are given a fixed number of learning trials. During the testing phase, subjects are presented with stimuli and told to decide which category a stimulus belongs in. There are three kinds of test stimuli: first, the old exemplars, second, new examplars generated by the same rules as the old ones, and, third, the prototype itself.

Experimental results vary, depending on details of the stimuli and task manipulations. However, in many cases and for many kinds of stimuli, experimental results show that the fastest response, the fewest errors in classification, and the strongest subjective certainty of correctness are found when the prototype is presented, which, of course, was never actually seen during the learning phase. Sometimes there is an advantage for the old examplars and sometimes old and new exemplars are apparently indistinguishable, depending on details of the experiment.

The model presented in Knapp and Anderson for these results is a straightforward realization of the linear model presented above. The coding of the state vectors was the key assumption for the model. It was assumed that there was a sensory surface with a topographic represention of visual space on it, inspired by the many such topographic maps in visual cortex. A dot in the world gives rise to a 'bump' of activity with exponential falloff on the surface. Nine dots (the traditional experimental number) gave rise to a pattern on the surface of nine bumps, arranged as the stimulus was in the world.

Since all the exemplars derived from a given prototype are associated together, outer products of stimulus and response are stored in the memory matrix, A, during the learning phase. Since we know the codings of the stimuli, we therefore know part of the outer product. The responses are somewhat arbitrary, since we do not know how the responses are represented, but the selectivity of the system (what we are observing experimentally) is primarily determined by the inputs, as long as the output patterns are equivalent in terms of the model (i.e., equal length, equal separation between the responses, etc.)

Since this is a linear model, the output will generally be a superposition of the learned responses, but the biggest response will usually be that associated with the appropriate prototype. The amplitude of the associated response will be a measure of response time, certainty, etc. In general this simple model gave a good account of itself in fitting a set of experimental data.

2. Error Correction and the BSB Model. A related but more powerful distributed, associative learning model takes the basic linear associator and applies two additional steps: error correction and a simple limiting non-linearity. By using an error correcting technique related to the Widrow-Hoff procedure we can force the system to give us correct associations. Suppose information is represented by associated vectors $f1 \rightarrow g1$, $f2 \rightarrow g2$ ... A vector, f, is selected at random. Then the matrix, A, is incremented according to the rule

$$\Delta A = \eta (g - Af) f^T$$

where ΔA is the change in the matrix A and where the learning coefficient, η, is chosen so as to maintain stability.

If f and g are identical, the system is called 'autoassociative'. Autoassociation reconstructs missing parts of stimuli. The averaging prototype extractor is strongly affected by error correction. Consider the autoassociative system. Suppose we are learning a number of examples as before. If the d terms are small, then $A p \propto n p$, that is, an eigenvector of A will be close to the prototype with eigenvalue proportional to n. There is a strong dependency of the eigenvalue (a measure of the strength of response) on the number of presentations. This is sensible for small n. However, if we consider the enormous relative frequencies of occurence of words, say, this result becomes unrealistic for large n. The more frequent items would completely dominate the system.

If error correction is perfect,

$$Af_i = f_i$$

In this case all the eigenvectors correspond to members of the stimulus set, and have eigenvalue 1.

2.1. A Prototype Region. Since all the eigenvalues associated with the stimlus set are 1, the set of eigenvectors in the stimulus set are *degenerate*. Any sum of exemplars (which includes the average, the prototype in the previous system) is an eigenvector, forming a subspace spanned by the exemplars.

Internal to the subspace, the response to exemplars is not 'peaky' but 'flat' with no effect of frequency of presentation. The 'prototype' in a psychological sense perhaps can be interpreted as a region, not as a single vector. The actual average, the prototype in the simpler system, will be toward the center of this subspace.

Figure 1.1

Magnitude (Length) of Response to a Test Vector

```
        /\                      /-------------\
       /  \                    /               \
      /    \                  /                 \
     /      \                /                   \

     Prototype              With Error Correction
```

Some concept structures will be hard to learn. If a and b are regions related to particular concepts, Figure 1.2 shows a difficult concept structure.

Figure 1.2

Recalcitrant Subspace Configuration

```
              ---------------
             /      ---      \
   a        (   b  ( a )  b   )    a
             \      ---      /
              ---------------
```

The geometry show in Figure 1.2 is related to the 'exclusive or' problem, which is known to be difficult to solve with distributed models of the type discussed here (see Minsky and Papert, 1969).

2.2. Retrieval Using a Non-linear Model. Assume we have an autoassociative matrix that has learned a set {f}. Starting information is represented by a vector constructed according to rules used to form the original vectors, except missing information is represented by zeros. The retrieval strategy will be to repeatedly pass the information through the matrix A and to reconstruct the missing information using the cross connections. The filled-in information will provide the required additional answers.

We will use the following nonlinear algorithm. Let x(i) be the current state vector of the system at quantized time (i). f(0) is the initial vector. Then, let x(i+1), the state vector at the next time quantum be given by

$$x(i+1) = \text{LIMIT} [\alpha A x(i) + \gamma x(i) + \delta f(0)].$$

The first term, (α A x(i)), passes the current state through the matrix and adds more information reconstructed from cross connections. The second term, γ x(i), causes the current state to decay slightly. The third term, δ f(0), can keep the initial information constantly present. Once the element values for x(i+1) are calculated, the element values are 'limited'. This process contains the state vector within a set of limits, and hence we have called this model the 'brain state in a box' or BSB model. (Anderson, Silverstein, Ritz, and Jones, 1977). After many iterations, the system becomes stable and will not change. The final state will be the output of the system. This is a discrete approximation to an underlying continuous system. The dynamics of this system are similar to those discussed in Hopfield (1984). It can be shown to be minimizing an energy function. (Golden, 1986).

2.3. <u>General Description of Simulations.</u> In the specific examples of state vector generation that we will use for the simulations to follow, letters, words and sets of words are coded as concatenations of the bytes in their ASCII representation. Zeros are replaced with minus ones. This is a highly arbitrary coding and is a 'distributed' coding because a single letter or word is determined by a pattern of many elements. In the outputs from the simulations the underline, '_', corresponds to to an uninterpretable character whose amplitude is below an interpretation threshold. All the BSB simulations described are 200 dimensional (25 characters). The autoassociative matrix is 50% connected, that is half the connections, randomly chosen, are set identically to zero. During learning, items to be associated are chosen randomly from the stimulus set.

3. <u>Simulation 1. A Simple Hierarchy.</u> The set of stimuli used has a 'hierarchical' structure in that segments of the state vector to the right determine the segments to their left. Two separate hierarchies are stored in the matrix.

Table 3.1

Stimuli in Learned Set

F[1]. VehiclesAirplaneJetPlane_	F[7]. FurniturChairs Ostuffed_
F[2]. VehiclesAirplaneX15Rockt_	F[8]. FurniturChairs Diningrm_
F[3]. VehiclesAirplanePropplne_	F[9]. FurniturChairs Swivelch_
F[5]. VehiclesNewcars Porsche _	F[10]. FurniturBgtablesWorkshop_
F[6]. VehiclesNewcars Mercedes_	F[11]. FurniturBgtablesKitchen _
F[6]. VehiclesNewcars Chevrolt_	F[12]. FurniturBgtablesClassrom_

This is an autoassociative system. The error correction procedure applied to a set of exemplars, will, in the limit, have eigenvalues with value 1. Without error correction, the eigenvalues can be large. We would expect the eigenvalue spectrum of the largest eigenvalues to contract toward 1 as error correction proceeds.

Table 3.2

Ratio of Largest to Tenth Largest Eigenvalue

Presentations	Largest e.v.	10th Largest	Ratio
12	3.830	0.331	11.57
50	1.000	0.374	2.67
100	1.000	0.555	1.80
200	1.000	0.765	1.31
400	1.000	0.928	1.08

3.1. Eigenvectors. The most frequently occuring patterns (Vehicles, Furniture) appear in the eigenvectors. More eigenvectors become meaningful as learning progesses. The system is gradually structuring its eigenvectors for most accurate recall or, one might say, developing a more useful set of vector valued 'features.' The eigenvectors are not quite the same as the stimuli.

Table 3.3

Interpreted Largest Eigenvector

Number Pres.	Interpretation	Eigenvalue
12	VuxmidesBetabpg!Cmtscdod_	3.83
50	VehiclesAirplanePropplne_	1.00
100	FurniturBfeir~!2GM‘kossm_	1.00
200	FurniturChairs Ostuffdd_	1.00
800	FurniturChairs 034Uffet_	1.00

3.2. Using the Hierarchy. Concepts are frequently arranged hierarchically, and we have structured the stimulus set to show this. Retrieval respects the hierarachy. If we put in one field, information to the left is usually reconstructed correctly, in whole or part, whereas ambiguous information to the right is slow to be reconstructed and sometimes incorrect. This can be interpreted as superordinate general information (to the left) can be retrieved, but it is often impossible to give a specific example (to the right).

If the eigenvectors representing the stimuli are degenerate, then sums of stimuli should give about the same response as the learned stimuli. If 'JetPlane' is input, the recall process reconstructs 'Vehicles' after 16 iterations and 'Airplane' after 15 iterations. The sum of 'JetPlane' and 'X15Rockt' gives 'Vehicles' after 17 iterations and 'Airplane' after 16 iterations. Suppose we want to know what 'JetPlane' and 'Chevrolt' have in common. They are neither 'Airplanes' or 'Newcars' but both are 'Vehicles'. If we put in the average of 'JetPlane' and 'Chevrolt', the system reconstructs 'Vehicle' after 20 iterations. Apropriate use of commonalities can automatically raise or lower the level of computation in this concept hierarchy. Note that this 'hierarchy' does not need to be traversed node by node but nodes can be leapfrogged.

Table 3.4
Computation with a Hierarchy

Input	Result			
JetPlane	Vehicles	(16)	Airplanes	(15)
JetPlane + X15Rockt	Vehicles	(17)	Airplanes	(16)
JetPlane + X15Rockt + Propplne	Vehicles	(16)	Airplanes	(15)
JetPlane + Chevrolt	Vehicles	(20)	--	(50)
Mercedes + X15Rockt	Vehicles	(23)	--	(50)

The number in parenthesis is the number of iterations required to completely reconstruct the characters shown. '--' means nothing was successfully reconstructed.

4. Simulation 2. Natural Categorization. We were interested in a more complex test of the concept forming abilities of the network. We would try to teach it a complex category structure with many examples. If the system was operating correctly, it would categorize new inputs correctly, and hopefully give a

psychologically 'reasonable' pattern of response. It is premature to look for exact correspondance with cognitive data but most of the qualitative statements made below are consistent with data from psycholinguistics and cognitive psychology, as well as with common sense.

For a training set, we used a set of input vectors based on animals. Animals are convenient because they are familiar and have a natural taxonomy. We used dogs, cats, *Limulus*, *Aplysia*, and leeches. Some were invertebrates and some were vertebrates. In a hierarchical concept tree, all these items were basic level. There were also several 'subordinate' level examples: Siamese cats, German Shepherd dogs, Alley cats, etc.

There were two simulations in this group. One was a straightforward attempt to verify the nested subspace notion. Figure 4.1 presents the hierarchy we wanted to construct and Table 4.1 gives the individual associations used to realize the hierarchy.

Figure 4.1
Simple Concept Hierarchy

```
              LTHNG
             /  |  \
        CANIS   |   APLYS
       / | |    |
  GSHEP / /     |
  CSPNL-- /   FELIN
  PEKNG---   /     \
        SIAME     ALLEY
```

To code this structure, we made the important assumption that there was a linguistic part of the coding and a 'sensory' part which contains a complex physically oriented code. This sensory code is multi-modal and conjectured to be very large. If we have two somewhat independent systems, it is possible to play them off against each other. Suppose they had differing time constants. Then we have we have many intriguing possibilities for forming associations between linguistic material and sensory material. A complex set of linguistic associations would be possible while the sensory code remains relatively unchanged. We make use of this technique in Simulation 3.

In the coding we used, the first five positions in the state vector were the considered to be the 'linguistic' part and are written using capital letters for convenience in interpreting the state vectors. Therefore, CANIS is the linguistic coding of 'dog', FELIN is the linguistic coding of 'cat', and so on. There is also a 'sensory' coding in the last 20 positions of the character string. Since dogs and cats are somewhat similar in appearance and habitat, we made the state vectors similar:

'Dog' = CANISMammlFamiliaris_____
'Cat' = FELINMammlDomestica _____
'Aplysia' = APLYSAplysiaCaliforn_____

There was also some miscellaneous information about other living things ('LTHNG') including turkey, birds, reindeer, and mice. The last five positions were specific information about subordinate category members. Any real creature has specific detail. Thus

'German Shepherd' = GSHEPCanisfamiliarisDaaba

but an example of a random dog is

'Random Dog' = CANISCanisfamiliarisAbbca

During learning, the system saw seven different examples of random dog, seven different examples of random cats, four random Aplysia, and several different subordinate examples of dogs and cats.

Correct operation of the system would be checked by its classification behavior:

a) Partial information should usually be categorized at the basic level This is 'quick look' categorization and should agree with human quick look categorization, which tends to be basic level.

b) Novel examples should be categorized at the basic level if they are similar to a basic level category (i.e. a new example of a 'dog').

c) Examples with very sketchy information should be categorized at the highest level, i.e. LTHNG in this simulation.

We were attempting to form a nested categorization structure in state space that looked something like Figure 4.2.

Figure 4.2

Desired Categorization Structure

```
|--------------------------------------------------------|
|                     |----------------------------|     |
|                     |        FELIN      () ALLEY |     |
|   LTHNG        |-----------------------|          |     |
|                |                       |          |     |
|  |----------|  |        () GSHEP       |  (   )   |     |
|  |          |  | CANIS  () CSPNL       |  SIAME   |     |
|  |  APLYS   |  |                       |          |     |
|  |          |  |           PEKNG  () -----------|     |
|  |----------|  |-----------------------|                |
|________________________________________________________|
```

The system learned using the state vectors given in Table 4.1, learning for 1000 learning trials using Widrow-Hoff error correction learning, BSB retrieval techniques, and a 50 percent connected matrix.

Table 4.1

Stimulus Vectors: Two Layer Concept Network

Basic Layer Plus Subordinate

F[1]. CANISMammlFamiliarisAaaaa	F[2]. CANISMammlFamiliarisBbbbb
F[3]. CANISMammlFamiliarisCcccc	F[4]. CANISMammlFamiliarisDdddd
F[5]. CANISMammlFamiliarisAbcda	F[6]. CANISMammlFamiliarisBcdab
F[7]. CANISMammlFamiliarisCdabc	F[8]. FELINMammlDomestica Ttttt
F[9]. FELINMammlDomestica Lllll	F[10]. FELINMammlDomestica Xxxxx
F[11]. FELINMammlDomestica Ooooo	F[12]. FELINMammlDomestica Xolto
F[13]. FELINMammlDomestica Lxotl	F[14]. FELINMammlDomestica Olxot
F[15]. APLYSAplysiaCaliforn12345	F[16]. APLYSAplysiaCaliforn89012
F[17]. APLYSAplysiaCaliforn56789	F[18]. APLYSAplysiaCaliforn23456
F[19]. CSPNLMammlFamiliarisCcabc	F[20]. GSHEPMammlFamiliarisDaaba
F[21]. PEKNGMammlXXmiliarisPekin	F[22]. ALLEYMammlDomestica Ollot
F[23]. SIAMEMammlDomestica Tloxx	F[24]. SIAMEMammlDomestica Tloxt
F[25]. LTHNGMammlReindeerw.horns	F[26]. LTHNGMamRattusvulgarisRtb
F[27]. LTHNGMamRattusvulgarisPlk	F[28]. LTHNGMamRodMusDomesticaGr
F[29]. LTHNGMamRodGerbilPet4kids	F[30]. LTHNGMamFlyingBatBBnhugnn
F[31]. LTHNGBirdRobinRedWormsSpr	F[32]. LTHNGBirdTurkeyStupidEdib
F[33]. LTHNGBirdHummingbirdSmall	

When the matrices were constructed, the categorization behavior was tested by putting in partial information and checking the final state. In Tables 4.2 to 4.5 the numbers in parentheses are the numbers of iterations required to reach the state given, usually the point at which the first five characters were completely reconstructed.

Table 4.2 gives the states with very sparse input information. Except for the first entry, LTHNG was the final state, and probably an appropriate classification given the ambiguity of the input. When an unambiguous determination is possible, as it is in Tables 4.3 and 4.4, the correct categorization is produced in almost all cases. (FELIN or CANIS). Note that 'reaction time' (number of iterations required to determine the first five characters) does not change much, even when more information is given. When information adequate to give a subordinate categorization is provided, an appropriate subordinate category is chosen. Note the modest amount of generalization to state vectors that have never been seen but which are very close to the learned vectors. (Table 4.5.) The learned example corresponding to GSHEP is given for reference in the Table.

Table 4.2

Response of System to Various Input Vectors

Living Things

Input		Output	
_____M___________________	⇒	GSHEPMammlFamiliarisFccbc	(144)
_____Ma__________________	⇒	LTHNGMamml_am_diera_ag_k_	(50)
_____Mam_________________	⇒	LTHNGMam_l__mcdi_barigcjc	(54)
_____Mamm________________	⇒	LTHNGMammlPmmmleer_.Hobno	(65)
_____Mamml_______________	⇒	LTHNGMammlPmmmde__s_Hmbno	(62)

Table 4.3

Response of System to Various Input Vectors

Canis Familiaris

Input		Output	
_____MammlF______________	⇒	CANISMammlFamiliaris@''ba	(54)
_____MammlFa_____________	⇒	CANISMammlFamiliaris@'_b_	(30)
_____MammlFam____________	⇒	CANISMammlFamiliaris@'ab_	(36)
_____MammlFami___________	⇒	CANISMammlFamiliarisB_abc	(28)
_____MammlFamil__________	⇒	CANISMammlFamiliaris@'ab_	(39)
_____MammlFamili_________	⇒	CANISMammlFamiliaris@'aba	(44)
_____MammlFamilia________	⇒	CANISMammlFamiliaris@''ba	(65)
_____MammlFamiliar_______	⇒	CANISMammlFamiliaris@''ba	(75)
_____MammlFamiliari______	⇒	CANISMammlFamiliarisA'aba	(58)
_____MammlFamiliaris_____	⇒	CANISMammlFamiliaris_'ab_	(44)

Table 4.4
Response of System to Various Input Vectors
Felis Domestica

Input		Output	
_____MammlD_______________	→	ALLEYMammlDomestica \lltt	(84)
_____MammlDo______________	→	FELINMammlDomestica \lll_	(72)
_____MammlDom_____________	→	FELINMammlDomestica _llll	(52)
_____MammlDome____________	→	FELINMammlDomestica \lll_	(48)
_____MammlDomes___________	→	FELINMammlDomestica \ll|_	(59)
_____MammlDomest__________	→	FELINMammlDomestica \llt_	(65)
_____MammlDomesti_________	→	FELINMammlDomestica \llt_	(67)
_____MammlDomestic________	→	FELINMammlDomestica \llt_	(70)
_____MammlDomestica_______	→	FELINMammlDomestica \llt_	(62)
_____MammlDomestica _____	→	FELINMammlDomestica \llt_	(58)

Table 4.5
Response of System to Various Input Vectors
Specific Examples: Canis

Input		Output	
_____MammlFamiliarisD____	→	GSHEPMammlFamiliarisD_a''	(33)
_____MammlFamiliarisDa___	→	GSHEPMammlFamiliarisDaa'a	(19)
_____MammlFamiliarisDaa__	→	GSHEPMammlFamiliarisDaa'a	(16)
_____MammlFamiliarisDaab_	→	GSHEPMammlFamiliarisDaaba	(12)
_____MammlFamiliarisDaaba	→	GSHEPMammlFamiliarisDaaba	(9)
_____MammlFamiliarisDaabb	→	GSHEPMammlFamiliarisDaabb	(14)
_____MammlFamiliarisDaabc	→	GSHEPMammlFamiliarisDaabc	(14)

Learned Example GSHEPMammlFamiliarisDaaba

5. Simulation 3: A More Complex System. We wanted to try to learn a more complex hierarchy, which might display some more interesting results than the simple autoassociative classifier. The details of the techniques used have been discussed in Anderson, 1986 and Anderson & Murphy, 1986. The more complex hierarchy is diagrammed in Figure 5.1.

Figure 5.1

A Structured Category Hierarchy

```
                        ANIML
                      /       \
                     /         \
                    /           \
                VERTB            INVRT
               /   \            /  |  \
              /     \      LEECH   |   LIMUL
             /       \           APLYS
        CANIS         FELIN
        / | \         /   \
       /  |  \    ALLEY SIAME
      /   |   \
 GSHEP  CSPNL  PEKNG
```

Two partially connected (50%) matrices were learned and used. An autoassociative matrix was used to couple together the 'sensory' and the basic and subordinate category names and a heteroassociative matrix coupling together the lower level names (LEECH, CANIS) and the higher level names (VERTB, INVRT). The autoassociative state vectors are given in Table 5.1 and the heteroassociative pairs given in Table 5.2. Learning was using the Widrow-Hoff technique. Notice the heteroassociative learning was done only at the 'linguistic level', which was then relatively labile with upward and downward associations. The sensory regions of the state vector were stable and did not contain upward or downward associations.

Table 5.1

Three Layer Network: Two Matrices

Autoassociative Stimulus Set

CANISMammlFamiliarisAaaaa	CANISMammlFamiliarisBbbbb
CANISMammlFamiliarisCcccc	CANISMammlFamiliarisDdddd
CANISMammlFamiliarisAbcda	CANISMammlFamiliarisBcdab
CANISMammlFamiliarisCdabc	FELINMammlDomestica Ttttt
FELINMammlDomestica Lllll	FELINMammlDomestica Xxxxx
FELINMammlDomestica Ooooo	FELINMammlDomestica Xolto
FELINMammlDomestica Lxotl	FELINMammlDomestica Olxot
APLYSAplysiaCaliforn12345	APLYSAplysiaCaliforn89012
APLYSAplysiaCaliforn56789	APLYSAplysiaCaliforn23456
CSPNLMammlFamiliarisCcabc	GSHEPMammlFamiliarisDaaba
PEKNGMammlXXmiliarisPekin	ALLEYMammlDomestica Ollot
SIAMEMammlDomestica Tloxx	SIAMEMammlDomestica Tloxt
LEECHHirudoMedicinalmmmmm	LEECHHirudoMedicinalqqqqq
LEECHHirudoMedicinalnnnnn	LEECHHirudoMedicinalmqnnq
LEECHHirudoMedicinalqmnmn	LEECHHirudoMedicinalnmqqm
LIMULLimulusPolyphem09890	LIMULLimulusPolyphem89090
LIMULLimulusPolyphem90898	LIMULLimulusPolyphem97653
VERTB____________________	INVRT____________________
ANIML____________________	

Table 5.2
Heteroassociative Stimulus Set

CANIS_	⇒	VERTB_
FELIN_	⇒	VERTB_
APLYS_	⇒	INVRT_
LEECH_	⇒	INVRT_
LIMUL_	⇒	INVRT_
INVRT_	⇒	ANIML_
VERTB_	⇒	ANIML_
SIAME_	⇒	FELIN_
ALLEY_	⇒	FELIN_
GSHEP_	⇒	CANIS_
PEKIN_	⇒	CANIS_
CSPNL_	⇒	CANIS_

The results showed similar initial categorization behavior to the two layer simulation, though with a little less precision, partly due to the larger number of different categories stored, 13 versus 9.

Because of their greater lability, strictly linguistic associations 'carry along' some of the sensory material. If we assume a weak learning process involved as well as reconstruction, we have a natural mechanism for high level (abstract) concept formation.

The system output state vector appears to 'ascend' the hierarchy as shown by the example in Table 5.3. This is due to the linguistic associations. Then, due to the fact than ANIML is associated with nothing and the sensory code is still present, the basic level category emerges again and the system cycles again through the network. Therefore, there is effectivly a backward association to the basic level of the system, but this backward association is not specifically taught.

The sensory part of the state vector may change due to brief periods of decay when the heteroassociative matrix is applied. In general, a more realistic system would incorporate adaptation of various types, which is currently largely an unexplored area. A paper by Kawamoto and Anderson (1985) used Hebbian anti-learning as an adaptation mechanism to explain shifts of interpretation for Necker cubes. In the current simulation, brief periods of 'adaptation' allowed the sensory part of the state vector to change. The system categorized inputs into a subordinate category only if the match was nearly exact. When the hierarchy was ascended up to

ANIML and then the system dropped back to the beginning, it dropped back to the *basic* level even though it initially categorized the input at the subordinate level because the match was no longer exact enough for a subordinate match. This selective distortion is an interesting potential categorization mechanism. Table 5.4 shows this effect.

Table 5.3

Selected Examples

Dogs and Cats Ascend and Descend a Hierarchy

Input		Step	Output
_____MammlFa_____________	⇒	44.	CANISMammlFamiliarisBaac'
		55.	VERTBMammlFamiliaris_aac'
		117.	ANIMLMamml__miliarisBa_c_
		138.	CANISMamml__miliarisBa_cb
		184.	VERTBMammlFamiliarisB__cb
_____MammlDo_____________	⇒	45.	FELINMammlDomestica Siosz
		73.	VERTBMammlDomestica Si_sz
		145.	ANIMLMammlDomestica_Shysz
		199.	FELINMammlDomestica C_y_j

Table 5.4

A Specific Dog Gets Distorted in Retrieval

Input		Step	Output
_____MammlFamiliarisDaabb	⇒	22.	GSHEPMammlFamiliarisD'_bb
		38.	CANISMammlFamiliarisD_cbb
		83.	VERTBMammlFamiliarisF____
		134.	ANIMLMamml_amiliarisF_c__
		165.	CANISMammlFamiliarisFaacb

The autoassociative system is capable of filling in missing sensory information from linguistic material. Table 5.5 shows this information retrieval. This is simple example of a 'data base' application, of which many other examples are available. (See Anderson, 1986).

Table 5.5

Answers to Queries

Query		Answer
FELIN____________________	⇒	50. FELINMammlDomestica [srsk
CANIS____________________	⇒	58. CANISMammlFamiliarisFrcb'
LEECH____________________	⇒	32. LEECHHirudoMedicinalyuvum

6. Specific Conclusions. These simulations suggest it is possible to develop a useful concept system based strictly on presenting specific examples and perhaps a little bit of abstract knowledge. It works robustly in computer simulation. The techniques we used are very simple and preliminary, but encouraging.

The initial representation of knowledge is critically important. The exact details of representation of information in the state vectors are at least as important as the details of the learning algorithms. Playing two systems off against each other, with slightly different properties, as in the last simulation, gives systems of great flexibility and power -- perhaps one reason for the success of language, which seems to be somewhat decoupled from other systems.

7. General Conclusions. In general, simple networks of this type are effective at categorization, data base applications, and disambiguation.

Such systems require low precision matrix multiplications, parallel computation, a degree of linearity, and high interconnectivity. They are excellent matches to optical computers or special purpose digital or analog hardware.

It is possible for even very simple, one layer systems to build quite complex cognitive structures. Knowledge networks and structures can be implemented so they evolve in time, as we showed in the last simulation and in Anderson (1986) where a 'semantic network' was traversed by having nodes in the network appear in succession.

A technological prediction is that this type of system (i.e. 'connectionist' or 'neural network') will be part of the hardware and organizational concepts used to build the brain like computers which will be constructed in the near future. The first practical applications of these network models are now starting to appear. However, the software will prove to be much tougher than hardware, as hardware builders are just starting to realize.

Such systems have some serious drawbacks. Some examples are: (1) Network systems make unexplainable mistakes. (2) Such systems have weak formal computing power, basically only associative or similarity operations. (3) They realize 'logic' and show precision only with great difficulty. (But then, so do we!) (4) And, most of the power of the system is in the representional details of the vector codings. The systems are strongly memory driven and are somewhat inflexible.

These neural network systems are direct, though now somewhat remote, descendents from brain models. An important claim of this discussion is that most well formulated models of the large scale organization of the nervous system will have significant behavioral implications. These behavioral implications are just as interesting and testable as lower level neuroscientific predictions.

This work is sponsored by the National Science Foundation under grant BNS-85-18675 administered by the Memory and Cognitive Processes section and by the Office of Naval Research under contract N-0014-86-K-0600.

References

Anderson, J.A. (1970). Two models for memory organization using interacting traces. *Mathematical Biosciences*, *8*, 137-160.

Anderson, J.A. (1986). Cognitive capabilities of a parallel system. In E. Bienenstock, F. Fogelmann, and G. Weisbuch (Eds.), *Disordered Systems and Biological Organization*, Berlin: Springer.

Anderson, J.A., Golden, R.M. & Murphy, G.L. (1986). Concepts in distributed systems. In H. Szu (Ed.), *Optical and Hybrid Computing*, Bellingham, WA: S.P.I.E.

Anderson, J.A. & Murphy, G.L. (1986). Psychological concepts in a parallel system. *Physica D*, (In Press).

Anderson, J.A., Silverstein, J.W., Ritz, S.A. & Jones, R.S. (1977). Distinctive features, categorical perception, and probability learning: Some applications of a neural model. *Psychological Review*, *84*, 413-451.

Collins, A. & Qullian, M. (1969). Retrieval times from semantic memory. *Journal of Verbal Learning and Verbal Behavior*, *8*, 241-248.

Golden, R. (1986) The 'Brain-State-in-a-Box' neural model is a gradient descent algorithm. Journal of Mathematical Psychology 30, 73-80.

Hinton, G.E. & Anderson, J.A. (Eds.), Parallel Models of Associative Memory, Hillsdale, NJ: Erlbaum Associates.

Hopfield, J.J. (1984). Neurons with graded response have collective computational properties like those of two-state neurons. Proc. Natl. Acad. Sci. U.S.A., 81, 3088-3092.

Kawamoto, A. & Anderson, J.A. (1985). A neural network model of multistable perception. Acta Psychologica, 59, 1-31.

Knapp, A.G. & Anderson, J.A. (1984). Theory of categorization based on distributed memory storage. Journal of Experimental Psychology: Learning, Memory, and Cognition, 10, 616-637.

Kohonen, T. (1977). Associative Memory, Berlin: Springer.

Kohonen, T. (1984). Self Organization and Associative Memory, Berlin: Springer.

McClelland, J.L. & Rumelhart, D.E. (1985). Distributed memory and the representation of general and specific memory. Journal of Experimental: General, 114, 159-188.

McClelland, J.L. & Rumelhart, D.E. Eds. (1986). Parallel, Distributed Processing, Volume 2, Cambridge, MA: MIT Press.

Minsky, M. & Papert, S. (1969) Perceptrons, Cambridge, MA: MIT Press.

Posner, M.I. & Keele, S.W. (1968). On the genesis of abstract ideas. Journal of Experimental Psychology, 77, 353-363.

Rips, L.J., Shoben, & Smith, E.E. (1973). Semantic distance and the verification of semantic relations. Journal of Verbal Learning and Verbal Behavior, 12, 1-20.

Rosch, E. (1978). Principles of categorization. In E. Rosch & B.B. Lloyd (Eds.), Cognition and Categorization, Hillsdale, NJ: Erlbaum.

Rumelhart, D.L. & McClelland, J.L. Eds.(1986) Parallel, Distributed Processing. Volume 1 Cambridge, MA: MIT Press.

Smith, E.E. & Medin, D.L. (1981). Categories and Concepts, Cambridge, MA: Harvard University Press.

Categorization in a Neural Network

Discussion
edited by *Hanspeter Mallot*

(1) From the theory of associative memory and statistical learning, limitations of the Widrow-Hoff learning rule were pointed out: First, although this rule is superior to the simple Hebb-rule in that it does not require an orthogonal set of concept state vectors, it still allows only a linear segmentation of the feature space; i.e., a segmentation into linear subspaces. Thus, the concept state vectors have to be linear-independent. The admissible number of concepts is therefore not increased by the use of this rule. Clearly, this problem cannot be overcome with a linear retrieval technique. - Second, the proof that the Widrow-Hoff learning rule actually results in the correlation matrix relies on the so-called pseudo-inverse of that matrix. This pseudo-inverse, however, is well defined only if the involved vectors are far from parallel. It runs into problems if the assumption of linear independence is poorly met.

(2) The model proposed here is similar to a perceptron; once the connectivity matrix is learned, it acts as a linear pattern recognition device. As such, it judges similarity from correlation and therefore runs into problems with shifted or rotated input vectors, since, in general, these are not collinear to each other. (They may, in fact, be orthogonal). A question raised in the discussion asked for the robustness of both the performing and the learning model against a positional scatter in the input vectors. As a possible remedy, a preprocessing layer could be added to the network which achieves

translation- and rotation-invariance by certain connectivity schemes used in the FFT, the R-transform, or the Walsh-Hadamard transform. These transforms have been used successfully for invariant pattern recognition. The networks required are natural to the proposed high connectivity system.

(3) The proposed model assumes that each unit is connected to each other. In the nervous system, this assumption does not generally hold, and the question was asked, what types of networks could be used alternatively and what implications that might have on the presented results. Possibilities for connectivity schemes include distinct modules, grids, "butterfly"-networks like those used with the FFT, and others. As to a closer modelling of the nervous system, distortions by continous ("topographical") maps may be considered as well. For some connectivity schems, such as local statistical connections, a continous approach might be more useful than the discrete one employed in the proposed model.

(4) Another remark dealt with the use of feedback in parallel networks. In stable linear systems, an equivalent feedforward realization for all feedback networks exists. In fact, the solution of a feedback equation results in just this equivalent feedforward system. Thus, it may be argued that feedback systems are not essentially different from feedforward ones. However, feedback may help to overcome constraints of the system such as limited connectivity or simple (first-order) temporal behaviour which may make the implementation of the equivalent feedforward system impossible. In general, what feedback actually adds to a linear system with given constraints is an interesting question.

(5) The combination of the (positive) feedback coupling with a saturation type nonlinearity links the proposed model to the theory of nonlinear dynamics. In this context, it is an important result that the system converges in time to a single point attractor. In biological systems, however, more complicated behaviour such as limit circles and strange or chaotic attractors have been found. It may be rewarding to study the memory storage, retrieval, and concept formation capabilities of these more general dynamic systems, thereby including their short-term temporal behaviour.

The Organization of Sequential Memory: Sparse Representations and the Targetting Problem

Graeme Mitchison

SUMMARY

Human memory seems particularly adapted to learning sequences of events or actions, as is shown by our ability to recite verse or perform music. This type of memory seems to have a large capacity, in the sense that a great many events in such sequences can be committed to long-term memory. I conjecture that there is a limit, equivalent to two bits per synapse, to the total amount of information a neuron can convey by giving learnt "0" or "1" responses. From this it follows that the mean information per event conveyed by a neuron which is involved in memory storage must be small, which in turn implies that only a small proportion of such neurons can be active during an event.

The fact that there is a considerable flexibility in organizing sequences composed of varied types of information suggests that the activity of one small set of neurons, which represents one event, must in principle be able to excite activity in any other set representing another event. Since an individual neuron makes synapses onto only a tiny fraction of all neurons in the brain, this poses what I call the "targetting problem". I shall suggest one possible solution to this problem

INTRODUCTION

Memory often has a sequential character, as when we recollect verse, music, a particular route, and so on. I shall be concerned here with this type of memory and shall use the term "event" for an item in a memorized sequence. The basic assumption I shall make is that events correspond to the activity of a given set of neurons, and the generation of the next event in the sequence is brought about by the activation, via the usual synaptic mechanisms, of a second set. There are of course other possible neuronal models; for instance, a sequence might be encoded as a single pattern of neural activity, components of which could be inspected or highlighted in some fashion. However, the sequential activation model seems to offer the most natural explanation for the strong order dependence of such memories. This order dependence, manifested as a need to return to a starting point after a slip of memory, is indeed ony too familiar to musicians and other performers.

The notion of coding a sequence as a series of events is nicely illustrated by the Place System (1) used in classical and mediaeval times, where items ("events") in a sequence were recorded as bizarre objects put into a set of locations in a building or some other piece of architecturally ornamented topography. The sequence was then retrieved by moving in a fixed order through the set of places. Highly trained exponents of the system were able to use it for learning large amounts of material; many thousands of lines of text could be rapidly memorized, in many cases with long-lasting recall.

The type of memory considered here can be roughly characterized as "knowing that" as opposed to "knowing how", and it is the former kind which is lost in the classical amnesic syndrome (2). Amnesia is thought to be caused by damage to the hippocampus, perhaps in conjunction with damage to the amygdala (3). However, the ability of amnesic patients to recollect events when given partial cues (4), and also the comparative sparing of memory for remote events compared to that for recent ones (5) argues against the hippocampus actually being the memory store. Instead, it is conjectured that it plays some role in consolidation of memories (6), or in linking cognitive with semantic memory operations (7), the latter perhaps involving pathways connecting temporal and frontal cortex. The hippocampus makes extensive connections with association cortex, the subiculum in particular making rich reciprocal connections with temporal association cortex (8). I shall assume that these cortical association areas are the likely site of storage for the sequential, data-based memory we are concerned with, though the arguments do not depend strongly on any anatomical assumptions.

It is not easy to arrive at a lower bound for the size of human long-term memory. The feats of mnemonists, for instance the exact memorization of the location of every word in the 12 volume Talmud by "Shass Pollaks" (9), certainly imply a vast capacity; but it is difficult to measure this in numbers of events (though it is likely that something resembling sequential memory is involved even in so-called "photographic" memory). The place system seems to provide more quantifiable information, and the performances of the Russian mnemonist Shereshevski are particulary helpful here, both because there are some detailed records of his technique, and also because his recall remained perfect over decades (10). One can estimate from examples given of the extraordinary elaborate constructs which he used for memorizing nonsense syllables that perhaps 1,000 steps might be involved in a night's performance as a mnemonist. If this seems unreasonably large, one should bear in mind Standing's experiments, showing that untrained subjects could distinguish, with 83% accuracy, seen from unseen pictures out of a set of 10^4 presented in a previous learning session (11).

Now, if we suppose that 1,000 events could be permanently learnt each day for 30 years we arrive at an event capacity of about 10^7. Though this is clearly an extremely crude calculation, it is interesting that it agrees quite well with a recent estimate. Landauer (12) argues that 10^9 bits may be encoded in human memory, and that a picture may require 10-15 bits for its encoding. If we translate pictures directly into events, this gives a capacity of 7×10^7 to 10^8 events. In what follows, I shall assume that the capacity exceeds 10^7 events. This figure may well be somewhat low.

MEMORY CAPACITY IN A SIMPLE NEURONAL MODEL: THE TWO BIT LIMIT

An event is assumed to correspond to the activity of a subset of neurons in some region of the brain which deals in highly processed, multi-modal information. The neurons $\mathfrak{E}_1$ active in an event e_1 are assumed to activate the set $\mathfrak{E}_2$ of those corresponding to a second event, e_2 , where $e_1 \rightarrow e_2$ is a step in a sequence.

(The existence of a reverse set of connections, corresponding to $e_2 \rightarrow e_1$, need not be postulated, and indeed the forward passage through a series of recollections is usually much easier than the reverse.) Unless otherwise stated, I assume here that this activation takes place in a single step, the axons of the neurons in $\mathcal{E}_1$ synapsing onto the dendrites of those in $\mathcal{E}_2$. That is, we assume the existence of of a set of recurrent connections $\mathcal{N} \rightarrow \mathcal{N}$, where $\mathcal{N}$ is the set of all neurons which participate in events.

Suppose that there are two states for a neuron, 0 or 1 (in other words we do not consider gradations of spiking rates), and suppose that a neuron in $\mathcal{N}$ takes the state 1 in some fraction f of events, and 0 otherwise. The average information it conveys by firing is given by the entropy $H(f)=-f.\log_2 f-(1-f).\log_2(1-f)$. Suppose furthermore that the synapses on its dendritic tree are either inactive ("0") or active ("1"). If there are N synapses on the dendritic tree, then the total information they can encode with their states is N bits. From this it follows that the neuron can learn to give the correct response for at most N/H(f) independent events, so, if M is the number of events which can be stored, $M \leq N/H(f)$.

A capacity of this order is attained by the associative net of Willshaw (13,14,15,16) in the case when $f=\log_2 N/N$, for then it can be shown that $\log_e 2(N/\log_2 N)^2$ events can be stored before significant errors occur. This is approximately .69 of our limit, when N is large enough to ignore $\log_2\log_2 N$ compared to $\log_2 N$. In the case of this network, the synapses take only two states, but in general synapses might take some number s of states. The total information in the dendritic tree is then $(\log_2 s)N$ bits, and the foregoing argument gives a limit $M \leq (\log_2 s)N/H(f)$. If the synapses can take arbitary integral or real values, as in the Hopfield network for instance (17), then this argument places no finite bound on M. However, if neurons can be treated as linear threshold devices, then for a neuron to give correct responses (0 or 1) to randomly chosen vectors is equivalent to being able to find a plane in N-dimensional space separating those vectors which require response 0 from those requiring response 1. It is known (18,19) that if the vectors are random and their associated responses, 0 or 1, are randomly and equiprobably assigned (so f=1/2), then the number M of vectors which can be separated this way is 2N. More precisely, the probability of finding a separating plane decreases with M, and reaches 50% when M=2N. The shape of this probability distribution is sigmoidal, the transition at M=2N becoming increasingly sharp as N increases.

In this instance $f=1/2$, so $H(f)=1$. If the synapses were binary the informational limit for M would be $N/H(f)=N$. Therefore the limit for binary synapses and that for arbitary real synaptic weights are in the ratio 1:2; in other words, there are at most two bits per synapse available for storing events. I conjecture that this two bit limit holds for all firing frequencies f, so that the number M of events which can be stored by a neuron with real-valued synaptic weights satisfies

$$M \le 2N/H(f) = 2N/\{-f.\log_2 f-(1-f).\log_2(1-f)\} \qquad (1)$$

Figure 1 gives some evidence for this from numerical simulations, obtained by applying the perceptron learning algorithm (20) to sets of M randomly chosen vectors $\mathbf{v}^1, \mathbf{v}^2, .., \mathbf{v}^M$. This algorithm passes through the complete set of input vectors $\{\mathbf{v}^i\}$ updating the weight vector $\mathbf{w}$ according to the rule $\mathbf{w}^+=\mathbf{w}+\mathbf{v}^i.(z^i-z)$, where $\mathbf{w}^+$ is the new value of $\mathbf{w}$, z^i is the desired output (0 or 1) for $\mathbf{v}^i$ and z is the actual output obtained by thresholding $\mathbf{w}.\mathbf{v}^i$. This is repeated until the algorithm either finds a weight vector $\mathbf{w}$ which gives the correct outputs or else cycles.

To obtain Figure 1a, sets of M random binary vectors were generated, each a string of N equiprobable 1's or 0's. To each set of M vectors, outputs were randomly assigned, with exactly f.M outputs of 1, the rest 0. The learning algorithm was then applied to this set to see whether a value of $\mathbf{w}$ which gave correct outputs could be found or whether a cycle occurred. This entire procedure was repeated for at least 100 such sets, and the fraction of successes found. This fraction is shown on the ordinate, while the abscissa shows M as a fraction of $N/H(f)$; in effect, the abscissa shows the number of bits per synapse which are available for storing events. N was taken to be 9,10 or 11, the precise value being chosen so that f.M would be an integer for values of M close to the point of 50% success rate (to locate this point as exactly as possible). For these values of N, when no correct $\mathbf{w}$ existed, cycles occured within a few hundred passes through the complete set of vectors $\{\mathbf{v}^i\}$. As can be seen, the 50% points on the curves for f=0.5, 0.1 and 0.04 all lie near or below the two bit limit, $M.H(f)/N=2$. There is a noticeable tendency for the 50% point to fall below this value as f decreases.

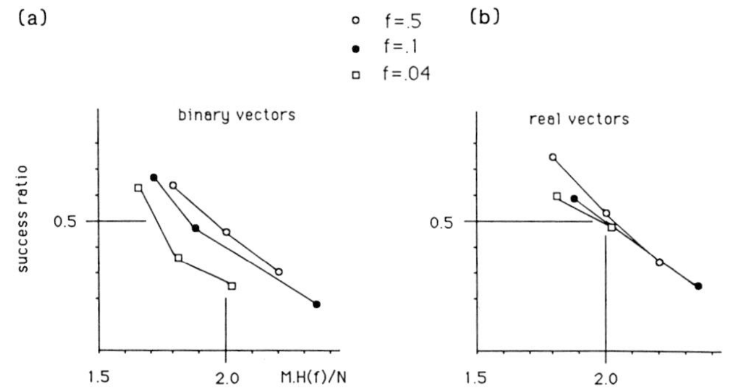

FIGURE 1

Figure 1b shows the same calculation applied to vectors whose components are real numbers x in the range $0 \leq x \leq 1$ instead of binary components, 0 or 1. Again, the outputs are randomly assigned 0's or 1's with exactly f.M of the latter. With real components to the input vectors, cycles of the learning algorithm are never found, presumably because they are very long. Instead, it is necessary to run the algorithm for a great many iterations, and even when this fails to find a correct **w** one can never be sure that it does not exist. It is reassuring, therefore, that the success ratios for f=0.5 closely approximate the theoretically expected values (18, 19). For vectors with real components, the 50% success rate for small values of f lies closer to M.H(f)/N=2 than with binary vectors. I conjecture that this limit is attained by random vectors with real components, so that M=2N/H(f) in this case.

I shall assume that the maximum event capacity of neurons is given by the two bit limit (inequality 1), even though a linear threshold device may be a poor model of a real neuron.

RANDOM CONNECTION MODEL

Let us now consider a model in which connections between neurons in $\mathcal{N}$ are made at random, and where an event is represented by randomly distributed activity in some fraction f of neurons in $\mathcal{N}$. Of course, this is not a very realistic model, but it helps to illustrate the basic ideas.

Assume that all cells have the same number N of synapses and the same probability f of firing. When the number of events which have been stored exceeds the two bit limit (inequality 1), a neuron will respond to inappropriate events, or fail to respond when expected, and so bring about malfunction of the network. Of course, a few mis-firing cells may be tolerated, especially when the number of active cells is large. A less stringent criterion would be that the output activity of one net, when treated as an input to another, must be able to regenerate with high probability the next event in a sequence. However, there turns out to be little difference between these various definitions of "malfunction", since the breakdown is rapid once it starts to occur.

It follows that the number M of events which the network can store is given by the two bit limit, and is independent of the total number of neurons. To calculate M we need to know both N and f. The average number of synapses on a human cortical cell has been estimated to be about 40,000 (21,22); a large cortical pyramidal cell must have many more. However, we must bear in mind that not all synapses on a given neuron will be involved directly in memory storage, and also that a given presynaptic cell may make several synaptic contacts. It is known, for instance, that a single Ia fibre may make from 2 to 7 synapses on the same motoneuron (23). I shall therefore assume that $N=10^5$, which is probably quite a generous estimate, even though a large pyramidal cell may have more synapses in total.

Suppose first that f=1/2, as in Hopfield's network (17), for instance. Then the two bit limit implies that at most 2N events can be stored, giving $M=2x10^5$ which is far below our bound of 10^7 events. Of course, more events can be stored when f is smaller. But there is a limit to how small f can be. Suppose f is so small that the expected number f.N of active synapses on a given neuron supplied by the recurrent pathway $\mathcal{N} \rightarrow \mathcal{N}$ is 1. Then, after a synapse has been activated, there is a 1/N chance that that synapse will be activated again by an input and the cell caused to fire, even when this is not desired. In other words, the combinatorial potential of the synapses on a cell is not being exploited.

There are more general reasons for expecting this combinatorial potential to be used. Human memory exhibits a capacity for completion: A fragment of an event enables us to recall the rest. Completion can be achieved naturally by an associative net if the partial information corresponds to a subset of the total event in question. This implies that there should be a combinatorial code in which the elements are (possibly quite high order) features. Since a given event will generally be an agglomerate of subevents which themselves have to be uniquely coded, it is clear that the representation must be quite rich in active neurons. It also suggests that the cells which are post-synaptic to the event-coding cells (ie other event-coding cells, by a recurrent pathway, or some intermediate layer) must receive many active inputs from a given event, since a subevent must have a high probability of exciting these cells and so completing the event.

Assuming a random distribution of connections, the expected number of active synapses on a post-synaptic cell with N synapses will be f.N. The physiology of two cell types, hippocampal granule cells and motoneurons, can help us make an educated guess as to how large f must be to ensure that a cell fires. In the case of the granule cell, a single EPSP produces on average a 0.1mV depolarization at the cell body (24), and a total depolarization of about 25 mV is needed to fire the cell. This implies that about 250 inputs must be active to fire the cell (McNaughton et al (24) reach a higher figure because they consider decay between incoming impulses, but we assume simultaneous inputs for reasons which appear later). Since the cell has about 10^4 synapses in all (25), this implies $f>2.5x10^{-2}$. In the case of the motoneuron, the size of an e.p.s.p. due to a single Ia axon lies in the range 0.2 - 0.4 mV (23), and the cell may receive as many as $2x10^4$ synapses (26). This suggests the somewhat lower bound $f>5x10^{-3}$. We take $f>10^{-2}$ as a compromise. If we can scale this up from these cells to our hypothetical pyramidal cell with about ten times as many synapses, then we obtain $f>10^{-2}$ for the fraction of neurons active in a subevent, which suggests that $f>>10^{-2}$ in a richly structured event. From inequality 1, this implies $M << 2.8x10^6$, which is considerably below our limit for event capacity.

Of course, these numbers are arrived at by many intuitive

steps, and are not to be taken too seriously. However, it does seem likely that a randomly connected network is not a very good kind of memory, because of the conflict between the requirement for sparse activity, so that the number of stored events can be large, and the need for quite rich projection onto post-synaptic cells. These joint requirements constitute what I shall call the *targetting problem*. A natural way of solving this problem is to have the activity clustered and to arrange connections so that neurons tend to receive synapses from localized clusters of neurons. We consider this next.

FUNCTIONAL SPECIALIZATION AND A MANY-UNIT MODEL

One reason why the randomly connected network is unattractive as a model for memory is that it does not allow for the regional specialization which characterizes processing in the cortex. In addition to the many separate sensory areas in the cortex (27, 28), there are also localized regions involved in higher processing, for example, the well-known speech areas (Wernicke's and Broca's), and certain regions of infero-temporal cortex which contain cells specifically responsive to faces (29). If we assume that our hypothetical region $\mathcal{N}$ possesses a similar kind of functional specialization, then an event will consist not of uniformly spread activity but of a small number of clusters of active cells, each cluster associated with some particular aspect of the event.

Let us therefore suppose that $\mathcal{N}$ can be decomposed into S disjoint *nets* $\mathcal{N}_1$, $\mathcal{N}_2$ . . $\mathcal{N}_S$, each of which stores some particular kind of information, and suppose for the moment that each net $\mathcal{N}_i$ makes recurrent connections with itself only, so that it is possible to store sequences of a given type but not transitions between these types. The two bit limit for M which applied originally to the whole of $\mathcal{N}$ now applies to each $\mathcal{N}_i$, and S times as many events can be stored. Of course, this gain in number of events has been paid for in the diminished size of the representation; but this is no loss if the restriction to a particular type of event enables it to be encoded efficiently in the appropriate net $\mathcal{N}_i$. What has been truly lost is the flexibility of being able to store arbitrary sequences.

Figure 2 shows how one might connect all the nets $\mathcal{N}_i$ together into a more richly intercommunicating global network $\mathcal{N}$. Each net (dotted box outline in Figure 2a) is assumed to contain N/L neurons. The dendrites of neurons in the net are assumed to be subdivided into L regions or <u>subunits</u>, each containing N/L of the N synapses on the dendrites of a neuron. Each net projects to L subunits in other nets $\mathbf{N}_j$ (possibly including itself), so that each cell makes LxN/L = N synapses in all, conserving the number which are made and received. Each net has a set of forcing fibrės which produce the desired output (see Figure 2b), these fibres making some pre-established set of synaptic connections. The assumption is that an event e_1 involves the activity of some set S_1 of nets. Provided that each net in the set S_2 of nets which are active in e_2 can be reached from some net in S_1, then the sequence e_2 -> e_1 can be coded by the network.

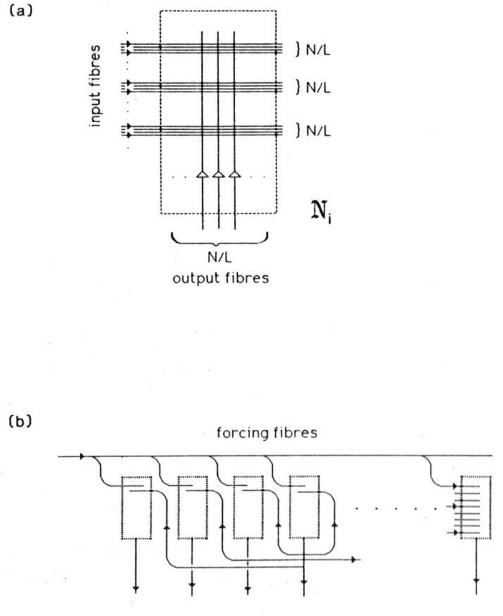

FIGURE 2

Before considering in any detail how the patterns of connections could be made between nets, let us consider how many events can be stored by the whole network $\mathcal{N}$. We may treat each subunit as a separate learning device which happens to share with other subunits an output through the axons of its net (assuming that the inputs from a subunit can activate a neuron). Suppose $\mathcal{N}_i$ is active in event e_1. Not all of the nets which receive inputs from $\mathcal{N}_i$ will be active in e_2; suppose that, on average, a fraction q of the nets supplied by the L branches from $\mathcal{N}_i$ are active in e_2, and suppose the frequency of an individual neuron being active in a net when that net as a whole is active is f_0. Then the frequency of that neuron being active when there is an input to a subunit is $f=q.f_0$. If we count as learnt events only those in which the subunit generates activity in its net, then for small f the event capacity for the subunit, M_u, is approximately given by $M_u=(2N/L).q/\{-q.f_0.\log_2(q.f_0)\}=(2N/L)/\{-f_0.\log(q.f_0)\}$, neglecting the term $(1-f).\log_2(1-f)$ in the denominator of inequality 1. The event capacity for the subunit regarded as an isolated storage device is given by $M_i=(2N/L)/\{-f_0\log(f_0)\}$ and clearly $M_u<M_i$.

This reduction in capacity is due to the need for the subunit not to activate nets except when they participate in the next event in sequence. If this were not the case, then a sequence of inputs would generate a mass of irrelevant and corrupting activity. This puts an added informational burden upon the subunit, for the requirement that it respond only to a given set of inputs means in effect that it learns those inputs as well as the appropriate outputs (for we can discover which inputs have been presented by measuring the size of the output appropriately). The associative net which uses two-state synapses (14) has the attractive property that it achieves a high specifity for the learnt inputs.

With this taken care of, the capacity of $\mathcal{N}$ as a whole is obtained by multiplying the event capacity per subunit, M_u, by L.S/K, where L.S is the total number of subunits in $\mathcal{N}$ and K is the average number active in an event (in the sense that there is both an active input and output). As an example, suppose $N=10^5$, L=30, so each net contains N/L=3000 neurons. Suppose there are 10^8 neurons in $\mathcal{N}$, which is about 1% of the neurons in the cortex (for comparison, there are less than 10^6 pyramidal cells in the hippocampus of man, since there are 1.2×10^6 fibres in the fornix (30), some of which come from other neurons). This means that $\mathcal{N}$

includes about 2.5% of the synapses in the cortex (since we are assuming that N is larger than Cragg's average (21) of $4x10^4$), perhaps not a bad estimate for the number directly involved in storing highly processed information. These 10^8 neurons will be divided into $S=3x10^4$ nets, giving $L.S=9x10^5$ subunits in all. Suppose 10 nets are active in an event, with q=.2. This means that if S_1 is the set of 10 nets active in e_1, S_2 the set of 10 active in e_2, then K=60 of the 300 subunits which receive inputs from in S_1 lie in S_2. Take $f_0=1/3$, so that the inputs to a given neuron activated by one subunit constitute a fraction 10^{-2} of all its N synapses, and can therefore fire the cell, according to the estimate derived earlier. Then we find $M_u=4.6x10^3$, and the total number of events which can be stored is $M_u.L.S/K = 8x10^7$.

The event capacity will be further increased if we can regard the formal subunits of the model as compartments of dendrites which are closely electrically coupled (31) and if summed activity in these compartments is particularly effective at firing the cells, perhaps by the initiation of a calcium spike (32, 33). In this case, rather few inputs would be able to fire a cell provided they converged upon a subunit. It is certainly not unreasonable to assume that few inputs in a small region could be highly effective; as an extreme example, the opening of a single channel can trigger a spike in a bovine chromaffin cell (34). If we assume that the same limit, $f=10^{-2}$, applies to a subunit, and otherwise assume the same numbers used in the above calculation of event capacity, we find that $1.2x10^9$ events can be stored.

As we shall see in the next section, some of this event capacity may be used for learning intermediate steps linking events, as a result of which the capacity in this last example will be reduced to about $2x10^8$. But this is still a comfortably large number. Note, however, that it has been achieved by having a small number of cells active in an event, with their activities all targetted upon an equally small number of postsynaptic cells, or even subunits of these cells. For the inputs to these postsynaptic cells to be maximally effective they should occur within a short time span. The two-state synapse model, with an irreversible Hebbian rule (so it is irreversibly switched from 0 to 1 when there is a conjunction of pre- and post-synaptic activity), has the property that a given input $\{I_i\}$ will produce the same sum $\Sigma w_{ij}.I_i$ in every postsynaptic cell that ought to fire (where w_{ij} are synaptic weights), no matter what other events that cell has participated in. If these cells have similar geometry, then the input can be expected to have similar effects on all of them and, other things being equal, to generate a nearly synchronous output. In this way targetting can also be achieved in the time domain as well as space, allowing sizeable and precise effects with even a small number of inputs.

SEQUENTIAL ROUTING AND OPTIMAL TOPOGRAPHY

The assumption we have made is that a given net projects directly to a number of other nets. Since on average each neuron can only make as many synapses as it receives, a typical net will only be able to map to a fraction of all the nets (if the total number of neurons exceeds N). One way to achieve a richer connectivity is to allow nets to be reached in several steps, so that every net can eventually be reached from any starting point.

Figure 3a shows a possible arrangement of this type in one dimension (where the rectangles represent nets). Suppose there are S nets in $\mathbf{N}$, where $S \leq n^r$ for some integers n and r. The nets are numbered, and each is connected to the next (n-1) in sequence, then to those numbered n, 2n, . . . (n-1)n further on in the sequence, then to those numbered n^2, $2n^2$, . . . $(n-1).n^2$ further on, and so on, till $(n-1)n^{r-1}$ is reached (the numbering is assumed to be cyclical; ie arithmetic mod S is used). Then every net can be reached in at most r steps, this being just the property that every integer less than n^r-1 can be expressed as a number of r digits in base n. The total number of connections made by a net is r(n-1), so L=r(n-1) in the notation of the previous section. In the example considered there, L=30 and $S=3x10^4$, and we can take n=6, r=6, so that every link can be made in 6 steps. Since these extra steps have to be learnt to link every pair of events, the total number of events which can be learnt is decreased by a factor of 6. The capacity of $1.2x10^9$ calculated in the previous section must therefore be reduced to $2x10^8$.

A more symmetrical pattern can be devised in which the connections are made, with equal spacing, on both sides of the source net (Figure 3b; here the nets are represented by dots, with links being made from the central heavy dot to surrounding heavy dots). An analog of this pattern in two dimensions is shown in Figure 3c: Each net makes links to a block of near neighbours, then to a set whose members are separated by the diameter of a block, and so on, in a manner quite reminiscent of neuronal connections in the cortex.

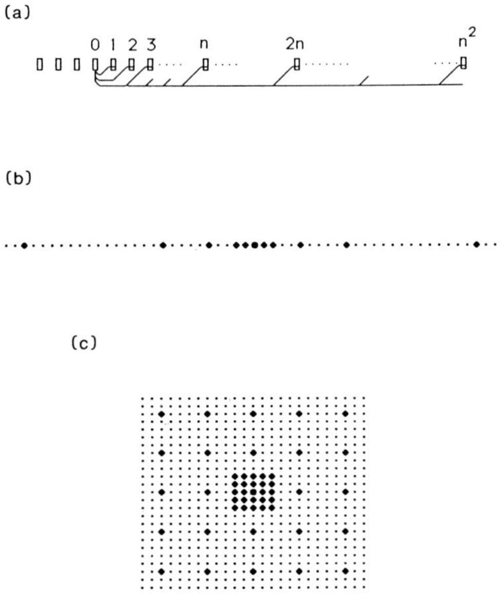

FIGURE 3

Clearly this kind of connection pattern will be most effective if the commonly occurring transitions correspond to links between topographically close nets. For instance, one might wish to arrange that all the most likely transitions occur in the block of near neighbours in the routing system of Figure 3c. However, this presupposes an essentially 2-dimensional structure for the set of possible events. Suppose instead that there is an m-dimensional "event space", in which the most probable transitions from a given event occur within some neighbourhood of that event. Then the natural routing structure is an m-dimensional analog of that in Figure 3c. To realize this as an efficient routing map on the cortex, with a minimum of long connections, we require a mapping from m to 2 dimensions which carries near points in m dimensions to near points in 2 dimensions. Since the number of neighbours increases with the dimension of a space (we think of both cortex and event spaces as discrete), it is not possible to preserve neighbourhood relations in a dimension-reducing map. However, one can seek a map f which minimizes the sum S^p of distances

$$S^p = \sum_{x} \sum_{y \in N(x)} |f(x)-f(y)|^p$$

where N(x) is the set of neighbours of x and $p>0$.

A simple model is provided by the case where m=2, the target space is one-dimensional, and where the set N(x) consists of the four (orthogonal) nearest neighbours of x. We can regard this as a map from a kxk array to the set of integers $1 \ldots k^2$. Figures 4a-c show some examples of maps which either minimize S^p or else come close to being minimal. The convention in each case is that the values of the map take consecutive (integral) values along the lines marked in the figure, increasing in the direction of the arrows. Thus, the map in Figure 4c starts with 1 in the top left-hand corner, fills in a small square from increasing columns of numbers, then fills in a larger square with several small squares made this way, and so on. The explicit numbering of the top left-hand corner of Figure 4c is shown in 4d, in the case where each small square is a 2x2 block.

The map for p=1 shown in Figure 4a is known to be minimal (35). Here, the numberings in the corners fill up a square along successive edges, thus packing the numbers close together. This is a good numbering until the square grows large, when the differences between numbers along the edges exceed those between the vertical stripes in the central region. The map can be regarded as a compromise between two modes: clustering and stripes. When $p>1$, the long distances $|f(x)-f(y)|$ are heavily penalized and this favours stripe patterns; for $p>2$ the minimal map is likely to consist of diagonal stripes (Figure 4d). When $p<1$ the long distances are more readily tolerated, and this gives rise to a pattern consisting of groups of local clusterings, as in Figure 4c. This last kind of map is particularly interesting because the mapping from blocks in the source space to regions in the target space may provide a model for the topography of cortical columnar structures, and even for the existence and arrangement of cortical areas themselves (36).

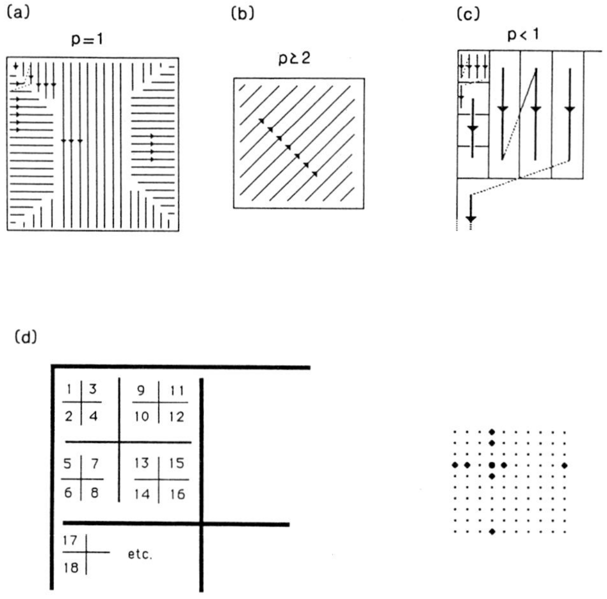

FIGURE 4

Returning to the routing problem, if we envisage such a map projecting from a high dimensional event space onto the cortex, some long connections will be needed, even for the most probable transitions (between neighbours in m-space). Figure 4d illustrates this. Let f be the map whose explicit numbering is given on the left-hand side of the figure. If we take our coordinates from the top left-hand corner, then f(2,3)=7, f(3,2)=10, and so on. We can define a map from 4 to 2 dimensions by g(i,j,m,n)={f(i,j),f(m,n)}. It has been conjectured (35) that if f is minimal then so is a map g defined in this way. The images under g of the point (2,3,3,2) and its 8 orthogonal neighbours are shown to the right in the figure. Notice that large steps are introduced between the images of (2,3,3,2) and its neighbours.

This not only shows the impossibility of preserving neighbourhood relations in dimension-reducing maps, but also gives a hint of the type of structure which is likely to emerge from the representation in memory of a multi-dimensional universe of events.

CONCLUSIONS

At first sight there is nothing difficult about storing many events in the brain. After all, there are about 10^{14} synapses in the human brain, so, if an event requires 25 bits, some 4×10^{12} events can be stored (equating bits with changes of two state synapses; for a more exact calculation along these lines see (16)). Observe, however, that encoding 25 bits corresponds to changing weights of synapses at the meeting of 5 axons with 5 dendrites. Thus, to activate the next event in a sequence, the activity of one set of 5 cells must be able to generate the activity of a second such set, and, in particular, the connections between these sets must exist. Since the average cortical cell receives about 4×10^4 synapses (21), and since there are surely many more than 4×10^4 cells involved in memory, the necessary connections are only likely to be available if there is some special topographical arrangement. This is what I have called the targetting problem.

My proposal is that events are represented by localized clusters of active neurons which map in a topographically organized fashion, so that their axons can make rich connections with the dendrites of another localized cluster of neurons. This assumption is not in itself sufficient, for some kind of routing, branching or spreading of connections would seem to be required to give reasonable flexibility in mapping. I propose a sequential mapping arrangement, but there are many other possibilities.

The line of reasoning which leads to the targetting problem is open to various objections. For instance, one could argue that if synaptic weights are not binary then memorizing could correspond to many small changes of synaptic weights in a large population of cells, so distributing the information more thinly and widely. Then the targetting problem might be avoided, since many more cells would be active in an event and the probability of making contacts between two populations representing consecutive events would be increased. Though this seems plausible and attractive, a more careful analysis suggests that there is in fact an upper bound on the event capacity, set by the "two bit limit". This implies that it is only possible to gain a factor of two over the informational limit for binary synapses by using arbitary real-valued synaptic weights.

Another objection can be directed against the notion of an event. As mentioned earlier, it is conceivable that a sequence is encoded by a single pattern of neural activity which is in some manner scanned or read out. If fewer and larger "events" are stored in this way then the targetting problem can be avoided. However, one has to explain how an item from the sequence is able to regenerate the rest of the sequence; the neural representation of the item must somehow be able to activate the whole pattern, and this would seem to imply an underlying ordered graph structure whose nodes are items and whose links are activating connections. The storage of the links in this structure is subject to the same informational constraints as the storage of the steps in a sequence.

The representation of events by few neurons can be most easily understood in terms of Barlow's "cardinal cell" schema (37), where individual neurons encode well-defined, high-order information. This raises the question of how such a representation could be generated, in itself a most challenging theoretical problem. In fact, the economical encoding of sensory information and the manipulation of the ensuing cognitive structures in memory are likely to be intimately related, and it can be hoped that the study of each will illuminate the other.

ACKNOWLEDGEMENTS

I thank Prof. H. B. Barlow and Dr. R. Durbin for helpful comments.

REFERENCES

(1) F. A. Yates, The Art of Memory, Ark Paperback, London, Melbourne and Henley 1965.

(2) N. J. Cohen & L. R. Squire, Science, 210, (1980), pp. 207-210.

(3) M. Mishkin, Nature 273, (1978), pp. 297-8.

(4) E. K. Warrington & L. Weiskrantz, Nature 217, (1968), pp. 972-4.

(5) M. S. Albert, N. Butters & J. Levin, Arch. Neurol., 36, (1979), pp. 211-216.

(6) B. Milner, in C. W. M. Whitty & O. L. Zangwill (eds), Amnesia, Butterworth , London1966, pp. 109-133.

(7) E. K. Warrington & L. Weiskrantz, Neuropsychologia, 20, (1982), pp. 233-248.

(8) G. W. Van Hoesen, D. L. Rosene & M.-M. Mesulam, Science 205 , (1979), pp. 608-610.

(9) G. M. Stratton, Psych. Rev. 24, (1917), pp. 244-7.

(10) A. R. Luria, The Mind of a Mnemonist, Jonathan Cape, London, (1969).

(11) L. Standing, Quart. J. Exp. Psych. 25, (1973), pp. 207-232.

(12) T. K. Landauer (manuscript in preparation).

(13) D. J. Willshaw, Proc. Roy. Soc. Lond. B 182, (1972), pp. 233-247.

(14) H. C. Longuet-Higgins, D. J. Willshaw & O. P. Buneman, Quart. Rev. Biophy. 3, (1970), pp. 223-244.

(15) G. Palm, Biol. Cyb. 35, (1980), pp. 1-13.

(16) G. Palm, in G. Palm & A. Aertsen (eds.), Brain Theory,

Springer, Berlin, Heidelberg 1986, pp. 211-228.

(17) J. J. Hopfield, Proc. Natl. Acad. Sci. U.S.A. 79, (1982), pp. 2554-8.

(18) T. M. Cover, IEEE Transact. Elect. Comp., 14, (1965), pp. 326-334.

(19) G. Meyer-Brötz & J. Schürmann, Methoden der automatischen Zeichenerkennung. Oldenbourg Verlag. Munich, Vienna (1970).

(20) M. Minsky & S. Papert, Perceptrons. MIT Press, Cambridge, Mass. & London, England, 1969.

(21) B. G. Cragg, Brain, 98, (1975), pp. 81-90.

(22) M. Colonnier in R.O. Schmitt, F.G. Worden, G. Adelman & S. G. Dennis (eds) The organization of the cerebral cortex, M.I.T. Press, Cambridge, Mass. and London 1981, pp. 125-152.

(23) S. Redman & B. Walmsley, J. Physiol., 343, (1983), pp. 117-133.

(24) B. L. McNaughton, C. A. Barnes & P. Andersen, J. Neurophysiol. 46, (1981), pp. 952-966.

(25) M. J. West & A. H. Andersen, Brain Res. Rev. 2, (1980), pp. 317-348.

(26) S. Gelfan, Nature 198, (1963), pp. 162-3.

(27) S. Zeki, Nature, 274, (1978), pp. 423-428.

(28) D. C. Van Essen, Ann. Rev. Neurosci, 2, (1979), pp. 227-263.

(29) D. I. Perrett, E. T. Rolls & W. Caan, Exp. Brain Res., 47, (1982), pp. 329-342.

(30) T. P. S. Powell, R. W. Guillery & W. M. Cowan, J. Anat. 91, (1957), pp. 419-437.

(31) C. Koch, T. Poggio & V. Torre, Phil. Trans. R. Soc. Lond. B

298, (1982), pp. 227-264.

(32) W. A. Spencer & E. R. Kandel, J. Neurophysiol., 24, (1961), pp. 272-285.

(33) R. D. Traub & R. Llinas, J. Neurophysiol. 42, (1979), pp. 476-498.

(34) E. M. Fenwick, A. Marty & E. Neher, J. Physiol. 331, (1982), pp. 577-597.

(35) G. J. Mitchison & R. Durbin, S.I.A.M. J. Alg. Disc. Meth., 7, (1986), pp. 571-582.

(36) R. Durbin & G. J. Mitchison (manuscript in preparation).

(37) H. B. Barlow, Perception, 1, (1972), pp. 371- 394.

The Organization of Sequential Memory: Sparse Representations and the Targetting Problem

Discussion
edited by *Ad Aertsen*

The discussion on Mitchison's presentation concentrated on the issue of finding a mapping from m-space to k-space (m>k), which optimally "carries near points in m dimensions to near points in k dimensions".

This mapping was invoked by Mitchison in connection to his proposal for an organizational design of the cortical neuronal network, which is claimed to be more optimal in terms of memory storage than a randomly connected network is. The problems with the latter type of arrangement, according to Mitchison, are caused by the conflicting requirements of 'sparse' neuronal activity on the one hand, and rich projection onto post-synaptic neurons on the other hand (the 'targetting problem'). As an alternative he proposed a scheme in which the cortical network is subdivided into a large number of disjoint sub-nets (each sub-net being a 'separate learning device'), the question now being how to find the appropriate organization of connections between the sub-nets, e.g. a connectivity such that, starting from any sub-net, all other sub-nets can be reached in several steps ('sequential routing').

The desire for a connectivity pattern of this type, especially one in which commonly used transitions correspond to connections between topographically neighbouring, or at least close sub-nets gives rise to the earlier mentioned problem of finding a mapping from the high-dimendional m-space ('event space') to a lower dimensional target space. Since in general such a mapping cannot be a achieved without certain sacrifices, this raises an optimization problem: find a mapping which most optimally fulfils a certain desired property, e.g. the preservation of small distances. For the case (m=2, k=1) Mitchison presented several such mappings, each particular solution depending on the optimization criterion used.

Questions and comments came in two flavours: those concerned with methodical issues of determining the 'optimal map', and those which relate the resulting map with findings on cortical anatomy and physiology.

How to find an 'optimal mapping' from m-space to k-space?

It was pointed out in the discussion that determining the 'optimal mapping' in fact involves two steps: (1) formalizing ones intuition about what should be optimized into the framework of a 'cost function', and (2), given a specific form of the 'cost function', how to find an optimizing solution. Important in this context was also the dependence of the solution on the specific form of the cost function (specifically on the parameter p in Mitchison's generalized cost function Cp).

Some doubt was expressed regarding the conjecture that, when mapping the m-space to a k-space in several stages (say 6-space to 3-space, followed by 3-space to 2-space), the overall optimal mapping from 6- to 2-space would be the direct product of the separately optimized mappings.

How does 'optimal mapping' relate to cortical anatomy and physiology?

The question was raised in which way observations on actual cortical connectivity (e.g. in the mouse cortex: strong local connectivity plus long ranging connections such that in 3 to 4 steps any location can be reached from any location (Palm)) may lead to possible clues regarding what has been 'optimized' in the 'design' of the cortex.

A related, but rather inverse question was whether the optimal mappings proposed by Mitchison, make any specific prediction regarding the geometrical arrangement one should be expected to observe in the functional organization of te cortical surface: Should one expect globular patches or rather stripes? What would be typical 'wavelengths' involved?

As an alternative design, Braitenberg (not being fully serious, and inspired by a mild form of number-juggling with actual observations on cortical anatomy) proposed the following arrangement. A natural way to subdivide the cortex would be into a number of units which is equal to the square root of the number of cortical elements: in that case each unit contains the same number of cells as there are such units, and there would be enough cells in each unit to interconnect all of them. For human cortex the square root is 10^{5} (corresponding to a unit of about one millimeter size which is about the size of the patches which occur in many contexts), in the mouse cortex the square root is down to $10^{3.5}$ (leading to a size which corresponds to the dendritic spread of the largest neuron in the mouse cortex, just like one millimeter corresponds to the dendritic spread of the largest neuron in the human cortex). Using Mitchison's numbers for the synapses on a pyramidal cell, such a 'square root arrangement' would allow each cell in a unit (at least the largest cells) to be informed about the state of all the other units. A necessary prerequisite for this argument to hold would be that the activity in each of the square root compartments is homogeneous, as indeed is claimed by some to hold

for the activity within a column.
This arrangement, which is more related to networks with random connectivity, according to Mitchison's argument would run into the earlier mentioned 'targetting problem' with regard to memory storage space. Still, "It's too good a coincidence to be completely accidental" (Braitenberg).

Finally, in view of the rather critical dependence of the solution to the 'optimal mapping problem' on the specific form chosen for the optimization criterion, some uneasiness was expressed regarding the robustness of conclusions with regard to the functional organization of the real cortical network.

Acknowledgement

This report was made on the basis of audio recordings of the discussion made by Ulrich Leinhos, which Shirley Wurth transformed into legible English, and on comments made by Guenther Palm and Gordon Shaw. I wish to thank Graeme Mitchison for kindly providing me with an early draft of his manuscript.

Neural Networks and Models for Learning and Memory

Gerard Toulouse, Jean-Pierre Changeux, Stanislas Dehaene and *Jean-Pierre Nadal*

SUMMARY

This text contains two parts. After an introduction on learning and memory, and generalities on the Hopfield model, the first part presents proposals for a different mechanism of learning, selective instead of instructive. In the second part, a specific neural network model, exhibiting short-term memory properties, is studied and discussed.

This paper has already been given in a similar form at the 'Fyssen Foundation Conference on 'Models of Visual Perception', March 1986, ed. by M.Imbert (Oxford U.Press).

1. Learning, memory and neural networks

1.1 Introduction on learning

Learning is adaptive change of an organism (called here the interior) interacting with an environment (the exterior). One of the goals of neural science is to discover the memory trace, the internal modifications subsequent to a learning event. Physicists are at their best when they find physical limitations (for instance, the Carnot principle in thermodynamics). One virtue of physical limitations is that they are valid for both natural and artificial objects (the Carnot principle holds true as well for an organism or a thermal machine). The limitations envisaged here are those due to the interference of distributed, superimposed, memory traces and bear on the storage capacity of a neural network.

In the ontogeny of the nervous system of an organism, there is a first distinction between innate, epigenetic and learning components. The innate component is genetic, i.e. under control of the genes; if this control is strict, it leads to hard-wired neural networks. I will reserve the term epigenetic for a process of autonomous maturation that occurs spontaneously, without influence of the exterior. Finally, learning is defined as a process of change in relation with the environment. Many subcategories of learning can be usefully distinguished such as learning during development versus adult learning, imprinting, short-term memory versus long-term memory, etc.

With such definition of learning, the immune system exhibits also something like learning and so does phylogeny as a whole, since evolution is adaptation to an environment. Indeed it is useful to think about these three systems (immune system, brain,phylogeny) jointly because they shed light on each other (1,2).

Now what are the general mechanisms of learning that have been contemplated? For purposes of comparison, I have drawn below an axis that runs from left (exterior-dominated) to right (interior-dominated).

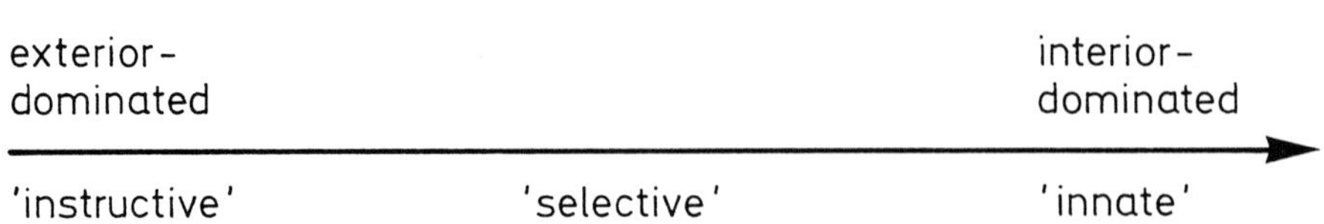

On the left are instructive mechanisms. They have been illustrated by the metaphor of the white paper on which experience draws endlessly. In its extreme form, the interior is supposed to be completely malleable, like a passive clay on which the exterior imprints its instructions. On the extreme right are innate mechanisms, where the interior knows everything a priori, and no real learning is taking place, since the role of the exterior is restricted to remind the interior of what it knew beforehand. This 'smart baby' viewpoint is nicely illustrated with a quotation from the Babylonian Talmud that I owe to my colleague Haim Sompolinsky:
"It (the embryo) is also taught all the Torah from beginning to end... As soon as it sees the light an angel approaches, slaps it on its mouth and causes it to forget all the Torah completely..."

In a somewhat intermediate position between these two opposite views stand the selective mechanisms, best known perhaps in the context of phylogeny, viz. the Darwinian theory. Selective mechanisms assume the existence of an internal generator of diversity, which is pruned by contact with the exterior. The Lamarck-Darwin opposition is one instance of the instructive-selective duality.

Twenty years ago, Niels Jerne (1), in a seminal paper on antibodies and learning, remarked:
"Looking back into the history of biology, it appears that wherever a phenomenon resembles learning, an instructive theory was first proposed to account for the underlying mechanisms. In every case, this was later replaced by a selective theory." So with Lamarck and Darwin, and so later in immunology.

The reason why instructive mechanisms are invoked at first is because of the open-endedness of the response. For instance, a higher vertebrate immune system is able to react toward molecules, that have been artificially synthesized and never existed on earth before, making hard to imagine how the genes could have somehow foreseen these artefacts. An attractive way out of this puzzle is to assume that the system's response is instructed by the stimulus.

But one reason why such theories have been abandoned after some time is because an instructive system is too much under the influence of the exterior and rapidly gets submerged. If the system is learning everything, it gets overwhelmed by the information load it receives. Indeed a neat illustration of this statement will appear below, in the overloading catastrophe of the Hopfield model.

1.2 Neural networks and Hopfield model

A few reminders. The theoretical conception of a neuron as an all-or-none device is often dated back to McCulloch and Pitts (3). The activity of a neuron, firing or resting, is described by a two-state spin, $S = \pm 1$. If the weighted sum of the inputs on neuron i, $\sum_j T_{ij} S_j$, where T_{ij} is the synaptic efficacy of neuron j on neuron i, exceeds a threshold θ_i, then neuron i fires and $S_i = +1$. If $\sum_j T_{ij} S_j < \theta_i$, $S_i = -1$. This is the simplest version of a formal neuron.

I do not know to whom credit should be given for introducing the spin-neuron analogy. As far back as three decades ago, Cragg and Temperley (4) have examined a possible analogy between the organisation of neurons and the collective interaction among compass needles or atomic magnets that leads to ferromagnetism. And they did raise good questions:

"It remains to be considered whether biological analogues can be found for the concepts of temperature and interaction energy in a physical system."

Indeed these two points were subsequently elaborated upon, twenty and thirty years later respectively, by Little and Hopfield, causing large waves of theoretical activity. Little (5) introduced the temperature-noise analogy and Hopfield (6) pointed out that an assumption of symmetric interactions $T_{ij} = T_{ji}$ allows one to define an 'energy' function

$$E = -\tfrac{1}{2} \sum_{i \neq j} T_{ij} S_i S_j - \sum_i \theta_i S_i \qquad (1)$$

such that the neuronal dynamics becomes a down hill motion on the energy landscape in configuration space (the space of all possible activity patterns of the neurons). The attractors of the dynamics have become therefore the local minima of an energy landscape.

In a perspective on neural networks theory, this Hopfield model appears as a clever step backwards. It is backwards because it is biologically implausible. Not only is it an untrue proposition that if neuron A synapses on neuron B, then neuron B does synapse on neuron A, but even if this happens to be the case, it is straight unlikely that the A → B and B → A efficacies will be equal in sign and magnitude. I think all opponents of the Hopfield model will accept without difficulty the notion that it is a step backwards. Now remains the harder task to convince them that it is a clever step nonetheless. Here are some arguments:

i) though implausible biologically, the symmetry assumption puts emphasis on the effects of feedback and reentrance, a feature that was clearly missing in the feedforward layered architectures of the perceptron models and one cause for their pitfalls;

ii) the revival of the analogy with statistical mechanics occurred at a time when the study of frustrated disordered systems, such as spin glasses, offered a much richer repertoire of collective effects than the rather poor ferromagnetic model, that was previously the dominant reference; and indeed the new theoretical tools such as the replica trick, applied to the Hopfield model, have led to a wealth of results and surprises: phase transitions, geometry of the basins of attraction, spurious memories, etc. (7,8); physicists know that they draw much benefit from theoretical surprises, and this model has opened a new frontier in statistical mechanics;

iii) although no result from such models can be taken as quantitatively realistic in itself for biology, the comparative analysis of various factors is probably pertinent; I mean for instance the relative difficulty in storing correlated versus uncorrelated memories, static patterns versus dynamic sequences; or the importance of using a local versus a non local learning rule; or, as discussed below, the question of large versus stationary memory capacity;

iv) symmetric interactions are perfectly legitimate restrictions for artificial intelligence and devices;

v) finally it is a happy, even if somewhat unexpected, fact that the results obtained assuming symmetric synapses appear remarkably robust to the introduction of dissymetries, so that the Hopfield model stands as a useful starting point for the study of more general network structures.

To end this discussion with a smile: All biologists know that it requires a rather highly evolved organism to be able to make a clever step backwards. Let me suggest that this ability is also uncommon among research scientists.

Some additional simplifying assumptions of the model concern the fully-connectedness (each neuron interacts with every other neuron), the registration of the inputs and the learning rule governing the synaptic modifications. The inputs are assumed to be complete, in the sense that an input pattern s clamps all the neurons in a particular configuration of activity $S_i = \{\mu_i^s\}$. The learning rule is the generalized Hebb rule (9).

$$\Delta T_{ij}(s) \sim \mu_i^s \mu_j^s \tag{2}$$

This is the optimal one-shot learning rule with locality in space, meaning that the modification of a synapse is only affected by the activity of adjacent neurons. It is easy to check that this synaptic modification tends to dig a hole in the energy landscape at the particular configuation $\{\mu_i^s\}$. After a sequential presentation of p patterns, the efficacies take the form

$$T_{ij} = \sum_{s=1}^{p} \Delta T_{ij}(s) \tag{3}$$

and the network functions as a distributed, content-addressable, fault-tolerant memory (6).

Theoretical effort so far has been mainly directed toward the question: What is the memory capacity? Simplest conditions (uncorrelated patterns, vanishing thresholds $\theta_i = 0$) are commonly assumed. How many patterns can then be faithfully stored on a network of given size (i.e. given number of neurons N)? A definition of the storage or retrieval quality of a memory is needed. Retrieval quality of a pattern will be measured by the following procedure: take as initial configuration of activity the nominal input pattern, let the system evolve with its neuronal dynamics and converge toward an attractor, then compute the overlap (scalar product) between the initial and final configurations

$$q^s = \frac{1}{N} \sum_i S_i\ (t = 0)\ S_i\ (t = \infty) \tag{4}$$

where $S_i(t=0) = \mu_i^s$. The sum is over all N neurons.

Of course, if the input pattern is perfectly stored, it has become the bottom of a valley and the retrieval quality is optimal, $q^s = 1$. This retrieval quality is indeed a measure of the accuracy of the positing of the valleys in the energy landscape.

The early numerical estimate by Hopfield (6) of the memory capacity turned out to be quite accurate

$$p \sim 0.15\ N \qquad (5)$$

In the Hopfield model, the number of neurons and the network connectivity are equal by construction. Now, roughly speaking, if the presentation and storage strategies are unchanged but the connectivity is smaller than the number of neurons, then the proper physical variable on the right hand side of (5) is the smaller of the two, i.e. the connectivity.

The next breakthrough was the discovery by the Israeli group (7,8) that the Hopfield model could be solved exactly, in the thermodynamic limit ($N \to 0$). As a result of this analytic tour de force, a phase diagram (temperature T, storage α) is obtained (the parameter T is a measure of noise level, the parameter α is defined as ratio of number of learnt patterns to number of neurons). The interested reader is referred to the original papers for a detailed discussion (7,8,10). For our purpose here, the main result is the existence of a sharp, discontinuous phase transition at $\alpha = \alpha_c \cong 0.14$ (for $T \cong 0$). When $\alpha < \alpha_c$, the retrieval quality is very good ($q > 0.97$) but it drops suddenly for $\alpha > \alpha_c$. Thus there is a catastrophic deterioration, due to overloading, as the learning process goes on. The more the system learns, the less it knows, and a state of total confusion is reached. All Frenchmen know this aphorism of Edouard Herriot, "culture is what remains when one has forgotten everything"; in the Hopfield model, a spin glass state is that what remains.

What is the cause of this catastrophic deterioration? The cause is interference of superimposed memory traces. Each stored pattern acts as noise on every other pattern and interferes destructively with its retrieval quality. With a non local or slow learning rule, the deterioration threshold can be pushed to higher values of α, but the overloading problem still exists.

The final diagnosis is that the model is instructive and the system gets submerged. Because it lacks resistance to exterior domination.

1.3 Biological critique

I leave aside the symmetry hypothesis (discussed above) and the critique of the generalized Hebb rule (because it would require too long developments).

The subsequent critiques are addressed to what may be called the standard version of the Hopfield model (6). Some critiques can be met by partial modifications of the model. Others would request a more drastic change in the approach. Actually Hopfield himself in a second paper (11) has proposed another conceptual framework (not considered here because it is further away from theoretical analysis).

i) Tabula rasa.
The tabula rasa hypothesis assumes that the interactions are negligible before learning, an assumption legitimate for artificial devices but not for biological networks.

ii) Absence of restrictions on synaptic modifications.
In the standard version, synaptic efficacies may grow without limits and change sign, from excitatory to inhibitory or vice versa.

iii) Uniform acquisition intensities.
There is a somewhat hidden assumption in (3) that the acquisition amplitudes (the prefactors in the synaptic modifications) are pattern-independent. Any input is thus assumed to make a similar impression, and there is no selectivity factor in the storage.

iv) Lack of hierarchical categorization.
It would be biologically desirable to obtain a memory with a hierarchical structure, so that recognition could proceed from class to sub-class, from crucial features to minor details.

v) Catastrophic deterioration.
Last but not least, the overloading problem aforementioned.

1.4 A spin glass model of learning by selection

Imagine an opposite strategy, that is, instead of starting from vanishing interactions, start from the opposite limit: the spin glass, with random excitatory and inhibitory synapses. As is now known, such a system has a remarkable energy landscape, with an abundance of attractors (number exponential in N). Let us call prerepresentions these attractors of the dynamics (i.e. local minima of the energy before learning). The distribution of the lowest local minima of a spin glass in configuration space has a property called ultrametricity (12), that means a form of hierarchical categorization. Namely the deepest energy valleys are not homogeneously distributed over the energy landscape, but there are valleys-within-valleys-within-valleys... and regions without valleys. The idea then is to prune selectively this excess of prepresentations with very small synaptic modifications associated with learning events (so that, among other things, no synaptic sign change need occur).

Now, in a realistic description of a biological network before learning, there should enter both specified (innate or epigenetic) features and random features. To know what are the relevant a priori constraints is a very hard task (it may require many further decades of work). So it is not the first thing to do in modeling. In contrast, complete randomness is easier to model and since interesting behaviour emerges already at that level, it may seem enough of an excuse for the spin glass model.

Some of you may detect a coming paradox, because earlier I told you that in the spin glass state the system is in a state of total confusion and unable to learn anything. And now I am on the way to claim that it is well possible to learn something in the spin glass state, and indeed to obtain a storage of arbitrary good quality with small synaptic modifications. More precisely, suppose I start with synaptic efficacies of random sign and of constant modulus, $T_{ij} = \pm A$. The claim is that the following synaptic modification:

$$\Delta T_{ij} = \varepsilon A \left(\frac{1}{\sqrt{N}}\right) \mu_i \mu_j \tag{6}$$

with $\varepsilon \gtrsim 2.5$, is sufficient to obtain quasi-perfect retrieval (Fig.1). Because of the factor $\frac{1}{\sqrt{N}}$, and N is supposed to be large, this synaptic modification is small (in particular, no synaptic sign change occurs in this learning event).

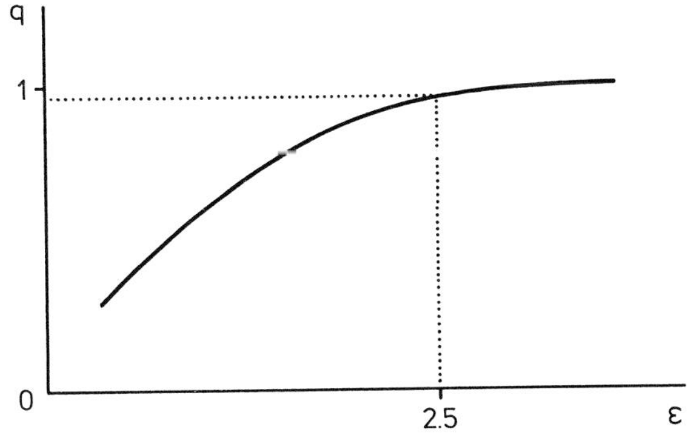

Fig.1 Sketch of the retrieval quality versus the acquisition amplitude factor ε, see formula (6), in the spin glass limit. For $\varepsilon \gtrsim 2.5$, the retrieval quality is better than 0.98.

The solution of the apparent paradox is that, in the framework of the Hopfield model, the acquisition amplitude is taken as a constant. As the number p of learnt patterns increases, $A \sim \sqrt{p} = \sqrt{\alpha}N$, and if the prefactor in (6) is fixed, $\varepsilon \sim \alpha^{-\frac{1}{2}}$, the coefficient ε is bound to fall below the safe storage limit, as learning proceeds and α increases.

However, if the acquisition amplitude is not fixed once for all, but tunable, it can be adjusted in order to get appropriate storage (of the last pattern). So the morality is that learning of a new pattern is always possible, provided that the acquisition amplitude is adjustable and properly adjusted.

Actually, in an ideal model of learning by selection, the system would learn with even smaller synaptic modifications, because the idea is not to store an arbitrary input pattern (i.e. to dig a hole anywhere in the energy landscape). Rather the idea is that after registration of an input stimulus the network activity evolves by its own dynamics toward one of its already existing attractors, i.e. a prepresentation. The attractor may be similar to the input stimulus (close in configuration space) or it may be far from it. If one considers that the learning process (the synaptic modifications) occurs over some learning time, two interesting limits appear. For a learning time smaller that the relaxation time of the neuronal dynamics, the input pattern is learnt; this is the Hopfield model. For a learning time longer than the relaxation time, it is the attractor (i.e. a prepresentation) that is learnt; and this then is a selective model. It is selective in a double sense, because the stimulus selects the attractor, but also because a stimulus that is very dissimilar to its attractor will be less well stored (because of less coherent synaptic modifications during the learning time) than a stimulus close to its attractor (this is one mechanism for the so-called 'resonance' effect).

These two limits have qualitatively different properties. As an example, in the instructive limit, learning is commutative, as visible on formula (3); since addition is commutative, the synaptic values are independent of the serial order of presentation of the patterns. That is clearly not the case in the selective limit, because each learning event changes the energy landscape and modifies the prepresentations available for the subsequent events.

Thus, despite its paucity of architecture and the lack of structural constraints (under genetic or epigenetic control), the spin glass model, with its rich energy landscape, offers a non trivial testbed for the implementation of some of the principles of learning by selection. This may come as a surprise for the biologists who would tend to equate tabula rasa and randomness, for lack of innate organisation. But the lesson drawn from the last decade in statistical physics is that there is considerable difference between vanishing and random interactions, with quite an interesting structure of the energy landscape in the latter case.

2. A model for a working (short-term) memory

2.1 Simple treatment of Hopfield model, and generalizations.

Here is a back-of-envelope derivation of the Hopfield model. It is not rigorous (some correlations are neglected), but it gives the essential results simply and rather accurately.

One defines two variables, that satisfy two equations. The first variable is the acquisition intensity for one pattern (an intensity is square of an amplitude)

$$k(s) = < [\Delta T_{ij}(s)]^2 > \tag{7}$$

where the brackets mean an average over the synapses. The second variable is the global intensity after learning p patterns

$$K(p) = < [T_{ij}]^2 > \tag{8}$$

The global intensity is modified after each new learning event, and the first equation just says that:

$$\frac{dK}{dp} = K'(p) = k(p) \tag{9}$$

For simplicity, the number of patterns p has been considered as a continuous variable (it is possible to write a difference equation, instead of a differential one, but it is more cumbersome, and largely irrelevant). Equation (9) is essentially a random walk formula, expressing additivity of the second cumulants.

The second equation gives the threshold signal-to-noise ratio for a pattern to be faithfully printed:

$$\frac{k}{K} = \frac{\varepsilon^2}{N} \tag{10}$$

As said earlier, in the discussion on equation (6), the acquisition intensity of a pattern has to be larger than some threshold, in order to ensure a good retrieval quality. The estimate for the numerical factor ε, obtained from the spin glass limit (Fig.1), is around $\varepsilon = 2.5$ (13).

Armed with these two equations (9) and (10), one can easily solve the Hopfield model, by introducing the corresponding hypotheses:

$$K(p=0) = 0 \quad \text{, tabula rasa,}$$
$$k(s) = k \quad \text{, uniform acquisition intensities.}$$

Then equation (9) tells that K(p) grows linearly:

$$K(p) = kp \tag{11}$$

When the global intensity K(p) gets larger than the critical value $\frac{Nk}{\varepsilon^2}$ given by equation (10), the signal-to-noise ratio falls below threshold for all patterns at the same time, and this is why a catastrophe occurs. Since all patterns have been printed with identical intensity, as this intensity gets below threshold, all memories sink together, each submerged by the noise created by all the others.

An estimate for the optimal capacity p_c is obtained from Fig.2

$$K(p_c) = kp_c = \frac{Nk}{\varepsilon^2} \tag{12}$$

Now if the Amit et al. (8) estimate for α_c is inserted in (12), one gets an estimate for ε, $\varepsilon \cong 2.7$, that compares very well to our spin glass estimate, $\varepsilon \cong 2.5$.

Note that k, the acquisition intensity, drops out from equation (12). There is no scale with which to compare it. In a more general treatment, where the initial values of the interactions are nonvanishing (tabula non rasa), (11) becomes

$$K(p) = K_o + kp \tag{13}$$

keeping the assumption of uniform acquisition intensities. A comparison of Fig.3 with Fig.2 reveals most simply the existence of two regimes. If $K_o < \frac{Nk}{\varepsilon^2}$, the optimal capacity is

$$p_c = \frac{N}{\varepsilon^2} - \frac{K_0}{k}$$

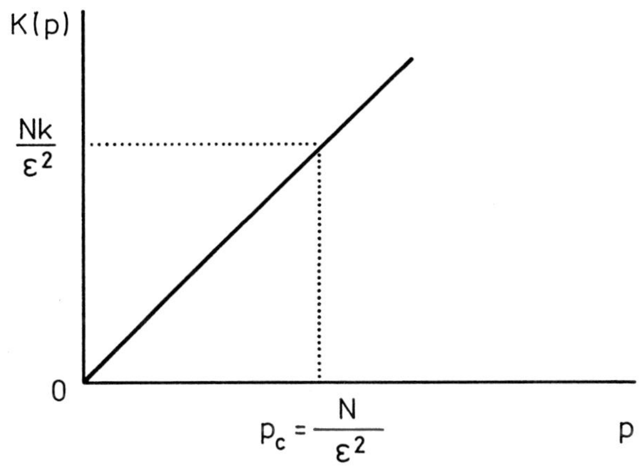

Fig.2 Variation of the global synaptic intensity, see definition (8), as a function of the number of learnt patterns, in the standard version of the Hopfield model (initially vanishing interactions). The optimal memory capacity p_c is attained when the critical signal-to-noise ratio is reached, see formula (10).

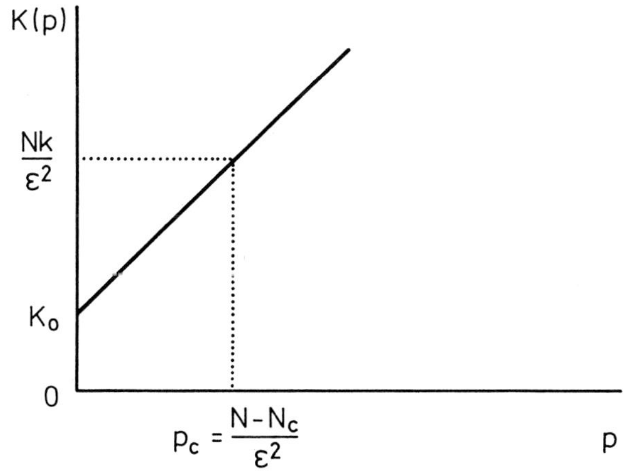

Fig.3 The difference with fig.2 is that the tabula rasa hypothesis is no longer assumed to hold. The interactions are not vanishing before learning and $K(p=0) = K_0$. See discussion of the consequences in the text.

whereas if $K_o > \frac{Nk}{\varepsilon^2}$, no faithful storage is possible. This is a rather neat manifestation of an emergence property. With a connectivity smaller than a critical value $N_c = \frac{\varepsilon^2 K_0}{k}$, the network cannot learn. Above this value (which depends on the initial value of the interactions and on the acquisition intensity), the network can function as a memory. The transition is continuous as a function of size, but it does separate two qualitatively different regimes.

2.2 A model with a stationary memory capacity

From now on, the definite aim is to find a learning process that avoids the memory ruin due to overloading. We want a system that keeps the ability to learn new patterns and, since there is no miracle, this will be at the expense of forgetting older items. In brief, a memory that keeps learning and forgetting, and is able to maintain a stationary capacity.

The simplest idea, that was called marginalist learning (13) consists in setting the acquisition intensity for new pattern learning just at the current threshold for good storage. Is not this a reasonable thing to do? In this case, equations (9) and (10) lead to

$$\frac{K(p)}{K_0} = \frac{k(p)}{k_0} = \exp\left(\frac{\varepsilon^2 p}{N}\right) \qquad (15)$$

As intuitively predictable, an exponential growth is obtained. The important point is that the memory reaches a stationary regime, with a running capacity that is about a third of the optimal storage capacity (14). Thus there is a price to pay for the ability to keep learning, and this price is a factor three reduction in the storage capacity. The forgetting process, i.e. the decay of the retrieval quality as a function of storage ancestry, has been studied in detail (13) and it does resemble the data on human short-term memory with its recency effect (for a review on the psychology literature, see Ref.15). A typical psychology experiment is done with a serial presentation of items belonging to some defined class (words, digits, musical tones, nonsense drawings, etc.) and a one second interval between successive presentations.

A drawback of the marginalist learning scheme is exponential growth, because in the long run exponential growth is not biologically plausible, nor is it convenient for artificial design. So we turn now to another scheme called learning within bounds.

2.3 Learning within bounds

Take the standard version of the Hopfield model and add the constraint (biologically reasonable) that the synaptic efficacies are bounded above and below

$$0 < T_{ij} < A \quad , \text{ for an excitatory synapse,}$$
$$-A < T_{ij} < 0 \quad , \text{ for an inhibitory synapse.}$$

It turns out that this constraint has two virtues: besides its plausibility, it produces a working memory without exponential growth. Acknowledgement is due to one paragraph in Hopfield (6) that presents similar hypothesis (without the sign constraint, though) and foresees several of its consequences. Parisi (16) has also considered a similar model. As a cautionary remark, because the confusion is often made, learning within bounds differs radically from the clipping process, also introduced by Hopfield (6).

Uniformity of the acquisition intensities will be assumed. This assumption requires justification, because it was one of the targets for critique in our discussion of the Hopfield model. The justification lies in the distinction between long-term and short-term memory. The studies of short-term memory are often done in a way that appeals to physicists, because of an effort to set the problem and the experiments as simply as possible (and thus setting aside complications, such as selectivity effects, that would require nonuniform acquisition intensities). As a result, a fairly universal figure, the magical number 7 ± 2 (17, 18), has emerged for the memory capacity. Universality is meant here with respect to sensory modalities, individual variations, etc. No such simplifications, and no such universal results, appear to be available in the more complex process of long-term memory, that seems at further reach for the theory of neural networks.

Now, if the acquisition intensity is much smaller than the bounds on the synaptic efficacies, an overloading catastrophe will occur before the bounds can play any role, and the situation remains qualitatively similar to the standard Hopfield model. However, in the opposite limit where the acquisition intensity is very large, at each learning event, all the synaptic efficacies will go bang on their bounds and the last pattern will be optimally stored, at the expense of all previously stored patterns. Thus, as first suggested by Parisi, the asymptotic memory capacity will be zero in the former case and one in the latter, and this suggests a phase transition between the two regimes, and indeed numerical observation shows that there is one (13).

This transition is intriguing for the physicist because the asymptotic memory capacity, that plays the role of the order parameter, has an unusual profile (fig.4): it goes through a maximum. This optimal capacity has been found numerically to vary linearly in N

$$C \sim 0.015\ N \tag{16}$$

Recently a whole family of working memory models (that includes marginalist learning, but not learning within bounds) has been solved analytically and exhibits a similar variation for the capacity (14).

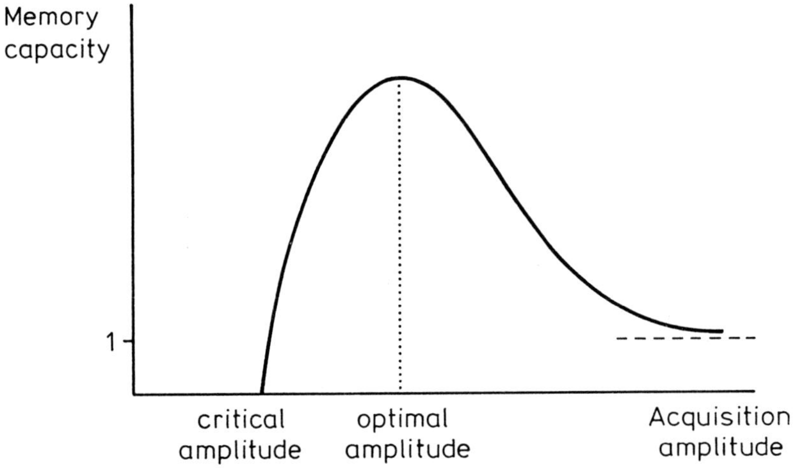

Fig.4 Sketch of the behaviour of the asymptotic memory capacity, in our model of short-term memory, as a function of the acquisition amplitude (measured relative to the bound on the synaptic efficacies).

Now, if one inserts into the l.h.s. of (16) the short-term memory span (i.e. the number 7), an estimate for the corresponding connectivity follows: N $\sim$ 500. Though smaller than the cortical connectivity (by and large, the fan in or fan out connectivity of a cortical neuron is around several thousands), it is not much smaller (specially in view of the crudeness of this simplified model) and this suggests the question:

Is the short-term memory span essentially determined by the cortex connectivity?

On one side, sits this fairly universal figure for the memory capacity, 7 ± 2, from psychology. On another side, sits another fairly universal estimate for the cortex connectivity, several thousands, from brain anatomy. A priori, the first number is much smaller than the second and they seem incommensurate. So the bold suggestion here concerns the existence of a direct and simple relation between these two numbers.

2.4 Discussion

i) Forgetting appears, in such models, as an active masking effect, distinct from a passive relaxational time decay (that might be envisaged also, as due to molecular turnover, or other cause). There has been a long-standing debate in the psychological literature between the interference and decay theories of forgetting. Massaro (15) has formulated the problem clearly:

"The critical difference between decay and interference theory is in the way people forget. Decay theory says that one forgets when one is unable to rehearse, or chooses not to do so. The memory trace fades automatically over time unless it is renewed. Interference theory assumes that forgetting occurs when a stimulus-response association is weakened by learning of another association. If no new material was learned, the subject would remember a given association permanently. This does not happen because we continuously learn new information that interferes with the old."

The review of Massaro (15) reports experiments that lend strong support to an interference mechanism. The question seems still open, though.

ii) It may seem that our relation between working memory span and connectivity is untestable, because of the difficulty of observing a change in the connectivity without simultaneous variation of many other brain parameters (like during development). There are however some cases like the song forebrain nuclei of the canari, where the neuronal connectivity can be modulated (in this case, by hormonal control) locally and accurately in an adult brain. This is just a passing remark and I shall not insist on the feasibility of short-term memory span measurements with birds...

iii) In the quantitative relation of equation (16), it would not be difficult to get a smaller coefficient by introducing stronger constraints on the synaptic plasticity or a more realistic (consequently less optimal) learning rule than the generalized Hebb. In contrast with most of our colleagues, our aim here is to find a smaller memory capacity!

iv) Time scales. There is a view on short-term memory, in the neuropsychological literature, that sees it as due to simple reverberation of the neuronal activity, without any change of the synaptic efficacies. Our model differs from this view and assumes that synaptic modifications can occur on a time scale of one second. This brings us close to the views of von der Malsburg and Bienenstock, who assume the possibility of even faster synaptic modifications (during perception time scales).

v) Optimality and fragility. The storage capacity of the Hopfield model is in some sense optimal (for a distributed memory with a local learning rule), but it is very fragile.

vi) Yerkes and Dodson rule. This classical (1908) law of experimental psychology predicts a maximum in a plot of performance versus motivation or attention. The curve of fig.4 offers a rather neat illustration of this phenomenon. It the acquisition amplitude or attention is set too low, the memory span vanishes (left part of the plot). If the attention is set too high (like students taking feverish notes), the memory span is poor also (right part of the plot). And the optimum is somewhere in between. This suggests an intriguing explanation for the primacy effect in serial recall (19). In a recall task, the subject will be initially tempted to set his attention high. Regulation toward the optimal acquisition amplitude is expected to occur as the task develops. Of course, this is assuming that the acquisition intensity is under control of a higher command center, an additional assumption on our model.

vii) In a lucid and entertaining little book (20), Simon reckons that the most striking limits on subjects capacities to employ efficient strategies, for problem solving, arise from the small capacity of short-term memory and from the time required to transfer a chunk of information from short-term to long-term memory. He contends that the contribution of neurophysiology, in the study of human behaviour, will be to explain such limits: 'Why is short-term memory limited to seven chunks; what is the physiological structure that corresponds to a "chunk"; what goes on during the five seconds that a chunk is being fixated; how are associational structures realized in the brain?' We have only begun to touch on the first question.

2.5 Conclusion and perspectives

The limitations due to interference effects of distributed superimposed memory traces are severe.

The introduction of inner constraints (such as bounds on synaptic efficacies) in the learning process can lead to improved memory performances (such as memory robustness).

These are the two main lessons.

Physicswise, an interesting phase transition has been found, with an unusual variation of the order parameter. And there are prospects for further exactly soluble models that will lead probably to further surprises. The theory of neural networks has become an active new frontier in statistical physics.

Biologywise, I shall be much more cautious. The claim is that perhaps we are raising a good question, about this very speculative relation between short-term memory span and cortex connectivity. Obviously the neural connectivity is one factor, the question is whether it is the dominant one. Perhaps one can go further and suggest that the organization of long-term memory aims precisely at fighting against this interference effect, that limits short-term memory. Another possible merit of this study is that the importance of tabula rasa, or the importance of constraints on synaptic modifications, have been dealt with at the most elementary level.

Acknowledgements

The authors are grateful to Kevin O'Regan for precious insights into experimental psychology, that have found their way into the written text. One of us (G.T.) thanks the Santa Barbara Institute for Theoretical Physics for providing a congenial setting for redaction and Miguel Virasoro and Christoph von der Malsburg for stimulating conversations.

REFERENCES

(1) Jerne,N. in The Neurosciences: A Study Program. eds.Quarton,G., Melnechuk,T. and Schmitt,F.O. (1967), The Rockefeller Univ.Press, New York, p.200.

(2) Changeux,J.P. L'homme neuronal, (1983), Fayard, Paris; Neuronal man, (1985), Pantheon, New York.

(3) McCulloch,W.S. and Pitts,W. Bull.Math.Biophys. 5, (1943), 115.

(4) Cragg,B.G. and Temperley,H.N.V. Electroencephalog.Clin.Neurophys. 6, (1954), 85.

(5) Little,W.A. Math.Biosci. 19, (1974), 101.

(6) Hopfield,J.J. Proc.Natl.Acad.Sci.USA. 79. (1982), 2554.

(7) Amit, D.J., Gutfreund,H. and Sompolinsky,H. Phys.Rev. A 32, (1985), 1007.

(8) Amit,D.J., Gutfreund,H. and Sompolinsky,H. Phys.Rev.Lett. 55, (1985), 1530.

(9) Hebb,D.O. The Organization of Behavior. (1949), Wiley, New York.

(10) Crisanti,A., Amit,D.J. and Gutfreund,H. Europhys.Letters, 2, (1986), 337.

(11) Hopfield,J.J. Proc.Natl.Acad.Sci.USA. 81. (1984), 3088.

(12) Rammal,R., Toulouse,G. and Virasoro,M.A. Rev.Mod.Phys. 58, (1986), 765.

(13) Nadal,J.P., Toulouse,G., Changeux,J.P. and Dehaene,S. Europhysics Letters, 1, (1986), 535.

(14) Mézard,M., Nadal,J.P. and Toulouse,G. J.Physique. 47, (1986), 1457.

(15) Massaro,D.W. Experimental Psychology and Information Processing. (1975), Rand McNally.

(16) Parisi,G. J.Phys. A 19, (1986) L 617.

(17) Miller,G.A. Psychological Review. 63, (1956), 81.

(18) Hitch,G.J. in Memory in animals and humans. ed. by Mayes,A. (1983), Van Nostrand Reinhold,UK. p.177.

(19) Murdock Jr. B.B. J.Exp.Psychology. 64, (1962), 482.

(20) Simon,H.A. The Sciences of the Artificial (1969), MIT Press, Cambridge.

Neural Networks and Models for Learning and Memory

Comments edited by *Shun-ichi Amari*

1. It is remarkable that many physicists are recently actively contributing to theoretical neuroscience. It indeed is welcome, because they provide with new theoretical methods of analysis. Physicists are apt at building such a neuron network model that can be treated by elaborated physical methods rather than it represents an intrinsic feature of neural information processing. However, their results are very useful if they are applicable to more general models. We can also expect that this approach will stimulate us to develop powerful methods which are useful to realistic models. The present talk is not only a good introduction to such an approach but elucidates interesting performances of a short-term memory aspect of an associative memory model by introducing a saturation characteristics of the magnitude of synatic efficacy.

2. The capacity of this model is shown to be 0.016N pattern, where N is the number of neurons in the network. Since the short-term memory capacity is proportional to N, this apparently shows a contradiction that the capacity becomes larger as memorized patterns becomes more complex. However, as the author replied, actual information patterns are coded in vectors of dimension N before entering the memory network, so that N does not directly represent the complexity of memory items.

There arises another problem common to all the associative memory network models: How patterns should be coded in an associative memory network? Is the random coding good enough? It can be shown theoretically that the random coding is good enough if each element takes on two values +1 and -1, but it is not so good when it takes on 0 and 1. We can prove that the capacity does not attain a level of order N in the latter case. It is possible to overcome this difficulty by using orthogonal learning or Widrow type learning whose ability was studies in(1).It was discussed by Palm that the capacity is more than N, if a low activity random coding is optimally chosen. However, as the author replied, this is possible only at the sacrifice of each attractor having a very narrow basin, i.e., by losing the stability.

3. The following questions are interesting in relation to a stochastic model vs. the deterministic model. It is assumed in a model analyzed by the physical method that each element behaves completely asynchronously with a special type of stochastic rule depending on the "tempreture". Are there any serious differences in its behaviour, if one uses a deterministic net with or without a synchronous change? Are there any differences if the state value of each element takes on analog values rather than takes on 1 and -1 (or 1 and 0). Moreover, in the spin glass chaotic situation, are there any relations betweeen behaviors of the present model and those of statistical neurodynamics of the randomly connected net model analyzed by Amari (2)? These seem to be very interesting theoretical questions.

Furthermore, the dynamics of the present net forms a Markov chain in the state space. There is another way to the theoretical study. The behaviour of the net can be directly treated by a dynamics of the space of all the probability distributions on the state space. This is connected with information geometry.

4. Historical remarks: Here, I would like to note some Japanese studies in early years on the associative memory models, because they are still not well known. The associative memory model by Nakano(3) (proposed in Japanese in 1969) used the synaptic efficacy of saturation type. The widths of the basins of attractors in the associative memory model were studied by Amari(4) by defining the stability number of an attractor. He also proposed an associative

memory model which memorizes temporal sequences of patterns and the stability numbers of such sequences (or cycles) were also studied. Uesaka and Ozeki (5) gave an exact analysis of the capacity of an associative memory net, where the random pattern encoder is used. The result is the same as that by Amit et al. The resolution of patterns in an associative memory model was shown in Amari (6), where "orthogonal learning" was introduced in order to eliminate the instability involved in the Widrow or generalized inverse learning method.

REFERENCES

(1) S. Amari, Biol. Cybern., 26, (1977), 175-185

(2) S. Amari, Kybernetik, 14 (1974), 201-215

(3) K. Nakano, IEEE Trans. SMC, SMC 2 (1972), 381-388

(4) S. Amari, IEEE Trans. Comp., C-21 (1972), 1197-1206

(5) Y. Uesaka and K. Ozeki, J. Inst. Electr. Comm. Eng. Japan (in Japanese), 55-D (1972), 323-330

Mathematical Theory of Self-Organization in Neural Nets

Shun-ichi Amari

SUMMARY

A mathematical framework is proposed for analyzing learning and self-organizing behaviors and capabilities of neural systems. A learning potential function is introduced to show the behaviors of synaptic modification in a single neuron, which modifies the synaptic efficacies according to input signals given from the outside information source. In order to analyze the learning and self-organizing behaviors of a neural net, we need to consider both the dynamics of neural excitations and the dynamics of synaptic modification, and also their interactions. A mathematical method of analysis is given for such a self-organizing network. The present method is applied to the mathematical analysis of the behaviors of a self-organizing neural encoder, which encodes signals given from the outside into neural excitation patterns in a neural field. It is proved that the encoder has two interesting properties. One is the amplification property by which frequently applied signals are encoded in patterns on wider regions having high resolution. The other is the categorization property by which similar signals in the outer world are automatically categorized and encoded in excitation patterns sharing the same region, so that the signal space and the neural field are automatically divided into blocks.

1. INTRODUCTION

It is helpful for understanding the brain functioning to build a neuron net model and to study its behaviors. Behaviors of a model may sometimes elucidate essential asspects of neural information processing. However, a model net is usually too much simplified compared with the actual brain. Therefore, it is sometimes possible, and important, to analyze a model net mathematically, thus elucidating possible information processing capabilities of neuron nets from a theoretical point of view. A mathematical analysis, if possible, gives us transparent understanding of possible mechanisms of neural information processing, and such mechanisms can in turn be used to build a more realistic model.

The present paper studies a unified method of mathematical analysis of self-organizing neural nets, based on the author's researches (1) $\sim$ (7). Section 2 treats learning or self-organizing behaviors of a single neuron by defining potential functions of learning. Various types of synaptic modifications are unifiedly treated in this framework. Section 3 studies dynamics of learning or self-organizing neural nets in which all the neurons cooperates to modify synaptic efficacies in the net. Section 4 is devoted to mathematical analysis of a self-organizing neural encoder of a neural field. By receiving signals from the outside, it automatically categorizes and encodes these information signals into neural excitation patterns in the neural field.

When a signal is encoded in a localized excitation pattern, an encoder defines a mapping from the signal space I to the neural field F, $m : I \rightarrow F$, such that a signal $\underline{x}$ arouses (i.e., is encoded in) an excitation pattern around the position $\xi = m\underline{x}$ in the neural field. The mapping or the encoder m is topological in some sense, and is formed by self-organization depending on the structures of I and F. We can prove some interesting properties of the encoder by mathematical analysis. 1) Amplification property : Those signals which appear frequently are mapped on a wider area of the field and have finer resolution. 2) Categrization property : Even when I is a continuous signal space, the neural encoder has an ability to categorize automatically signals into discrete structures. These properties are important when a neural encoder is used as a preprocessor to more complicated information processing systems.

2. LEARNING POTENTIAL OF A MODIFIABLE NEURON

Let us consider a mathematical model of a neuron, which receives a bundle of n input signals $x_1, x_2, \ldots, x_n$ or a vector input $\underline{x} = (x_1, x_2, \ldots, x_n)$ and emits an output signal z. These signals are considered to take on analog values representing pulse rates (firing rates) of respective axons. Let $s_1, \ldots, s_n$ be the synaptic efficacies of a neuron to the respective input axons. Then, the input-output behavior of a neuron is described by the dynamical equation of a potential u(t) of the neuron

$$(1)\quad \tau \dot{u} = -u + \underline{s} \cdot \underline{x} - h$$

where $\dot{u}$ denotes the time derivative du/dt, and τ is a time constant with which u decreases to a resting potential - h when no inputs are applied to and with which u increases in proportion to the inner product $\underline{s} \cdot \underline{x}$ of two vectors $\underline{s} = (s_1, \ldots, s_n)$, $\underline{x} = (x_1, \ldots, x_n)$, i.e., a weighted sum $\underline{s} \cdot \underline{x} = \sum s_i x_i$ of inputs. The potential u(t) represents a leaky integration of the weighted inputs, and it is assumed that an output firing rate z of the model neuron is determined as a function of the potential,

$$(2)\quad z(t) = f[u(t)],$$

where f is a monotonically increasing non-linear output function having a saturation characteristic. The equations (1) and (2) together represent the input-output behavior of a model neuron. When input signals $x_i(t)$ change very slowly compared with the time constant τ, the potential u(t) of (1) is approximated by

$$u(t) = \underline{s} \cdot \underline{x}(t) - h,$$

so that we have a simple input-output relation

$$(3)\quad z(t) = f[\underline{s} \cdot \underline{x}(t) - h].$$

This is an analog version of the McCulloch-Pitts formal neuron, which latter assumes that f is the unit step function.

The input-output behavior of a neuron changes by modifying the synaptic efficacy vector $\underline{s}$. In most learning models, the following dynamical equation is used to describe changes in the synaptic efficacies,

$$(4)\quad \tau' \dot{\underline{s}} = -\varepsilon \underline{s} + cr\underline{x},$$

where τ' is a time constant much larger than τ, $\dot{\underline{s}}$ is the time derivative $d\underline{s}/dt$, c is a constant, ε is a small decay constant, and r is a learning signal which specifies when the synaptic efficacy should be modified. In words, $\underline{s}$ increases in proportion to an input signal $\underline{x}$ when the learning signal r is positive, and decays very slowly when r = 0. The problem is to know what the learning signal is and how it is generated. Since there are many types of neurons, it is quite possible that different types of neurons use different learning signals. Some neurons receive a special input signal y in addition to the ordinary inputs $\underline{x}$, and this y plays a role of a teacher suggesting the neuron when the synaptic efficacies should be modified. For example, it is proved that the synaptic efficacies $\underline{s}$ of a Purkinje cell in the cerebellum to the parallel fibre input $\underline{x}$ are modified when a climbing fibre input y is arrived (8). When there are no such special signal as y, r may be generated by $\underline{s}$ and $\underline{x}$, for example, r could be equal to the output $z = f(\underline{s} \cdot \underline{x} - h)$. This type of modification is said to be Hebbian. More generally, r may depend on concentration of chemical neuromodifiers or activities of surrounding neurons. Here, we simply assume that r is a function of $\underline{s} \cdot \underline{x}$ and y,

$$r = r(\underline{s} \cdot \underline{x}, y).$$

When y is absent, it depends only on $\underline{s} \cdot \underline{x}$. Most modifiable synapses proposed so far belong to this class. For example, when $r = z$ or $1(z)$, where $1(z)$ is a unit step function equal to 1 when $z > 0$ and equal to 0 otherwise, learning is Hebbian. When a signal y exists and $r = y$, the learning scheme is called correlation learning. If we put $r = y - z$, Perceptron learning is obtained. The learning rule with $r = y - \underline{s} \cdot \underline{x}$ is called orthogonal or Widrow-Hoff learning.

Let us define a function R(u, y) by

$$R(u, y) = (\varepsilon/2)\ |\underline{s}|^2 - c\int_0^u r(v, y)dv.$$

Then, the learning equation (4) is rewritten as

$$\tau'\dot{\underline{s}} = -\partial R(\underline{s} \cdot \underline{x}, y)/\partial \underline{s}.$$

Since the present value $\underline{s}(t)$ at time t of synaptic efficacy is determined from the above equation, it depends on all the past history $\underline{x}(t)$ of input signals and of y(t) when it exists. It is difficult to know all the past history of a neuron. However, an input signal sequence $\underline{x}(t)$ is often regarded as a stationary stochastic process, which reflect the information structure of the outer world. In this case, it is expected that $\underline{s}(t)$ converges in some sense a value $\underline{s}$ which is determined from the stochastic property of the signals irrespective of a specific realized sample sequence $\underline{x}(t)$. To show this, we introduce an environment information source I, which produces an input signal $\underline{x}$ (and jointly y when it exists) with a joint probability distribution $p(\underline{x}, y)$. When y does not exist, only the frequency $p(\underline{x})$ of appearing $\underline{x}$ is specified.

Let $\langle\ \rangle$ be an ensemble average with respect to I, i.e.,

$$\langle a(\underline{s}, \underline{x}, y)\rangle = \int a(\underline{s}, \underline{x}, y)\ p(\underline{x}, y)d\underline{x}dy.$$

Then, it is expected that the asymptotic behavior of $\underline{s}(t)$ is very close to the solution of the averaged equation

(5) $$\tau'\dot{\underline{s}} = -\varepsilon\,\underline{s} + c\,\langle \underline{r}\ \underline{x}\rangle,$$

provided that ε is sufficiently small, τ' is sufficiently large, and $\underline{x}(t)$ is an i.i.d. sequence from I where each signal pair $(\underline{x}, y)$ is selected with probability $p(\underline{x}, y)$ and continues for a fixed time duration and then another pair is selected independently. See, e.g., Geman (9) for detailed mathematical analysis of such an equation.

The averaged learning equation can be rewritten as

(6) $$\tau'\dot{\underline{s}} = \partial L(\underline{s})/\partial \underline{s},$$

where

$$L(\underline{s}) = < R(\underline{s} \cdot \underline{x}, y) >_I.$$

We call this $L(\underline{s})$ the learning potential of a neuron under the environmental information source I. The synaptic efficacy $\underline{s}$ converges to one of the local minima of L, so that its learning behavior is understood from the shape of L determined from the learning signal r and the information source I. The local minima are determined from $\partial L/\partial \underline{s} = 0$.

We show some examples. When a special signal y exists, it usually designates the associated desired output for an input signal $\underline{x}$, so that a neuron modifies its behavior to output $y(\underline{x})$ when $\underline{x}$ is applied. This is learning of an input-output behavior. There have been proposed many types of learning schemes. For example, $r = y - z$ gives the Perceptron learning scheme, which works well when two categories of signals are li/nearly separabale. A simple learning scheme is given by putting $r = y$, and is called correlation learning, because the synaptic efficacy $\underline{s}$ converges to

$$\underline{s} = (c/\varepsilon)<y\underline{x}>_I,$$

the correlations of y and x_i. When I includes only a finite number of signals $\underline{x}_1, \ldots, \underline{x}_k$ with equal probabilities, $\underline{s}$ converges to

$$\underline{s} = (c/k\varepsilon)\sum y_\alpha \underline{x}_\alpha,$$

where y_α is the desired output to $\underline{x}_\alpha$ ($\alpha = 1, \ldots, k$). This scheme works very well when $\underline{x}_\alpha$ and $\underline{x}_\beta$ ($\alpha \neq \beta$) are orthogonal (or nearly orthogonal), and has been used as an underlying mechanism of associative memory models (10) $\sim$ (13).

However, the orthogonality of input signals is too strong a restriction. The least square error type learning is more flexible. This is approximately achieved by putting $r = y - \underline{s} \cdot \underline{x}$. We can prove that, when ε is small, $\underline{s}$ converges to

$$\underline{s} = <y\underline{x}^*> + O(\varepsilon) = \sum_{\alpha=1}^{k} y_\alpha \underline{x}_\alpha^* + O(\varepsilon),$$

where $O(\varepsilon)$ is a term of order ε and $\{\underline{x}_1^*, \ldots, \underline{x}_k^*\}$ is the reciprocal or dual set to the original set $\{\underline{x}_1, \ldots, \underline{x}_k\}$ of the input vectors. The dual vectors $\underline{x}_\alpha^*$ are linear combinations of $\underline{x}_\beta$'s which satisfy

$$\underline{x}_\alpha^* \cdot \underline{x}_\beta = \delta_{\alpha\beta},$$

i.e., equal to 1 when $\alpha = \beta$ and equal to 0 otherwise. The matrix composed of $\underline{x}_\alpha^*$'s is the generalized inverse of the matrix composed of $\underline{x}_\alpha$'s, and obviously

$$\underline{s} \cdot \underline{x}_\alpha = y_\alpha + O(\varepsilon)$$

holds, even if x_β's are not mutually orthogonal. It should be noted that the generalized inverse is unstable, i.e., it is not continuous in the mathematical sense. Here, a small constant ε plays an important role. It determines the degree of resolution of an associative memory. When two vectors, say $\underline{x}_1$ and $\underline{x}_2$, are very close, $\underline{x}_1^*$ and $\underline{x}_2^*$ become very large, thus causing instability. If we take the orthogonal learning rule of $r = y - \underline{s} \cdot \underline{x}$, two colse vectors $\underline{x}_1$ and $\underline{x}_2$ whose inner product is smaller than ε is automatically regarded as versions of one and the same signal, and no instability appears in this learning scheme. It is better than the ordinary least square or generalized inverse scheme in this sense. It is also proved in (1), (5) that, if input signals are noisy, a role of ε is played by this noise, stabilizing learning.

When there is no specific teacher signal, a neuron should generate a learning signal r by itself, as is the case with Hebbian learning. A neuron modifies in this case its synaptic efficacy $\underline{s}$ so as to adapt itself to the stochastic property of the information source I from which it receives signals. However, the learning potential L of Hebbian learning $r = z$ has a large number of local minima, and its behavior is rather confusing. If a neuron has modifiable inhibitory synapses in addition to excitatory ones, it responds to the structure of the information source I very nicely. It is plausible that the neural system uses this mechanism in forming feature extractors and encoders to the signals of the outer world, see e.g., (2), (7), (12), (14), (15). We show a mathematical analysis of this mechanism in the following.

Let us consider a neuron which receives an inhibitory input x_0 of a fixed magnitude with a modifiable symaptic efficacy s_0 in addition to excitatory inputs $\underline{x}$. Then, its input-output behavior is written as

(7) $z = f(\underline{s} \cdot \underline{x} - s_0 x_0)$,

where threshold h is omitted. When $\underline{s}$ and s_0 are modifiable with $r = z$, the averaged learning equation is written as

(8) $\tau'\dot{\underline{s}} = -\varepsilon \underline{s} + c\langle f(\underline{s} \cdot \underline{x} - s_0 x_0)\underline{x}\rangle$,

(9) $\tau'\dot{s}_0 = -\varepsilon s_0 + c'\langle x_0 f(\underline{s} \cdot \underline{x} - s_0 x_0)\rangle$,

where c' is a constant different from c.

Let us consider in the beginning a simple case where the information source I includes k signals $\underline{x}_1, \ldots, \underline{x}_k$. It is assumed that

$$|\underline{x}_i| = 1, \ |\underline{x}_i \cdot \underline{x}_j| \leqq b, \ (i \neq j).$$

A neuron with synaptic efficacies $(\underline{s}, s_0)$ is said to be a detector or representative cell of signal $\underline{x}_i$, when it is excited in response to $\underline{x}_i$ but is not excited by any other signals $\underline{x}_j$ in I. In other words, a neuron is a detector of $\underline{x}_i$, when

$$\underline{s} \cdot \underline{x}_i - s_0 x_0 > 0,$$

$$\underline{s} \cdot \underline{x}_j - s_0 x_0 < 0, \ (j \neq i).$$

We search for the condition which guarantees that a detector of any $\underline{x}_i$ in I can be formed by learning under I as a stable equilibrium solution of (8), (9). In this case, the equations are multi-stable, having k stable equilibria corresponding to each $\underline{x}_i$. If this is the case, and if there are a number of neurons which receive common signals from the same I, some neurons become detectors of $\underline{x}_1$, other neurons become those of $\underline{x}_2$, and so on, so that detectors of every signal are formed by self-organization. We may regard this set of neurons as an self-organizing encoder for an information source I given from the outside. We show a result of mathematical analysis without proof (see (2), (6), etc.).

Theorem 1. The equations (8), (9) have equilibrium solutions corresponding to all the $\underline{x}_i$'s in I but no other equilibrium solutions, when

$$1 > c'x_0^2/c > (1 + b)/2.$$

It is interesting that the constant $\lambda = c'x_0^2/c$ represents the resolution of signal detectors, because a detector of a signal $\underline{x}_i$ is responsive to those signals $\underline{x}$ satisfying

$$\underline{x}_i \cdot \underline{x} > \lambda,$$

where $|\underline{x}| = 1$. Hence, if two signals $\underline{x}_i$ and $\underline{x}_j$ are very close in I such that $\underline{x}_i \cdot \underline{x}_j > \lambda$, then a detector which is responsive to $\underline{x}_i$ but not to $\underline{x}_j$ cannot be formed in this case, but a common detecotor responsive to both of $\underline{x}_i$ and $\underline{x}_j$ can be formed by self-organization.

The mathematical analysis predicts that the resolution depends on c, c', x_0 in the above manner. Therefore, we can test and prove this resolution mechanism by experiments, if it is possible to change c, c' or x_0 by the use of effective chemicals.

3. SELF ORGANIZATION AND LEARNING OF NEURAL NETS

We have so far studied the learning behavior of a single neuron by introducing a learning signal r. However, a neuron is connected with many other neurons to form a network, and its dynamical behavior as well as learning behavior depends strongly on other neurons. This is because the value of the learning signal r of a neuron is influenced from the activities of other neurons. In order to understand the learning or self-organization behavior of a neural net or system, we need to study the dynamical behavior of neural activities, the dynamical behavior of synaptic efficacies, and their interactions at the same time.

Let N be a net consisting of n neurons $E_1, \ldots, E_n$, which are connected to one another with synaptic efficacy w_{ij} from E_j to E_i. When an excitatory vector signal $\underline{x}$ and an inhibitory signal x_0 are applied in common to all the neurons from an information source I, the dynamic behavior of N is written as

(10) $\tau \dot{u}_i = - u_i + \underline{s}_i \cdot \underline{x} - s_{0i}x_0 + \sum_j w_{ij}z_j$,

where u_i is the membrane potential of E_i, $\underline{s}_i$ is the synaptic efficacy vector of E_i receiving $\underline{x}$, s_{0i} is the inhibitory synaptic efficacy of E_i, and

$$z_j = f(u_j)$$

is the output pulse rate of E_j. Let $U_j(\underline{x})$ be the equilibrium solution of the above equation, when $\underline{x}$ is applied from the outside. Then, it shows the excitation pattern of the net in response to $\underline{x}$, and is given by the solution of the non-linear equation,

(11) $U_i(\underline{x}) = \underline{s}_i \cdot \underline{x} - s_{0i}x_0 + \sum_j w_{ij} f(U_j)$,

obtained from $\dot{u}_i = 0$.

All of the synaptic efficacies $\underline{s}_i$, s_{0i}, w_{ij} are eligible for modification by learning or self-organization. However, their roles are different in information processing. It may be possible to assign two prototypes of information processing to them as follows. One is a network in which $\underline{s}_i$ and s_{0i} are modified but w_{ij} are fixed. Such a network plays a role of an information encoder by which signals $\underline{x}$ are encoded in activity patterns $\underline{z} = (z_i)$ or potential patterns $U_i(\underline{x})$ of N. The encoding is modified by self-organization such that it adapts to the stochastic property of information signals in I. We analyze this type of self-organization in the following. There exists another type of learning network in which synaptic efficacies w_{ij} of mutual connections are modified, but its encoding mechanism is fixed. The correlation type associative memory network is such an example, and the programming mechanism of a parallel information processor (16) might be realized in such a network.

The dynamic behavior of synaptic modification is written as

(12) $\tau' \dot{\underline{s}}_i = - \varepsilon \underline{s}_i + c\langle r_i \underline{x} \rangle$,

(13) $\tau' \dot{s}_{0i} = - \varepsilon' s_{0i} + c'\langle r_i x_0 \rangle$,

(14) $\tau' \dot{w}_{ij} = - \varepsilon'' w_{ij} + c''\langle r'_i z_j \rangle$,

where r_i and r'_i are possibly different learning signals of the synapses of E_i. Since r_i and r'_i in general depend on the output $z_i = f(u_i)$ which is subject to the dynamic equation (10), two sets of equations are mutually related. However, when the time constant τ of dynamics of excitation is much smaller than the time constant τ' of synaptic modification, it is possible to separate the two sets of equations by using the so-called adiabatic approximation or the slaving principle. This is to assume that the synaptic efficacies are fixed, when one solves the dynamic equation (10), and to use its equilibrium solution $U_i(\underline{x})$ when one solves (12) $\sim$ (14). Obviously $U_i(\underline{x})$ depends on all the synaptic efficacies so that the two types of dynamics are still coupled.

In the following we study the ability of a neural net to form an information encoder by self-organization, we assume that w_{ij} is not modifiable but fixed and that the learning signal r_i is equal to $1(z_i)$, i.e., is equal to 1 when $U_i(\underline{x}) > 0$ and is equal to 0 otherwise. Then, a signal $\underline{x}$ is encoded in, or represented by, an activity pattern vector $\underline{U} = \{U_1(\underline{x}), \ldots, U_n(\underline{x})\}$ in the network. This representation or encoding $U_i(\underline{x})$ is the equilibrium solution of (11) and it obviously depends on $\underline{s}_i$, etc., so that it changes as self-organization takes place. In order to show the changes in the representation patterns $\underline{U}(\underline{x})$, we put $\underline{V} = (V_i(\underline{x}))$,

$$V_i(\underline{x}) = \underline{s}_i \cdot \underline{x} - s_{0i}x_0$$

which obviously changes as $\underline{s}_i$ and s_{0i} do. By differentiating V_i with respect to t, we have

$$(15)\quad \tau'\dot{V}_i(\underline{x}) = \tau'(\dot{\underline{s}}_i \cdot \underline{x} - \dot{s}_{0i}x_0).$$

Hence, by substituting (12), (13), we have the equations

$$(16)\quad \tau'\dot{V}_i(\underline{x}) = - V_i(\underline{x}) + k(\underline{x}, \underline{x}') * f[U_i(\underline{x}')],$$

$$(17)\quad U_i(\underline{x}) = V_i(\underline{x}) + \sum_j w_{ij} f[U_j(\underline{x})],$$

which determine how the neural encoding changes by self-organization, where we put

$$(18)\quad k(\underline{x}, \underline{x}') = c\underline{x} \cdot \underline{x}' - c'x_0^2$$

$$k(\underline{x}, \underline{x}') * f[U(\underline{x}')] = \langle k(\underline{x}, \underline{x}')\, f[U(\underline{x}')]\rangle_I = \int p(x')k(\underline{x}, \underline{x}')f[U(\underline{x}')]d\underline{x}'.$$

By solving (16) and (17), we know how the representation $\underline{U}(\underline{x})$ of $\underline{x}$ in the net changes. We analyze the equations in more detail in the next section, where we treat a net of homogeneous spatial structure, that is a neural field.

4. CHARACTERISTICS OF SELF-ORGANIZING NEURAL ENCODER

In order to analyze characteristics of a neural encoder, we consider a self-organizing neural field F in which neurons are arranged to form a continuum. Let $\xi = (\xi_1, \xi_2)$ be a coordinate system of the neural field, and let $u(\xi)$ be the average mebrane potential of neurons at spatial position ξ. Then, the dynamical equation of neural excitation corresponding to (10) is written as

$$(19)\quad \tau\frac{\partial u(\xi, t)}{\partial t} = - u(\xi, t) + \underline{s}(\xi) \cdot \underline{x} - s_0(\xi)x_0 + \int w(\xi, \xi')f[u(\xi', t)]d\xi'$$

when $\underline{x}$ is applied to the field. Here $\underline{s}(\xi)$ is the synaptic efficacy vector to input $\underline{x}$ of the neurons at position ξ, $s_0(\xi)$ is the inhibitory synaptic efficacy to x_0 of the neurons at ξ, $w(\xi, \xi')$ is the synaptic efficacy of the recurrent connection within F from the neurons at ξ' to the neurons at ξ(Fig. 1). Let $U(\xi, \underline{x})$ be the equilibrium solution of (19), satisfying

$$(20)\quad U(\xi, \underline{x}) = V(\xi, \underline{x}) + w \circ f[U(\xi, \underline{x})],$$

where

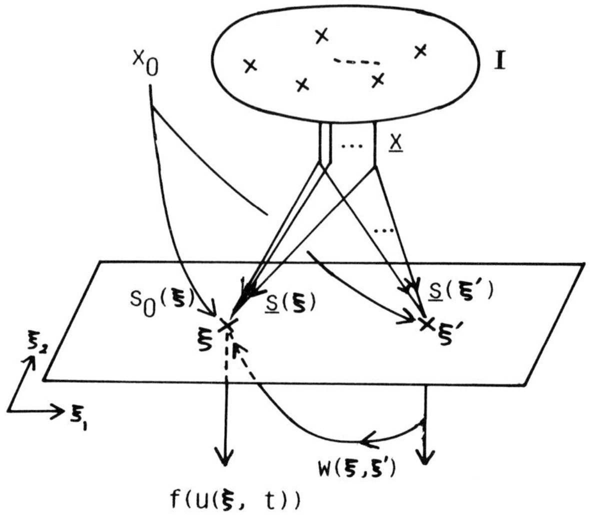

Fig.1

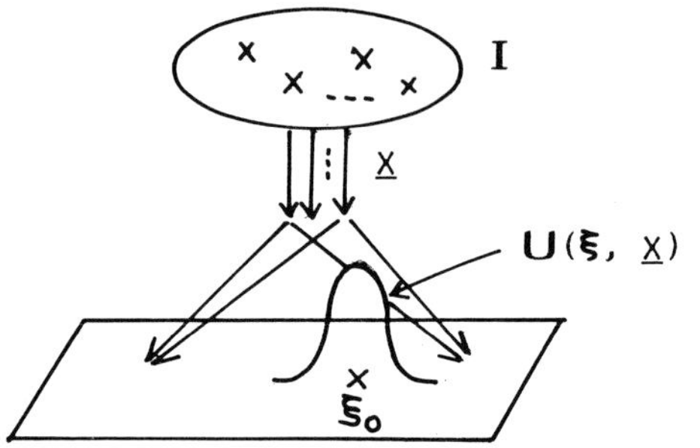

Fig.2

$$V(\xi, \underline{x}) = \underline{s}(\xi) \cdot \underline{x} - s_0(\xi)x_0,$$

$$w \circ f[U] = \int w(\xi, \xi')f[U(\xi', \underline{x})]d\xi'.$$

The quantity $V(\xi, \underline{x})$ denotes the weighted sum of stimuli which a neuron at position ξ receives when $\underline{x}$ is applied to from the outside. Hence, it changes as the synaptic efficacies $\underline{s}$ and s_0 are modified. On the other hand, $U(\xi, \underline{x})$ represents the equilibrium activity pattern aroused in the field when $\underline{x}$ is applied. If $U(\xi, \underline{x})$ is positive in a small neighborhood of ξ_0, we may say that a signal $\underline{x}$ is represented by, or encoded in, a position at (or more procisely a region around) ξ_0, because only the neurons around ξ_0 are excited in response to $\underline{x}$ (Fig. 2).

The present paper studies how signals $\underline{x}$ of a given information source I are to be represented in a neural field and to be encoded in its activity patterns by self-organization. The equations of synaptic modification are rewritten as follows, where $w(\xi, \xi')$ is assumed to be fixed,

$$(21) \quad \tau' \partial \underline{s}(\xi, t)/\partial t = - \underline{s}(\xi, t) + c < f[U(\xi, \underline{x}')]x' >_I,$$

$$(22) \quad \tau' \partial s_0(\xi, t)/\partial t = - s_0(\xi, t) + c' < f[U(\xi, x')] > x_0.$$

From the above equations, we derive the following equations which more directly represent how the encoding scheme changes by self-organization. By differentiating V with respect to t and by substituting (21), (22), we derive

$$(23) \quad \tau' \partial V(\xi, \underline{x})/\partial t = - V(\xi, \underline{x}) + k * f[U(\xi, \underline{x})],$$

where

$$k(\underline{x}, \underline{x}') = c\underline{x} \cdot \underline{x}' - c_0 x_0^2$$

and

$$k * f[U] = \int k(\underline{x}, \underline{x}') f[U(\ , \underline{x}')] p(\underline{x}') d\underline{x}'.$$

This equation together with

$$(24) \quad U(\xi, \underline{x}) = V(\xi, \underline{x}) + w \circ f[U]$$

represents how the neural representation or encoding changes. Its final equilibrium solution $\bar{U}(\xi, \underline{x})$ satisfies

$$(25) \quad \bar{U}(\xi, \underline{x}) = w \circ f[\bar{U}] + k * f[\bar{U}].$$

The connectivity function w represents the inner structure of recurrent connections in the field F. When $w(\xi, \xi')$ is of an on-center off-surround type such that neurons are connected excitatorily, i.e., $w(\xi, \xi') > 0$ when $|\xi - \xi'|$ is small, and are connected inhibitorily, i.e., $w(\xi, \xi') < 0$ when $|\xi - \xi'|$ is large, this w naturally represents the topology of F. We assume this type of connections in order to guarantee that each $\underline{x}$ is encoded in a local excitation in F. On the other hand, the correlation function $k(\underline{x}, \underline{x}')$ represents how two signals $\underline{x}$ and $\underline{x}'$ are correlated, representing the closeness or the topology of signals in I. Therefore, it is interesting to know how the signals in I are mapped to F such that the topology of I is compatible with that of F, which in general is different from that of I. In other words, the problem is how two neighboring signals $\underline{x}$ and $\underline{x}'$ in I can be represented by neighboring positions in F. Such a property is important in the formation of retino-tectal connections (4), (17), the formation of feature extractors (2), (12), (14) in an early stage of development, and the neural representation in the somato-sensory cortex (18).

In order to study such topological properties of self-organizing neural encoders, we use a simple mathematical model in which signals $\underline{x}$ are arranged in a one-dimensional region in I, i.e., they are concentrated on a curve in I. Then, by introducing a coordinate (a scalar parameter) θ on the curve, the signals are specified by θ as $\underline{x}(\theta)$. We further assume that the neural field F is one-dimensional, i.e., ξ is a scalar. Then, I is essentially one-dimensional and so is F, so that the encoding should be very simple in this case. Let us put

$$U(\xi, \theta) = U(\xi, \underline{x}(\theta)),$$

which shows that a signal at θ (i.e., $\underline{x}(\theta)$) is represented at ξ_0 in F around which $U(\xi, \theta)$ is positive (Fig. 3). The correlation function becomes

$$k(\theta, \theta') = k[\underline{x}(\theta), \underline{x}(\theta')],$$

and

$$k * f[U] = \int p(\theta')k(\theta, \theta')f[U(\xi, \theta')]d\theta',$$

where $p(\theta) = p(\underline{x}(\theta))$.

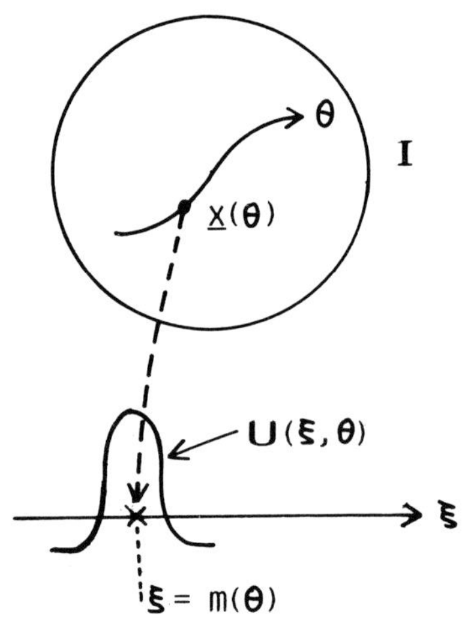

Fig. 3

We analyze the following two interesting problems by using the equations (24) and (25). 1) When some signals appear more frequently than others, does the encoding have such an amplification property that those frequently applied signals are represented in a wider regions in F having accurate resolutions? 2) When $\underline{x}(\theta)$ is continuous, is its map $U(\xi, \theta)$ continuous in ξ, too? Or does a neural encoding have such a categorization property that similar signals are categorized into one class and that each class of signals are encoded in the same region in the field F? If this is the case, both the field F and the signal space I are divided into regions or blocks, as can be seen in the columnar microstructures in the brain. We show a possible mathematical mechanism responsible for keeping such a neural structure stably.

In order to prove that the neural field has these properties, we assume that the information signals are homogeneously arranged such that

$$k(\theta, \theta') = k(|\theta - \theta'|)$$

holds and that $k(\theta)$ is a unimodal function. Let a signal $\underline{x}(\theta)$ be encoded in an excitation pattern whose center is at $\xi = m(\theta)$ and whose width is $a(\theta)$, i.e., $\bar{U}(\xi, \theta)$ is positive in the interval $(m(\theta) - a/2, m(\theta) + a/2)$. The function $\xi = m(\theta)$ shows the position in F at which $\underline{x}(\theta)$ is encoded, and its derivative

$$d\xi/d\theta = m'(\theta)$$

shows the amplification factor that $d\theta$ in the signal space is enlarged by a factor $m'(\theta)$ in the neural representation. However, if the width $a(\theta)$ of excitation is large, the excitation patterns of signals $\underline{x}(\theta)$ and $\underline{x}(\theta + d\theta)$ may overlap so that their separation is not good. Therefore, we define the resolution of a signal $\underline{x}(\theta)$ in the neural encoding by

(26) $\mathrm{res}(\theta) = m'(\theta)/a(\theta)$.

Under the assumption that the correlation function is very narrow and may be approximated by the delta function minus a constant

$$k(\theta) = \delta(\theta) - c,$$

we can prove the following theorem (4), which shows that the amplification property exists in our self-organizing neural encoder.

Theorem 2. The resolution function $\mathrm{res}(\theta)$ is in proportion to the relative frequency $p(\theta)$ of signals,

$$\mathrm{res}(\theta) \propto p(\theta).$$

This result may be compared with physiological experiments on the somato-sensory cortex, where the neural representation is shown to be modified in some degree by self-organization (18).

We next prove that our neural encoding system has the categorization property. To show this, we assume that $p(\theta)$ is constant over the signal space, and $k(\theta)$ cannot be approximated by (27). Then, by adopting an adequate scaling, the field equation has a uniform equilibrium solution

$$\xi = m(\theta) = \theta ,$$

$$a(\theta) = \text{const.}$$

Signals $x(\theta)$ in a curve in F are mapped continuously in positions $\xi = \theta$ of F by this solution (Fig. 4). However, this continuous representation cannot be realized, unless the solution is stable. When this is not stable, computer simulated experiment show that the region on which $U(\xi, \theta) > 0$ is divided as is shown in Fig. 5. This shows the categorization property, and each block is a category of signals which are represented in the same region in the field F. If

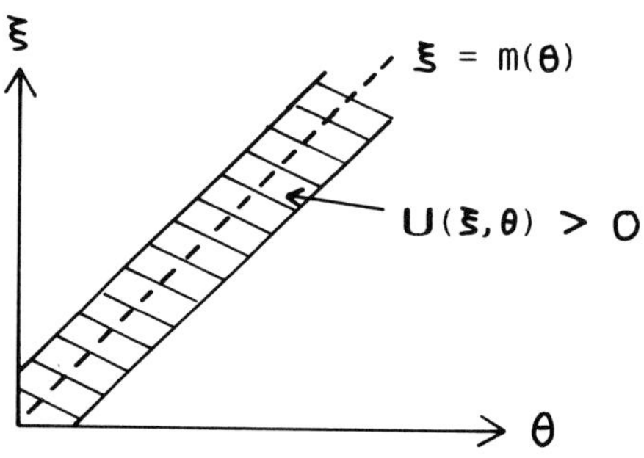

Fig.4

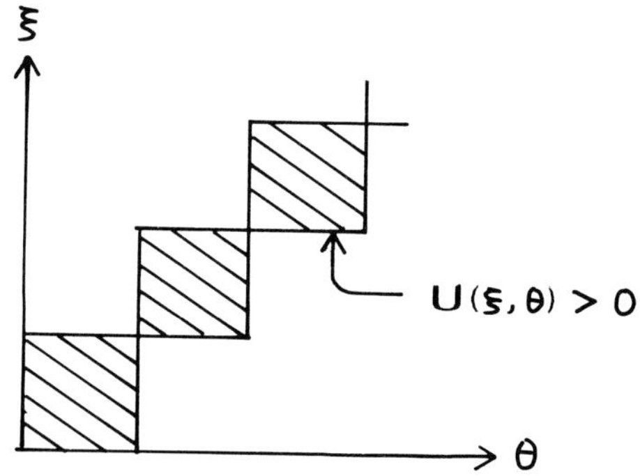

Fig.5

there are other types of neurons in the field, it is possible that each region has a further microstructure in which the same category of signals are encoded in different micropatterns. This is a way to compromise the different topologies of I and F, as is seen in the microcolumnar structure of the visual cortex. We study the variational equations of (23) and (24) around this continuous uniform solution. Since it needs a complicated mathematical analysis, we do not show here its exact mathematical analysis (5), (6), but only the result. Let $K(\theta)$ and $W(\xi)$ be, respectively,

$$K(\theta) = \int_0^{\theta} k(t)dt, \quad W(\xi) = \int_0^{\xi} w(\eta)d\eta.$$

Theorem 3. The continuous uniform solution is stable when $k(a) < 0$, and is unstable when $k(a) > 0$, where a is the width of the excitation in the field F aroused by a signal and is determined from

$$W(a) + K(a) = 0.$$

Roughly speaking, the theorem shows that the neural encoder has a categorization property when the range of effective correlations of signals is larger than the effective on-center range in F. This categorization is observed in various neural fields in visual cortices, somato-sensory cortices, and others. This categorization is an excellent way of compromising two different topologies of I and F. We have proved it in a very simple model that both F and I are one-dimensional, but it is plausible that the categorization property generally holds in neural self-organization.

5. CONCLUSION

We have shown a method of mathematical analysis of self-organizing neural systems. The bahaviors of a self-organizing neural encoder are analyzed by this method. The amplification property and the categorization property are proved to exist in such an encoder. A neural encoder is used as a preprocessor to a more complex association network or parallel processor, whose behaviors can be also analyzed by extending the present method. It is important to study a neural memory system in which both the modifiable encoding ability and association mechanism exist at the same time. It has an addressing system related to a modifiable neural encoder and an association mechanism at the same time.

6. REFERENCES

(1) S. Amari, Biol. Cybern. 26 (1977) 175-185

(2) S. Amari, A. Takeuchi, Biol. Cybern. 29 (1978) 127-136

(3) S. Amari, Biol. Cybern. 27 (1977) 77-87

(4) S. Amari, Bull. Math. Biol. 42 (1980) 339-364

(5) A. Takeuchi, S. Amari, Biol. Cybern. 35 (1979) 63-72

(6) S. Amari, IEEE Trans. Systems, Man & Cybern. SMC-13 (1983) 741-748

(7) S. Amari, Mathematical Theory of Neural Nets (in Japanese), Sangyo-Tosho, Tokyo 1978

(8) M. Ito, The Cerebellum and Neural Control, Raven, New York 1984

(9) S. German, SIAM J. App. Math. 36 (1979) 86-105

(10) K. Nakano, IEEE Trans. Syst. Man Cybern. SMC-2 (1972)

(11) S. Amari, IEEE Trans. Comp. C-21 (1972) 1197-1206

(12) T. Kohonen, Self-Organization and Associative Memory, Springer, Berlin 1984

(13) J. A. Anderson, Math. Biosci. 14 (1972) 197-220

(14) Ch. von der Malsburg, Kybernetik 14 (1973) 85-100

(15) E. L. Bienenstock, L.N. Cooper, P.W. Munro, J. Neurosci. 2 (1982) 32-48

(16) D. E. Rumelhart, J. L. McClelland (ed.), Parallel Distributed Processing, Vols. 1, 2., MIT Press, Cambridge, Mass. 1986

(17) D. J. Willshaw, Ch. von der Nalsburg, Proc. Roy. Soc. B-194 (1976) 431-445

(18) M. M. Merzenich, this volume (1987)

Mathematical Theory of Self-Organization in Neural Nets

Comments
edited by *Gordon Shaw*

The self-organization of Amari's network of neurons into small sub-assemblies as a result of input experience is a major result. The dependence on the various details of the calculations need to be carefully examined since these results are so important. Specifically, some sort of comparison could actually be made with Mountcastle's minicolumns as described in his organizational principle of cortex. Furthermore, the results presented here by Merzenich on the many striking features observed in his laboratory on the reorganization of somatosensory cortex should be addressed by such models as presented by Amari.

Remaining questions are:

- Can some theoretical statements be made about the relative number and strength of the connections of the neurons in a sub-assembly versus the corresponding quantities between neurons in different sub-assemblies in a more general situation? In other words, are there more connections within a sub-assembly than between them? If so, then one could use this feature to help establish an anatomical basis for a sub-assembly.

- Are these self-organizing networks primarily to be considered as "processors" or do they also involve storage and recall of associative memories?

It would be of interest to be able to assess the relative merits of plasticity in the input versus the internal synapses in establishing Amari's results. Similarly, we would like to know the number of input units as compared to the number of internal units. These quantities should perhaps be related to the magnitude of the processing capabilities or of the memory storage for the network. These questions were of large interest in the talks of Mitchison and Toulouse.

Learning Control Structures with Neuron-Like Associative Memory Systems

Enis Ersü and *Henning Tolle*

SUMMARY

The study of the human brain and its theoretical and experimental outcomes are sources from which principles can be derived to design control loops with information processing structures similar to those of the the biological system. The paper discusses a learning control system concept which combines in its macro-structure the human problem solving strategy as a basic control scheme and associative memories as models of the information processing in neuronal networks in its micro-structure. The latter are used for system descriptions where the information is memorized as a nonlinear input-output mapping of stimulus-response type. After the presentation of the basic control concept some structural variations are also given which show the similarity to the biological system.

---

* : This work was performed at the Technical University Darmstadt, Institute of Control Engineering, Dept. Control Systems Theory. Financial support has been provided in part by Deutsche Forschungsgemeinschaft and Stiftung Volkswagenwerk.

I. INTRODUCTION

The application of the principles of biological information processing to real technical tasks has not been attempted very often except in a few cases of without promising any generality. However, the results of psychological and neurological research on the human brain are not only interesting from the point of view of providing physical explanations to mental effects, but also from a cybernetical viewpoint of providing design principles for man-made systems.

Optimized by evolution, the human brain - which can be regarded as a highly sophisticated, self-organizing system with tremendous abilities to learn, adapt, associate and influence or change its environment - displays properties needed for the control of complex systems.

In recent years the application of learning, adaptive or self-tuning methods to automatic control systems has become an important area of research (8,14,15). This is due to the fact that most of the complex industrial processes are of nonlinear nature, but control theory deals mostly with linear systems. The problem lies not only in the difficulty of modelling nonlinear systems but also in finding some classes of nonlinear systems to which then a somehow standard control design algorithm can be applied. The learning or adaptive control schemes have been proposed to bridge this gap of missing mathematically prestructured description of nonlinear complex systems.

This paper describes a learning control method motivated by neurobiological and psychological research on the human brain, trying to make some of its tremendous abilities to learn, to adapt, to associate and to prepare signals to influence and/or change its environment applicable to man-made systems. Actually the method is a synthesis of the macro-intelligence given by human problem solving capabilities and the micro-intelligence to be found in the basic working structure of neuronal networks. The main features of the concept which has been developed and implemented by E. Ersü (see 3,4,5) are that a predictive model of the unknown environment and the control strategy evaluated via a goal oriented performance index working with the predictive model are both treated as general mathematical mappings which are represented and stored on-line in associative memory systems similar to information storage in neuronal networks.

II. HUMAN PROBLEM SOLVING: AN INTELLIGENT CONTROL STRATEGY

The human problem solving behaviour as a higher mental process is an intelligent connection from <u>perception</u> to <u>action</u> via <u>reasoning</u> on the basis of past and present experiences. The most important properties that contribute to such an intelligent behaviour are see (6,9,12) :

- modifiable predictive model of the unknown environment
- the ability to plan actions and action sequences which represent solutions to upcoming problems by using the predictive model
- the ability of generalizing the past experience during perception, action and, of course, planning for similar problem situations.

The act of perception and the act of planning result in memorization of what is perceived and planned, that means in two corresponding memories: one for the predictive model, one for the strategy selected by the planning process. The ability of generalization is to be handled by the way of memorization and will be dealt with in section III.

In terms of control theory the perception corresponds to learning of the predictive model of the process at hand, the reasoning to weighting possible actions by using the predictive model together with a goal oriented perfomance index, the generalization to suppressing of planning activities when for a certain situation the best action is already learnt and now a very similar situation arises.

Fig.II.1. illustrates this basic concept in form of a control system. The short term memory is needed for the environmental information of the near past as unique characterization of the actual situation. The shaded associative memories (long term memories LTM) store the predictive model for planning and the selected good actions for reaction in "similar" situations without reference to the planning level, " similar" being dependent on the generalizing capabilities in the information storage cycle.

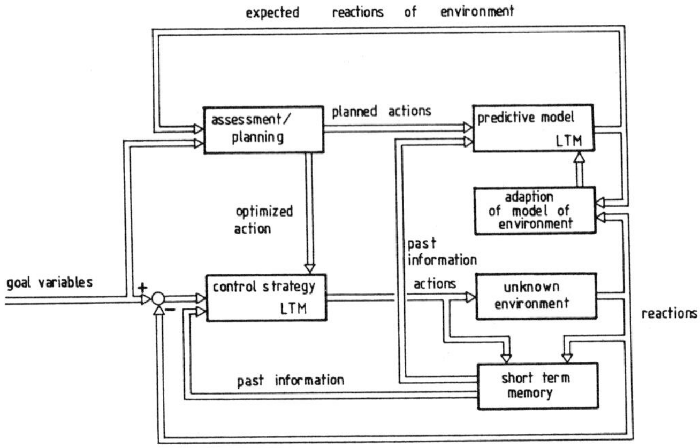

Figure II.1. Human problem solving procedure as learning control system concept.

III. INFORMATION STORAGE IN NEURONAL NETWORKS: ASSOCIATIVE STIMULUS-RESPONSE MAPPING AS GENERAL SYSTEM REPRESENTATION

Fig.II.1. is an assumption on the macrostructure of learning in the human brain. The supporting microstructure is given by the neuronal networks, and huge progress has been made since the beginning of this century - especially in the last two decades - in understanding their functioning by theoretical and experimental research. The general result is that incoming information pattern represented by dendritic activity of the specific neuronal network is mapped through weighted synaptic connections as a coding onto an output information pattern, which is the axonal activity of the network. This mapping of afferent information (stimulus) onto an efferent one (response) can be modelled from a technical point of view by an associative memory system (10,17,18) and is also one of the basic elements of the psychological association theory.

An associative memory system is by the way an information storage in which the location of the output information is addressed by the content of the input information only (10). This type of information retrieval - which is an outstanding feature of the associative neuronal network models - avoids long searching procedures as they are necessary in general in the conventional memory systems.

In classical system theory input- output descriptions are based on an assumed or predetermined mathematical structure, normally a set of linear differential equations. Replacement of these predetermined structures by learned stimulus-response type of associative memory mappings lead to more general system representations by (in general nonlinear) mathematical mappings between n-dimensional input vectors $\underline{I}$ and m-dimensional output-vectors $\underline{O}$:

$$S : \underline{I} \rightarrow \underline{O} \quad \underline{I} \in R^n , \underline{O} \in R^m \qquad \text{III.1}$$

The utilization of such mappings for system representation is the most fundamental idea underlying the learning control concept introduced in this paper.

Given a time-discrete representation of a time-invariant deterministic system

$$\underline{y}(k+1) = \underline{f}_s [\underline{\Phi}(k) , \underline{u}(k) , \underline{v}(k)] \qquad \text{III.2}$$

with input vector $\underline{u} \in U$

output vector $\underline{y} \in Y$

measurable disturbance vector $\underline{v} \in V$

vector of the "inner state" $\underline{\Phi}^T(k) = (\underline{y}^T(k),\ldots,\underline{y}^T(k-\alpha_s),\ \underline{u}^T(k-1),\ldots,\underline{u}^T(k-\beta_s),\ \underline{v}^T(k-1),\ldots,\underline{v}^T(k-\delta_s))$

scalars $\alpha_s , \beta_s , \delta_s$

sampled time $t = k \cdot T_s$

and sampling period T_s

a general mathematical mapping for III.2 can be defined by

$$S : [\ \underline{\Phi}(k),\ \underline{u}(k),\ \underline{v}(k)] \rightarrow \underline{y}(k+1) \qquad \text{III.3}$$

The mathematical expression III.3 memorized in an associative neural network model, i.e. an associative memory system, does not require any a-priori structural information about the system at hand except the scalars α, β, δ representing the amount of history to be used in the model.

IV. AMS: A SOFTWARE-IMPLEMENTED NEURONAL MODEL

As a result of a comparative study of various neuronal network models (5,10,19) the associative memory system called AMS and discussed below was found to be a suitable system for real-time control applications . AMS is conceptually based on CMAC (Cerebellar Model Articulation Controller) conceived by Albus as a neuronal network model of the human cerebellar cortex (1,2). CMAC itself is a modified version of the Perceptron, a trainable pattern classifier proposed by Rosenblatt as a model of the human visual system (13).

In terms of computer science AMS is a distributed associative memory system by which a function or a mapping of several variables can be evaluated via simple content addressing mechanisms rather than by complex mathematical operations.

The AMS information processing structure is illustrated in Fig.IV.1. Mathematically AMS can be presented by the overall mapping

$$H : \underline{s} \rightarrow \underline{p} \qquad \text{(IV.1)}$$

by which according to the neurophysiological notation (1) the sensory input $\underline{s} = (s_1, ..., s_n)^T$ of n variables s_i may have R_i distinguishable values, is mapped onto the m-dimensional output vector $\underline{p}$ representing the axons of Purkinje cells leaving the cerebellar cortex.

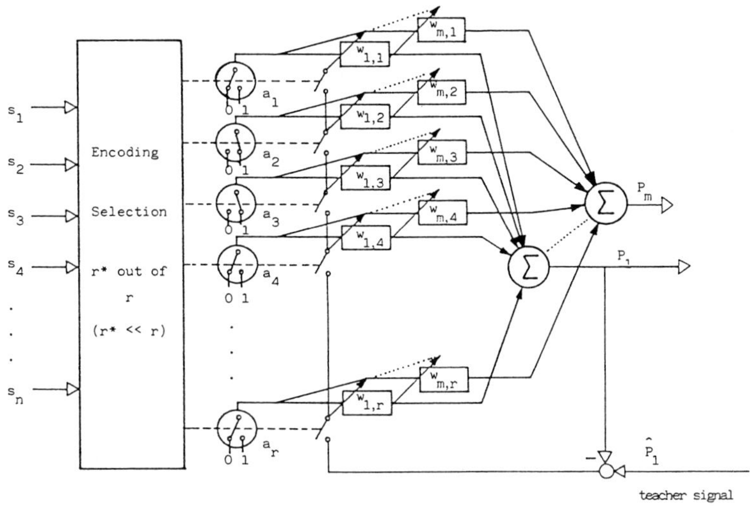

Figure IV.1. AMS : Information processing structure.

Depending on the contents of the input information s a constant number r* of association cells (pointers for locations in the physical memory marked by * in Fig.IV.1.) are activated by a special encoding procedure out of r memory locations. Thus the incoming information is subdivided onto r* information elements which are linked together to form the output information. The concatenation of these information elements, i.e. the sum of weights w_{ij} of these selected memory locations, represents the value of each output channel p_j.

Learning, i.e. training occurs by adjusting the selected weights w for an input s depending on the error e between the desired $\hat{p}$ and the actual output p

$$\underline{e} = \hat{\underline{p}} - \underline{p} \qquad \text{(IV.2)}$$

which is distributed over the selected weights.

The encoding procedure by which similar inputs share some of the r* memory locations - the number of shared locations being dependent on the degree of similarity - is responsible for the AMS's most fundamental feature of generalization, i.e. similar inputs tend to produce (for details see 5) similar outputs even if they are not trained before. This ability - which is of great importance during learning - means mathematically that for untrained input variables

$\underline{s}_x$ the memory system evaluates an output $\underline{p}_x$ the value of which depends on the similarity of the corresponding input $\underline{s}_x$ to all other already trained $\underline{s}_i$ in a specific nearest neighbourhood. For a single $\underline{s}_i$ a generalization expression can be given by

$$|\underline{p}_i - \underline{p}_x| = \alpha \cdot ||\underline{s}_i - \underline{s}_x|| \qquad \text{(IV.3)}$$

The scalar α, the degree of generalization can be pretuned by the generalization variable r^*.

This effect of local generalization is illustrated in Figure IV.2, which is great if the number of locations used for the storage of the weighting factors r^* is big and which is small, if r^* is small. With $r^* = 1$, there is no generalization and the associative memory is just a look-up-table.

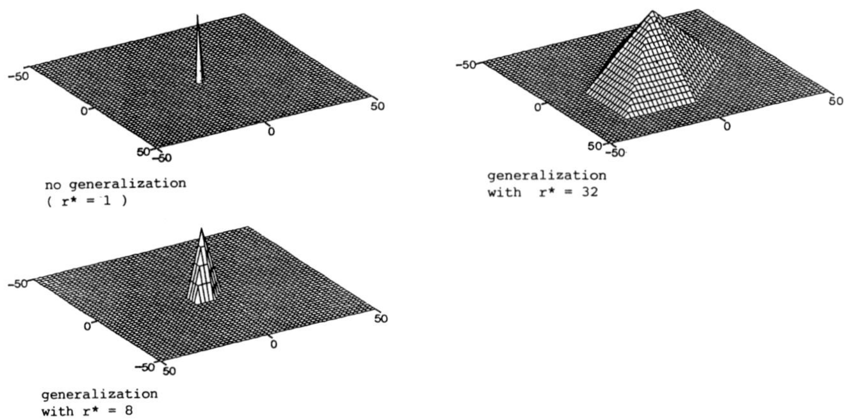

Figure IV.2. Response of three AMS-configurations with different generalization after training at the origin of the input plane.

The distribution of information into r^* locations instead of 1 location has, by the way, besides the effect of generalization also the advantageous effect of reduction of necessary memory size and training effort, and furtheron - if some averaging procedure is built in for the calculation of the w_{ji} - the effect of filtering regarding measurement noise.

Since each additional element of the input vector $\underline{s}$ extends the memory size in one more dimension and in general only rough estimates of the range of each s_i are easily obtainable one easily get into the problem of huge memory sizes ("course of dimensionality"); but on the other hand it is wellknown, that very seldom all possible s_{ij} - combinations are to be expected in reality, a further reduction of memory space by hash-coding is wise and possible. One estimates a realistic size of the memory and distributes the information onto this physical memory by randomly generating the addresses of the r^* memory places. If a high number of inputs is concentrated on a limited amount of memory space, also different inputs $\underline{s}$ could lead to the same memory addresses, the well known effect of hash collisions arises. These are avoided to a certain extent by generating out of the input information a further pseudo-random number, an input pattern identifier. This identifier and an additional information indicating the number of training frequency for a certain memory location are used also

- for noise filtering by indicating the quantity of information repetitions
- for finding out, how many of the r^* memory locations connected to a certain input $\underline{s}$ are trained to a certain extent already.

The latter information is an indicator, how reliable the output p_j for a not trained input $\underline{s}$ is; for further calculations normally only outputs with a sufficient training level are to be taken into account, that means only specific parts of the results of generalizations are considered to be reliable information.

The following example demonstrates some of the properties of the memory concept :

A reference function

$$p^d = f(s_1, s_2) = 300 \cdot \sin(2\Pi \cdot s_1/200) \cdot \sin(2\Pi \cdot s_2/200)$$

shown in Fig. IV.3.a with s_1, s_2 defined both in the interval [0,200] with a resolution of unity ($s_{1,2} = \sigma$; $\sigma = 0,1...$) is trained into the memory with $r^* = 8$ by a single training cycle, and the filtering algorithm $w_k = w_{k-1}+(p^d-p)/k+1$ in 2500 randomly chosen points of the possible 40401 points (6,25%). During training a measurement noise of 7% is assumed. The result is shown in

Fig.(IV.3.b.). One gets with the training of 6,25% of the possible points by the property of local generalization the desired function with a mean square error of 8% and no visible effect of the measurement noise, this being due to the multiple addressing of the same memory cells via generalization even in the case of one training cycle in a low percentage of the possible points.

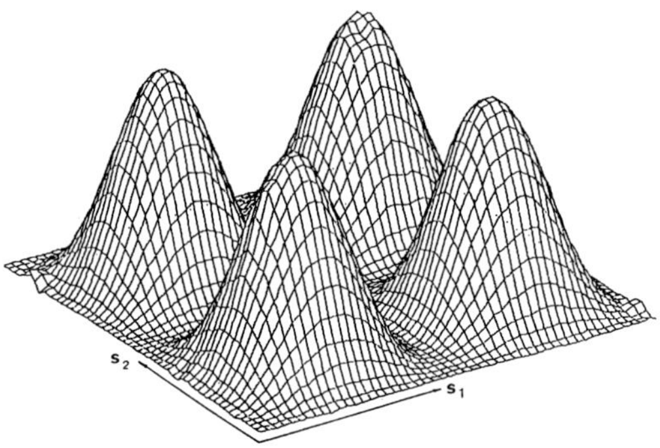

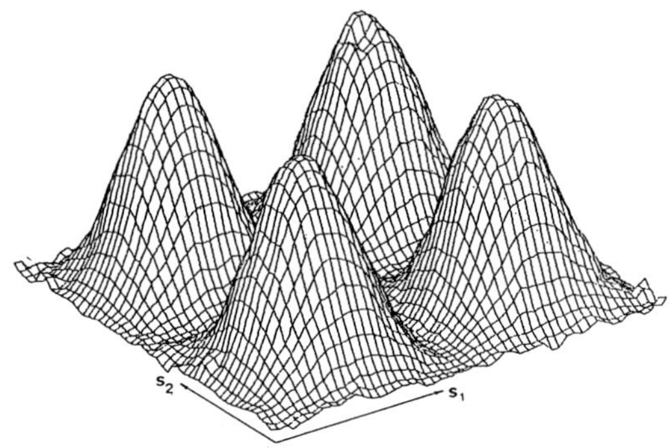

Figure IV.3.a. Reference function. Figure IV.3.b. AMS - response.

V. THE IMPLEMENTED LEARNING CONTROL SYSTEM

Based on the basic ideas of the two preceding sections a control system has been proposed and designed for technical control tasks. As mentioned before the concept incorporates two associative memory systems of type AMS, one for the predictive model of the unknown process and one for the control strategy. Fig. V.1 shows the control loop scheme - called LERNAS (Learning Control With Neuronlike Associative Memory System).

For better understanding some definitions and assumptions are given:

a) The unknown multivariable process at hand is deterministic, time-invariant or weakly time-variant.

b) Time is sampled with a sampling period T_s ($k=kT_s$) to allow time for learning and planning.

c) The control input $\underline{u} \varepsilon U$ is quantizied and the control input space U is finite.

d) The process output $\underline{y} \in Y$ is quantizied and out the output space Y is finite.

e) The measurable disturbances $\underline{v} \in V$ are quantizied and out of a finite space V.

f) The process can be described mathematically by equation (III.3) with some α_s , β_s and δ_s .

The concept based on AMS uses an output predictive algorithm scheme with a 1-step ahead control strategy where the predictive process model

$$M : [\underline{\Phi}_M(k), \underline{u}(k), \underline{v}(k)] \rightarrow \hat{\underline{y}}(k+1) \qquad (V.1.)$$

with some α_m , β_m and δ_m

and the control strategy

$$C: [\underline{\Phi}_c(k), \underline{v}(k), \underline{w}(k)] \rightarrow \underline{u}(k) \qquad (V.2.)$$

are represented by two general mappings stored in two different AMSs.

A performance index represented by a 1-step ahead subgoal

$$I_s(k) = \sum_{i=1}^{1} L_s[\underline{y}(k+i), \underline{w}(k+i), \underline{u}(k-1+i)] \qquad (V.3.)$$

with $\underline{w} \in W$ desired goal variable

used to evaluate possible actions $\underline{u}^i$, in order to assess them and to choose the optimal one.

The algorithmic scheme in each time cycle is as follows (Fig.V.1.) :

i. the predictive model is updated by the measured prediction error

$$\underline{e}(k) = \underline{y}(k) - \underline{\hat{y}}(k) \qquad (V.4)$$

where $\underline{\hat{y}}$ is obtained by :

$$M: [\underline{\Phi}_M(k-1), \underline{u}(k-1), \underline{v}(k-1)] \rightarrow \underline{y}(k) \qquad (V.5)$$

ii. an optimization (decision or planning) scheme is activated, if necessary, for calculating an optimal control decision $\underline{u}^*(k)$ for the subgoal (V.3).

To speed up the optimization a starting approximation for evaluating the optimal action $\underline{u}^*(k)$ can be obtained by the past decision experience

$$C: [\underline{\Phi}_C(k), \underline{v}(k), \underline{w}(k)] \rightarrow \underline{u}^o(k) \qquad (V.6)$$

iii. the control decision $\underline{u}^*(k)$ optimized by (ii.) is memorized in control memory to be used as the best decision making (ii.) superfluous in the long range, and giving either an excellent optimization starting point (see V.6.) or being used from a time on (to be decided by the user) without further inclusion of the predictive learning loop.

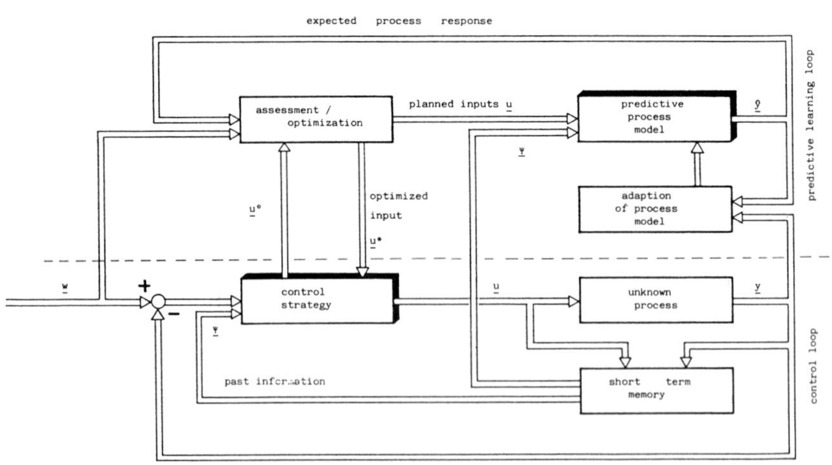

Figure V.1.Functional organization of LERNAS.

IV. the last operation before applying a control input to the process is to look for a suboptimal control input $\underline{\hat{u}}(k) \not\equiv \underline{u}^*(k)$ which is not yet trained in the process model memory.
$\underline{\hat{u}}(k)$ applied to the process excites it to further untrained information for the model AMS and, so, enlarges the information entropy.
This exploratory procedure called <u>active learning</u> is essential for the learning speed.

The concept was successfully tested by simulations on several nonlinear examples, single-input-single output, as well as multi-input-multi-output. The principle learning behaviour will be demonstrated by a real-time implemented first order single-input-single-output example.

<u>Example</u> : For the first order process

$$z(t) = [u(t) - z(t)]/10$$

the input and output variables for the AMS's are defined as:

$$u(k) = u(k) \quad ; \quad y(k) = z(k)$$
$$\Phi_M(k) = \Phi_C(k) = y(k) \quad ; \quad w = 5 \; .$$

The one-step ahead performance index is of the form

$$I_S(k) = (y(k) - w)^2 \; .$$

The control input is bounded

$$0 \leq u(k) \leq 10 \; .$$

With a sampling period of T_s = 6 sec Fig. V.1.a and Fig. V.1.b demonstrate the evolution of learning by successive learning trials. Neither the model nor the control-AMS had any a-priori information before the first run. Fig. V.1.b illustrates the learning curves which are the performance indexes for each learning trial plotted down over the number of trials. Three learning curves with different generalizations for the predictive process model show the learning performance of the control system and the influence of the generalization on the learning speed. The learning curves are similar to those obtained by the biological experiments. In the

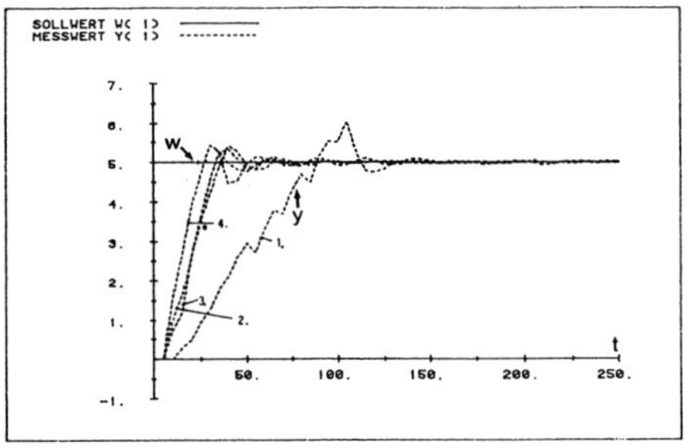

Figure V.1.a. Learning control.
(four trials)

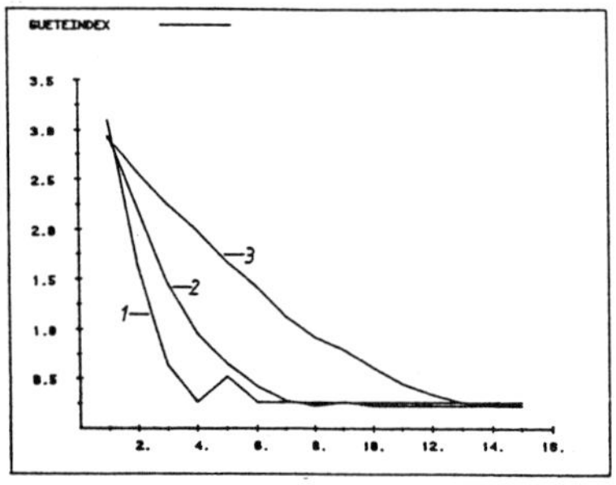

Figure V.1.b. Learning curves.
(1 : r* = 32 ;
2 : r* = 16 ;
3 : r* = 8)

first trials the learning progress is huge and slows down with the growing number of trials. The optimum is approached asymptotically. Fig.V.2. shows the response of the model-AMS as a two-dimensional function. The amount of information trained in the memory is illustrated for different time points. In Fig.V.3. the learning behaviour for changes of the desired goal variable is demonstrated.

The examples above discuss only some principal features of the concept. (4) shows a more detailed example of a waste-water neutralization process and (6) demonstrates the real-time example for a distillation column.

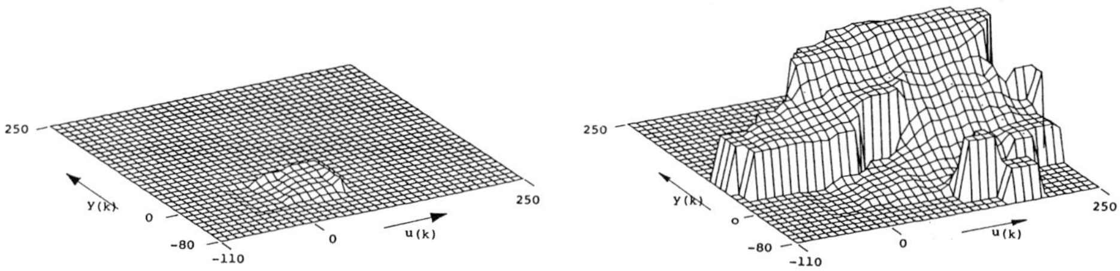

a. b.

Figure V.2. Response of the model-AMS after a.: 20 and b.: 150 sampling periods.

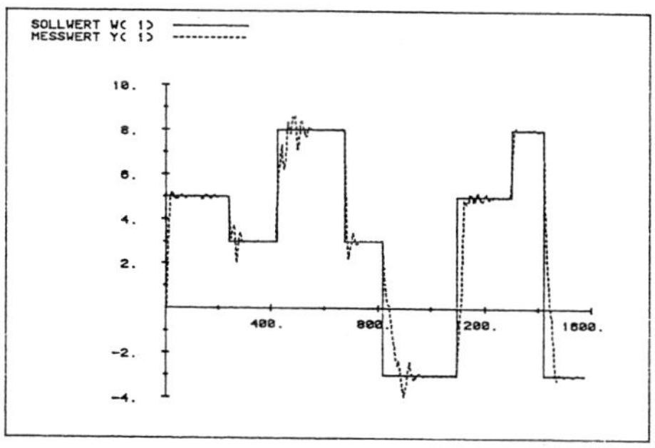

Figure V.3. Learning behaviour for changes in the desired output value w.

VI. STRUCTURAL VARIATIONS

VI.1. MINLERNAS

In human behaviour not always intelligent reasoning on the basis of an explicit predictive model takes place. Especially during the initial phase of human learning and for some simple situations implicit process knowledge acquired by basic learning strategies - like as "learning-by-doing" or "trial-and-error" - is used and planning is performed by assessing the memorized performance index values of past actions. This is modelled with the structure given in

Fig. VI.1. In this structure which is called MINLERNAS the assosiative memory in the planning level is used to store the values of an performance index like (V.3) instead of an explicit process model. Especially for complex processes with many outputs this saves storage, since the associative memory system described in section IV. needs the same amount of memory places for each output. In all other aspects MINLERNAS is in principle similar to LERNAS.

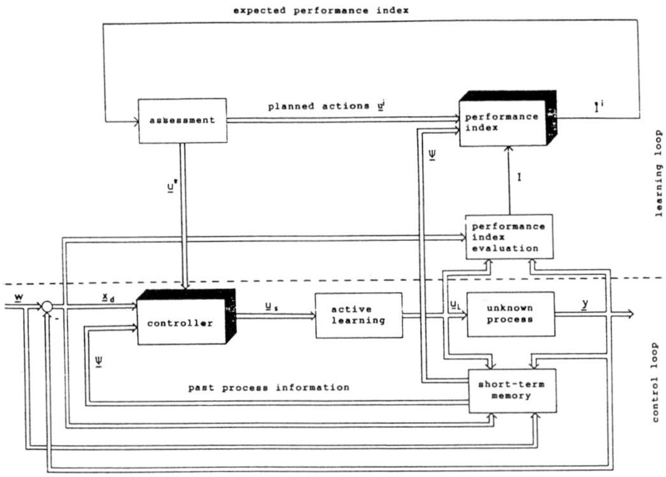

Figure VI.1. Basic structure of MINLERNAS.

The concept of MINLERNAS is of course more elementary and the expected performance depends only on the action oriented experience of the control system. This can lead to slower learning behaviour due to missing predictive planning with the process model.

The simulation results verify this expected slower performance. This handicap can be overcome by a modified memory configuration strategy resulting in variable generalization, i.e. tuning up the generalization depending on the information entropy in the specific associative memory.

VI.2. VARIABLE GENERALIZATION

In the preceding sections the generalization ability of the AMS is used with a constant r^*, i.e. no changes during learning in generalization. The determination of r^* for a specific control task has to consider the required accuracy with respect to the stored information and the learning speed. A broader generalization, i.e. r^* is big, provides fast learning behaviour, but in the long range the information stored can become inaccurate, as the cross-talk between stored patterns in a broader neighbourhood influences the obtainable accuracy. A strategy which could be used by the biological system is to tune the generalization depending on the learning progress, thus working with a broader generalization at the beginning for fast learning and decreasing it gradually with the on-going information acquisition to smaller sizes which then results in more accurate information recall in the long range.

The easiest method to reach variable generalization is the idea to train not just one associative memory with $r^* = r^*_1$, but to train in parallel a number of associative memories with $r^* = r^*_1, r^*_2, ...$ $(r^*_1 > r^*_2 > ...)$ and to take as the output a linear combination of the output of the different memories with weighting factors a_i, which are changed during learning.

Instead another strategy for variable generalization shall be discussed. With the above strategy the problem is to select the weighting factors. However, this can be automated by taking into account, that an indication of the training status for a certain input can be gained. This leads to the idea to search through the different associative memories with different r^* values, to find out, whether for the actual input the output can be regarded as a trainded value. The best would be, if one would get already an answer for the smallest local generalization. Therefore this memory is searched first. If one gains no satisfactory answer here, one looks at the memory with the next broader local generalization, and so on.

This scheme (see Fig. VI.2.) was implemented on MINLERNAS and tested for the control of a first order element again (Example 1). However, in this case both associative memory systems, the memory for the performance index and the memory for the controller were equipped with a variable generalization procedure of the above-mentioned type.

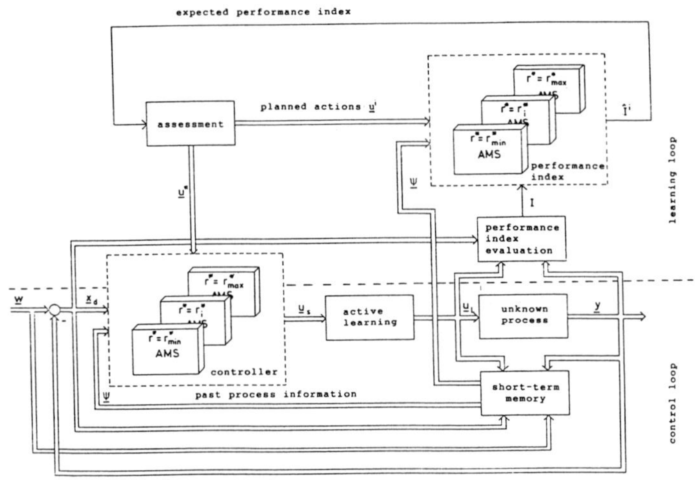

Figure VI.2. MINLERNAS with variable generalization.

Fig. VI.3.a and b give results of MINLERNAS with fixed and variable generalization, respectively. In Fig. VI.3.a the local generalization factor was for the controller memory r* = 4 and for the performance index memory r* = 8. One sees a good learning success for the first changes of the required output, but only a fair overall control performance in the long run. Since this is not the case with LERNAS, this indicates - as to be expected - that the implicit modelling of the process through the performance index is less effective as working with a direct predictive process model in the planning level.

Fig. VI.3.b shows on the other hand, that excellent results can be reached by using variable local generalization than with fixed local generalization. This gives rise to the statement that variable local generalization is an important improvement for real-time applications of the learning control loops studied in this paper. Four values of r* were implemented: r* = 16; r* = 8; r* = 4; r* = 1.

Fig. VI.4. shows for the performance index memory how the usage of the different memories shifts during learning from finding well trained inputs in the r* = 16 memory only to the r* = 8, r* = 4 and, finally also, the r* = 1 memory. This result encourages the use of MINLERNAS with variable generalization for heavily non-linear processes like the pH-neutralization process studied already with LERNAS and fixed local generalization in (4). In simulations similar good learning behaviour is obtained with MINLERNAS due to the variable generalization.

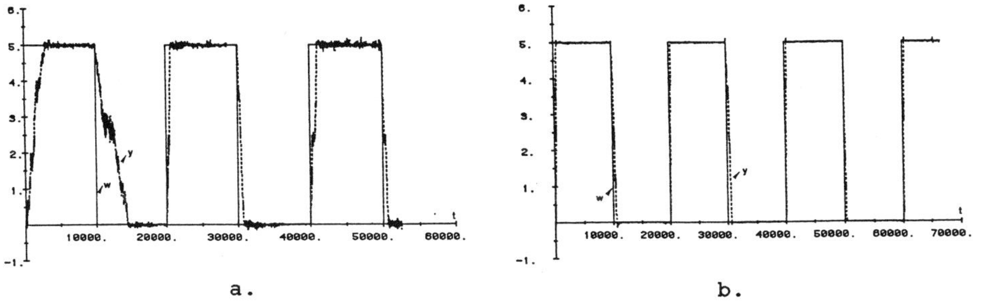

Figure VI.3. MINLERNAS - Learning behaviour a.: with and b.: without generalization.

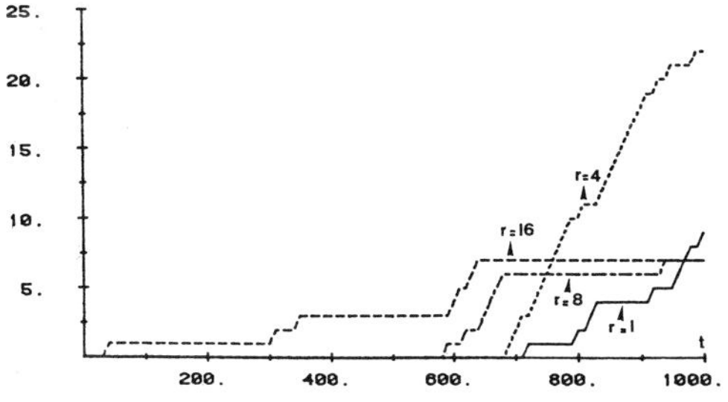

Figure VI.4. Switching of memories of different generalization during learning.

VII. CONCLUSION

The careful study of learning control loops during a number of years - starting from LERNAS in its most simple conception and making use of experimental experiences and deeper consideration for modifications of the microintelligence and the macrointelligence level - have led to the result that different macrointelligence structures can successfully handle even complicated control tasks with very little a priori knowledge about the process. However, there are limits to the productive power of the different schemes and the organization of the macrointelligence level and the microintelligence level is not independent. Therefore it seems advantageous for an efficient scheme without an explicit process model to have the feature of variable local generalization. There may be other connections and/or problems which we have not discovered up to now. But already those indications reached up to now may give rise to considerations and/or experiments by other researchers to clear the situation furtheron.

We are going to concentrate, in our further research, on the superposition of such control loop structures like LERNAS or MINLERNAS in hierarchical systems as it may be the case in the brain and on the possible cooperation between such control loop structures which learn from input and output signals of the process in an "unconscious" way with some explicit "conscious" (abstract) reasoning as it has been developed in the so-called "artificial intelligence" methods.

Actually we have to admit that although our research was initiated and supported by an interest in the behaviour of the brain and its possible structures, there is on the other hand a striving force after searching new methods for the control of technical processes. A further application of the AMS is its use for learning and interpolation of multidimensional nonlinear relationships, so-called "characteristic fields". By changing input and output one can especially learn the inverse nonlinear characteristic and by using this as a feed-forward filter for linearizing and decoupling complex processes, making them more handable for control considerations.
Nowadays the implementation of an AMS-like concept on a VLSI-basis with a highly parallel cellular processing architecture is technologically no severe difficulty. Such an implementation would result in an neuronal associative processor with huge application potential in industrial control problems.

REFERENCES

(1) Albus, J.S. (1972) Theoretical and experimental aspects of a cerebellar model. PhD Thesis, University of Maryland, USA.

(2) Albus, J.S. (1975) The Cerebellar Model Articulation Controller, Transact. ASME, series G, Vol 97, NO. 3.

(3) Ersü, E. (1984) On the application of associative neuronal network models to technical control problems in "Localization and Orientation in Biology and Engineering, ed. by Varjü/Schnitzler-Springer Verlag pg. 90-93

(4) Ersü, E., Mao, X. (1983) Control of pH using a self-organizing concept with associative memories. Proceed. Int. INSTED Conf. on "Applied Control and Identification, Copenhagen, Denmark.

(5) Ersü, E., Militzer, J. (1982) Software implementation of a neuron-like associative memory system for control applications. Proc. of the ISMM 8th Int. Symp. MIMI, Davos, Switzerland.

(6) Ersü, E., Militzer, J. (1984) Real-time implementation of an associative memory based learning control scheme for nonlinear multi variable control; Proceerd. of the Symp."Application of Multivariable System Technique" Plymouth, U.K.

(7) Ersü, E., Tolle, H. (1984) A new concept for learning control inspired by brain Theory, Proceed. 9th IFA World Congres, Budapest, Hungaria.

(8) Aström, K.J., Borrison, K.; Ljung, L. and Wittenberg, B. (1977). Theory and applications of selftuning regulators. Automatica,

(9) Dörner, D. (1974). Die kognitive Organisation beim Problemlösen. Hans Huber Verlag, Bern.

(10) Kohonen, T. (1977). Associative Memory. Springer, Berlin, Heidelberg, New York.

(12) Newell, A. and Simon, H.A. (1972). Human problem solving. Englewood Cliffs, N.J. Prentice Hall.

(13) Rosenblatt, F. (1962). Principles of neurodynamics. Spartan

Books, Washington D.C.

(14) Saridis, G.N. (1979). Self-organizing control of stochastic systems. Marcel Dekker, New York.

(15) Tsypkin, Ya.Z. (1973). Foundations of the theory of learning systems, Academic Press, New York.

(16) Arbib, M.A. (1972). The metaphorical brain. Wiley & Sons, New York.

(17) Amari, S.I. (1977). Neural theory of association and concept formation. Biol. Cybernetics, Vol.26, pp. 175 - 186.

(18) Palm, G. (1982). Neural assemblies. Springer, Berlin, Heidelberg, New York.

(19) Ersü, E. (1981). A learning mechanism for an associative storage system. Proc. of Int. IEEE Conference on Systems, Man and Cybernetic, 1981, Atlanta.

Learning Control Structures with Neuron-Like Associative Memory Systems

Discussion edited by *Werner von Seelen*

The paper was based on a neural strategy used for a complex control task. So naturally a number of questions were directed towards understanding the procedural steps necessitated by the constraints of the application. More general questions referred to the form of the learning rule and its consequences for the systems's capability of generalization. The authors reported here on experiences with special cases. The speaker considered the proposed introduction of a "forgetting function" to make sense only for time variant systems. Questions on the control quality were answered in terms of the equations and examples in the paper. The kind of functions that are employed in the learning process were more broadly discussed. The authors answered questions about the extent to which impulse shaped signals are sensible, whether continuity conditions should be introduced and whether point by point estimation should not be replaced by parameter estimation with a given approximation function. They called their method a compromise between the various requirements, in which a priori knowledge on the process and computation time play a decisive role. The discussion revealed that the size of the knowledge base at the start of the learning process is most important for further essential subproblems. This holds for the learning rate of the system as well as for its stability behavior.

The authors pointed out several times, in response to questions, that an essential advantage of the proposed procedure was that it could be scaled and that the superposition of several control loop structures in a hierarchical system was a necessary extension. To solve this problem, it seemed that close contact between neuroscience and control theory would be extremely helpful.

Summary Statement:

An Epigenetic Landscape of Brain Theory

by *Walter J. Freeman*

Abstract

A wide range of approaches to brain theory was offered at this meeting on The Organization of Structure and Function in the Brain. These approaches are summarized in terms of four research traditions with historical roots in the 19th century. An "epigenetic landscape" is sketched within which the work of each of the speakers is located, and new options for future research in theory are outlined in terms of recent work undertaken by conference participants.

Historical Trajectories

"Brain Theory" is a good example of a fuzzy set. Some theories are about brains, others derive from or explain brain studies, still others simulate what brains purportedly do, and so on. A current idiom in the United States is "neural models", which displays a becoming modesty in disclaiming to be theory and in restricting consideration to nets of model neurons that make up small parts of brains. Within even this limited part of brain theory there is a bewildering variety of approaches. The papers presented at this Conference constitute virtually a random sample from a broad array of topics and investigators. So was the selection at Trieste two years ago, and so will be the next selection two years hence.

A summary of a Conference such as this might take the form of a medieval bestiary comprising a brief description of each in the order of discovery or presentation. This does little to help participants and readers who wish to form a contextual overview, to remember the salient features of each, and especially to establish relations or distances among topics, that is, similarities and differences that are critical for understanding and evaluation.

An alternative view is Darwinian, laying out an evolutionary tree in which each brain model has a geneology (Figure 1). A useful conceptual tool has been provided for this by C.H. Waddington (1977) in the form of the epigenetic landscape. Brain theory is conceived as a living community, growing, reproducing, evolving, and spreading out over a terrain of basins separated by ridges or watersheds. Each basin is populated by members of a species that cross-fertilizes itself, and that has connecting flow lines or "chreods" for both ancestors and progeny. This is at once a mnemonic device, a historical overview, and an editorial commentary of who is related to whom and for what reasons. It also provides a technique for discerning significant differences between topics that might otherwise seem superficially to be quite similar.

Lines can readily be drawn back to Aristotle and Hippocrates but more cogent wellsprings or chreods lie in the 19th century with four outstanding figures. The pre-eminent French physiologist Claude Bernard introduced the fundamental concept of the constancy of the internal milieu, that is, of the chemical environment bathing the cells of the organs of the body and par excellence the brain. Walter Cannon elaborated brain physiology in terms of dynamic processes operating by negative feedback to ensure homeostasis. Norbert Wiener generalized these concepts into cybernetics and laid the mathematical foundations for automatic control systems. E. Ersu's presentation at this meeting on learning control structures represents a continuation of this approach using linear time-invariant equations with adaptive coefficents.

The second approach to brain theory has its roots in the cytoarchi-tectural work in Spain of Santiago Ramon y Cajal, which was advanced by Rafael Lorente de No' in the present century by his introduction into neuroanatomy of the notion of feedback: the reverberating circuit. Three related lines of research were developed within this cellular approach by theorists, who relied heavily on the network properties of cortex suggested by the Golgi technique. Warren McCulloch and Walter Pitts developed a calculus of neural nets able to solve problems in Boolean algebra with two-valued devices that expressed their states in nerve impulses, thereby introducing a whole theory of finite state automata. John von Neumann drew heavily on their neural metaphors in his development of the modern digital computer. From this stem has developed artificial intelligence exemplified by David Marr and represented at this meeting in the reports by T. Poggio and N. Grzywacz on the computation of motion by the visual system, by E. Bienenstock on the problem of visual recognition of invariant shapes, and by G. Mitchison on the organization of sequential memory.

A second line in parallel with the digital computer was the Perceptron of Frank Rosenblatt, which sired a family of imitators and competitors until the species was terminated by the criticisms of Minsky and Pappert. A third line also emanating explicitly from Lorente de No's anatomical work comprised Donald Hebb's concepts of the nerve cell assembly as the organ of the memory trace and the modifiable synapse as the basis of learning. These concepts served to emphasize again the notion of feedback in the guise of re-entrant excitation.

From these lines has emerged in the past 10 years the concept of interactive sheets of neurons. One of the first investigators to realize its potential for pattern recognition was James Anderson, who conceived the "Brain State in a Box" metaphor as described in his report here. Hopfield rediscovered this phenomenon in the properties of spin glasses and introduced the descriptive language of basins, attractors, and potential wells. This view is elaborated here by G. Toulouse, who shows how under certain simplifying assumptions of universal connectivity, complete input, and a generalized Hebb rule, it is possible to predict the memory capacity, retrieval quality, and the point of onset of interference effects leading to catastrophic breakdowns of neural nets. Amari extends this line by comparing four classes of associational memory (Pavlovian, Perceptron-like, orthogonal, and Hebbian), the last providing for "learning without a teacher" by self-organizing processes of neural interaction. Such systems again have multiple potential wells, each corresponding to a learned signal and accessed by presentation of an example of that signal.

Interactive nerve cell assemblies are fundamentally akin to diffusion-coupled chemical reaction systems and to self-organizing embryological systems, as described here by A. Gierer. The obvious implication is that the structures of brains are in part shaped by dynamic patterns of neural activity that exist at intermediate stage of growth, and that the process of creation of new spatial structure in neural nets must continue on into adult life by the auto-creation of nerve cell assembles during learning.

A third chreod was from masterful work of the German psychophysicist Hermann von Helmholtz. One set of descendents led through Nernst and Bernstein into cellular biophysics. Studies of the cochlea by von Bekesy that were previously thought to have disproven the resonance theory of Helmholtz are re-evaluated in the report by E. de Boer, who shows that Helenholtz was correct after all. Another set of descendents includes Sechenov and his pupil I.P. Pavlov; the American school of mathematical psychology exemplified by Clark Hull and W. Estes and presently by Stephen Grossberg; and by the German school of reflex physiology exemplified here in the work reported by K.P. Hoffmann on the retinal image stabilization system and its dynamic genesis of optokinetic nystagmus.

The main stream of brain theory flows from work of Charles Sherrington on the physiology of the nerve cell, the synapse, the receptor field, the hierarchy of reflexes, and the integrative action of the nervous system. By means of microelectrode technology these concepts were elaborated at the cellular level by scientists such as Bernhard Katz, John Eccles and Stephen Kuffler. The study of axonal potentials led to two schools of thought. One, exemplified by the work of Horace Barlow (the "cardinal cell"). J.Y. Lettvin (the "feature detector") and Donald Maynard (the "command neuron") led to the now dying tradition of the "grandmother cell" hypothesis. The other school led by Mountcastle, Hubel and Wiesel was more strictly descriptive. It is expressed mathematically by J.J. Koenderink at two levels, first in showing how nerve cell assemblies can grow by means of pair-wise correlation as required by the Hebb rule in its differential form (common rates of change in firing), and how features can be extracted by a calculus of receptor fields in the descriptive equations of differential geometry. Thereby the sensorium is treated as ROM with universal format in abstract linked data structures, easily implemented by known neural structures.

Recognition of the importance of the distributed character of information-bearing neural activity over assemblies of nerve cells has encouraged the development of multineuronal techniques for observation, including the study of multicellular dendritic potentials by use of the EEG, MEG, NMR optical probes with dyes, metabolic markers such as 2-DOG, and cerebral blood flow, and of multiple units by optical probes and arrays of microelectrodes. Two examples reported here by G. Gerstein and M. Merzenich both emphasize the variation over time and with the state of the animals in the receptor fields and correlational structures of active neurons. The brain is not a rigidly hard-wired network; it operates more like a fluid dynamic system.

Current Trends

The papers by J. Anderson on concept use by distributed systems, by A Gierer on the generation of spatial order in the developing nervous system, and by S. Amari and G. Toulouse on self-organizing neural nets, have several elements in common in their approach to brain theory. This approach focuses on interactive systems that generate ordered states from less ordered initial conditions, while interacting with their environment via a process that can be topologically modelled in terms of bifurcations and subsequent convergence to an attractor within a basin. The explanatory models used by this approach are influenced by information on how brains operate, rather than by information about man-made machine procedures. While the models produced by this approach thus far do not pretend to represent accurately the details of brain function, they offer insights into the kinds of distributed processes that brains might use to produce certain classes of behavior.

Expanding the metaphor of "attractors" to apply to the landscape just sketched, we can distinguish several basins of attraction into which brain theories fall. The central basin which we will call "NEURONLAND" is flanked by neighboring basins. Two of these basins represent incompatible theoretical approaches currently available to brain theorists. One of these basins we shall call "TURINGLAND"; it includes work on robotics, automatic control systems, digital computers, and AI. Research within this basin of attraction tends to concentrate on producing machine (digital computer) models that are functional equivalents of brains with respect to output, but that use mechanisms quite unlike those used by brains. A second basin of attraction on the epigenetic landscape we shall call "BRAINLAND". It includes research on interactive systems as applied to brain theory. As mentioned above, this approach is distinguished by its attention to the role of distributed, self-organizing processes in the production of behavior, resembling those that brains employ.

The recent multiunit recording approach, including the work of Merzenich and Gerstein, lies outside of both theory basins and well within the NEURONLAND basin of attraction. The evidence both found for plasticity suggests that self-organized functional changes in time occur in the brain with the emergence of neural patterns that can be modelled in ways similar to processes modelled by Toulouse, Amari and Anderson. It seems reasonable to expect that multiunit recording research will soon be swept into BRAINLAND.

The Future of Brain Theory?

Looking ahead we ask where we might go from here? What future chreods will be found by theorists about the structure and function of the brain? It seems likely that recent theories of interactive systems, theories in BRAINLAND, will be significantly expanded with cooperation among mathematicians, neuroscientists, physicists, and computer specialists, but in what ways?

Some insights can be gained by asking, what is the crucial property of the watershed that separates BRAINLAND from its neighbors? It is the property of self-organization, which in turn is endowed by a configuration of properties that are emphasized in all four reports. Their systems are composed of large numbers of elements widely if not universally interconnected and interactive, constituting dynamic regenerative feedback. They are nonlinear. They incorporate delays in the feedback paths. And they are open, having throughputs of energy, materials and activation, which drive them far from equilibrium. When driven sufficiently far from a previously dominating attractor, a system falls into the basin of a new attractor and bifurcates to a new stable state. Each attractor in formed by a learning process involving modification of synaptic connection strengths. Information is not localized; it is widely disseminated. "Meaning" is not attached to any part or activity level, as it is in the reflex arc, the grandmother cell, and the Turing machine; it is inherent in the function of the entire system in its relation to the environment. The question cannot arise at the local level. Information is created within the network, not merely processed or taught as in reflex, feed forward and AI models.

We can expect that brain theory will increasingly incorporate the mathematics and computer simulations of nonlinear dynamics. We can expect further that the needs for spatial pattern analysis of dynamic systems will create an insatiable demand for description and analysis of spatial patterns of neural activity, thereby accelerating the pace of research into recording from arrays of microelectrodes, EEG electrodes, magnetometers and optical probes. Bifurcation theory will become a standard tool in the workshop of brain theory.

A more fine-grained view of BRAINLAND shows that it is not homogeneous, and that several different models of neural activity are contained within its boundaries. Current interactive models can be classified in terms of a hierarchy of the kinds of attractors they employ, as revealed in the forms of behavior exhibited by the neural nets that they model. Amari, Toulouse, and Anderson model the neural network along the lines also adhered to by Hopfield and Grossberg: system dynamics is represented in the phase portrait as a set of point attractors. Gierer's model pictures system dynamics additionally in terms of limit cycle attractors. According to this model, an essential feature of brain function may be oscillatory behavior as exemplified by the pulsed behavior he describes of slime molds that allows escape from rigid forms.

At a yet more complex level exist strange attractors. Work reported 2 years ago in Trieste by Freeman and Viana Di Prisco

(1986) indicates that the brain exhibits behavior that can be modelled with chaotic attractors, thus introducing another basin on the landscape of BRAINLAND. Data on olfactory dynamics tell us that the EEG activity that can be modelled using a point attractor is only present under deep anesthesia or death, while genuine oscillatory behavior is exhibited in motivated animals upon recognition of an odor. Learning consists in convergence to a limit cycle attractor. But the data also indicate the presence of a deterministic, low-dimensional chaotic state. This chaotic attractor is dominant from the low level state of rest to the high level energy state of seizure (Freeman, 1986).

The only way to detect these kinds of state and the transitions betweem them is to observe the time and space dependent patterns of neural activity. The only way to model them is to set up dynamical systems of equations and study their time and space dependent solutions as trajectories under the appropriate initial and boundary conditions. Hence the near future is likely to bring widespread interest in the spatiotemporal nonlinear dynamics of brain models, in which the overt time courses of neural actions are simulated by solutions to equations, and in which spatial patterns of neural activity are measured and simulated, thereby superceding the present use of binary logic, clock-driven synchronized elements, and restrictions to nearest-neighbor interactions or back propagation. The bases for these trends are clearly evident in the papers presented here.

References

W.J. Freeman. Petit mal seizure spikes in olfactory bulb and cortex caused by runaway inhibition after exhaustion of excitation, Brain Res. Reviews 11, (1986), pp. 259-284.

W.J. Freeman and G. Viana Di Prisco. In G. Palm and A. Aertsen (eds.), EEG spatial pattern difference with discriminated odors manifest chaotic and limit cycle attractors in olfactory bulb of rabbits. In: Brain Theory, Springer-Verlag, Berlin (1986), pp. 97-119.

C.H. Waddington, Tools for Thought, Basic Books, Inc., New York (1977), 250 pp.

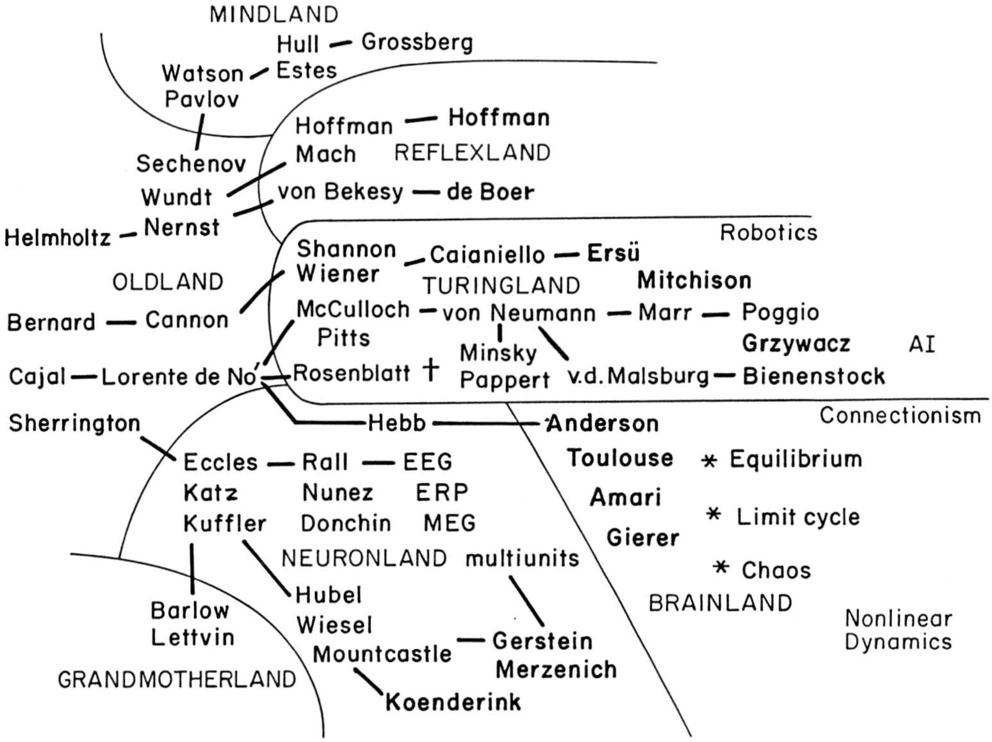

Figure 1.

The historical flow of ideas in brain theory is abstracted and reduced to two dimensions. The surface is a cone with its apex in Hippocrates. The cut is between mind and brain, so that Mindland and Brainland are nearest neighbors. The line segments are flow channels or chreods in Waddington's terminology. The light curves are the watersheds or separatrices of domains in the landscape. The asterisks are three classes of attractors in the domain of nonlinear dynamics, which serve to predict a fruitful direction of new growth. Abbreviations: EEG, electroencephalography; MEG, magnetoencephalo-graphy; ERP, event related potential.

Index